# FROM THE LABORATORY TO THE MOON

Lemelson Center Studies in Invention and Innovation

Joyce Bedi, Arthur Daemmrich, and Eric S. Hintz, general editors

Arthur P. Molella and Joyce Bedi, editors, *Inventing for the Environment*

Paul E. Ceruzzi, *Internet Alley: High Technology in Tysons Corner, 1945–2005*

Robert H. Kargon and Arthur P. Molella, *Invented Edens: Techno-Cities of the Twentieth Century*

Kurt Beyer, *Grace Hopper and the Invention of the Information Age*

Michael Brian Schiffer, *Power Struggles: Scientific Authority and the Creation of Practical Electricity before Edison*

Regina Lee Blaszczyk, *The Color Revolution*

Sarah Kate Gillespie, *The Early American Daguerreotype: Cross-Currents in Art and Technology*

Matthew Wisnioski, Eric S. Hintz, and Marie Stettler Kleine, editors, *Does America Need More Innovators?*

Kathryn D. Sullivan, *Handprints on Hubble: An Astronaut's Story of Invention*

Joris Mercelis, *Beyond Bakelite: Leo Baekeland and the Business of Science and Invention*

Eric S. Hintz, *American Independent Inventors in an Era of Corporate R&D*

David H. DeVorkin, *From the Laboratory to the Moon: The Quiet Genius of George R. Carruthers*

# FROM THE LABORATORY TO THE MOON

## THE QUIET GENIUS OF GEORGE R. CARRUTHERS

DAVID H. DEVORKIN

The MIT Press
Cambridge, Massachusetts
London, England

The MIT Press
Massachusetts Institute of Technology
77 Massachusetts Avenue, Cambridge, MA 02139
mitpress.mit.edu

The MIT Press would like to thank the anonymous peer reviewers who provided comments on drafts of this book. The generous work of academic experts is essential for establishing the authority and quality of our publications. We acknowledge with gratitude the contributions of these otherwise uncredited readers.

This book was set in Bembo Book MT Pro by Westchester Publishing Services. Printed and bound in the United States of America.

Library of Congress Cataloging-in-Publication Data
Names: DeVorkin, David H., 1944- author.
Title: From the laboratory to the moon : the quiet genius of George R. Carruthers / David H. DeVorkin.
Description: Cambridge, Massachusetts : The MIT Press, [2025] | Series: Lemelson center studies in invention and innovation series | Includes bibliographical references and index.
Identifiers: LCCN 2024022810 (print) | LCCN 2024022811 (ebook) | ISBN 9780262551397 (paperback) | ISBN 9780262381802 (epub) | ISBN 9780262381819 (pdf)
Subjects: LCSH: Carruthers, George R. | Astrophysicists—United States—Biography.
Classification: LCC QB460.72.C37 D48 2025 (print) | LCC QB460.72.C37 (ebook) | DDC 523.01092 [B]—dc23/eng20240924
LC record available at https://lccn.loc.gov/2024022810
LC ebook record available at https://lccn.loc.gov/2024022811

10 9 8 7 6 5 4 3 2 1

EU product safety and compliance information contact is: mitp-eu-gpsr@mit.edu

This book is dedicated to
The family of George R. Carruthers
and
To those who find inspiration in his work

# CONTENTS

# SERIES FOREWORD

About two decades before astronaut Neil Armstrong made his "one giant leap for mankind," a young boy on an Ohio farm stared up at the night sky, pretending that his soda bottle was a telescope. He wrote stories about space travel. His teachers reported that he was drawing spaceships instead of doing his schoolwork. At age twelve, he built a small telescope and recalled, "I went out and looked at the stars, the Moon, and the planets and that sort of thing."

This fascination with the universe stuck with George R. Carruthers throughout his life. He became the visionary inventor of a new type of small but powerful telescopic camera, with applications ranging from studying the Earth's outermost atmosphere to mapping the universe. In 1972, one of his instruments became the first sent to the Moon, onboard *Apollo 16*. He was, as author David DeVorkin notes, "a highly successful scientist who achieved notoriety as a Black man in science, and then true fame for his spectacular achievements."

Carruthers eschewed the spotlight, however. The biography of this extraordinarily talented space scientist is also the story of, as DeVorkin characterizes him, a quiet genius. Although embedded in the tumultuous 1960s through 1980s in Washington, DC, Carruthers stayed intensely focused on his work, creating new and more powerful ways to explore the universe. He also dedicated himself in his later years to mentoring the next generation, becoming a role model to aspiring Black STEM students.

Invention and innovation have long been recognized as transformational forces in American history, not only in technological realms but also in politics, society, and culture. Since 1995, the Smithsonian's Lemelson Center has been investigating the history of invention and innovation from an interdisciplinary perspective. Books in the Lemelson Center Studies in Invention and Innovation extend this work to enhance public understanding of humanity's inventive impulse. Authors in the series raise new questions and offer new insights about the work of inventors and the technologies they create, while stimulating cross-disciplinary dialogue. By opening channels of communication between the various disciplines and sectors of society concerned with technological innovation, the Lemelson Center Studies advance scholarship in the history of technology, engineering, science, architecture, the arts, and related fields and disseminate it to a general interest audience.

Joyce Bedi, Arthur Daemmrich, and Eric S. Hintz
Series editors, Lemelson Center Studies in Invention and Innovation

## AUTHOR'S AFFINITY

When I first interviewed George Carruthers in 1992, I felt an immediate affinity as he recounted experiences and passions he felt were responsible for shaping his life. I shared those influences growing up a few years after he did, from space travel comics, astronomy articles, and encyclopedias; the landmark *Collier's* series in the early 1950s envisioning spaceflight; and my father's astronomy books, to telescope building at home and at a local planetarium. There were many parallels in our early years. Our fathers were professionally trained engineers who encouraged science, mathematics, and building things. We were both the eldest child. Carruthers built rockets and engines; my father tested them at the Jet Propulsion Laboratory, and in my dreams and my cartooning, flew them into space. Carruthers was a Black man; I am Jewish. As children, we both experienced mild discrimination: teasing and bullying. Carruthers was less bothered by it than I was.

Carruthers and I have very different personalities and capabilities. He had built and done things I had only imagined. Exploring his life, therefore, was a window into a world I perceived and cherished, but never fully entered. Hence my fascination with the subject. I hope the readers of this book will take this into account.

# ACKNOWLEDGMENTS

This book could not have been written without the advice and aid of many people and institutions. The single most important stimulant was contact with George Carruthers starting in the 1980s and 1990s—interviewing him and his colleagues, touring his and others' laboratories at the Naval Research Laboratory (NRL), and working with him on educational activities at the Smithsonian's National Air and Space Museum (NASM). We stayed in touch over the years as his outreach efforts increased. Equally important to making the scientific and institutional side of the story available was the fact that his former boss, Herbert Friedman, had deposited his research papers at the American Philosophical Society, where they were beautifully organized and made electronically available to me in the 2020s pandemic. They contained a great deal of information and insight about Carruthers's professional life and about NRL's relationship with NASA from the 1960s through the early 1990s.

Carruthers's family—his wife, Deborah ("Debra"), and brother Gerald Carruthers—have been most supportive, providing insights, images, and archival records available nowhere else in the universe. Their images of the family have greatly enhanced my appreciation of their life and times, their heritage, and, most of all, Carruthers's personality and, at times, antics. Gerald, Carruthers's youngest brother, opened up memories of their early lives in Ohio and then in Chicago. Debra, Carruthers's widow, provided aid and insight, including biographical and professional materials from their home that now reside at NASM's archives. Suzanne Phillip, cousin of Carruthers's first wife, Sandra Redhead, shared memories of their life together.

In his later career, Carruthers became very active in several organizations devoted to improving educational and career opportunities for all children, especially the disadvantaged. His prominence in science made him a symbol, and he parlayed that by his activism, which made him invaluable to the cause. Helping me appreciate this aspect of his career and life were Hattie Carwell and Valerie L. Thomas, both leaders of organizations like S.M.A.R.T. Inc. (Science, Mathematics, Aerospace, Research and Technology) and the NTA (National Technical Association).

One of the many surprises in my research efforts was to discover a partial family history in the papers of Benjamin Carruthers, George's uncle, held by the Manuscripts, Archives and Rare Books Division, Schomburg Center for Research in Black Culture. Bridgett Pride kindly sent me valuable material that raised more questions and required more detailed scrutiny. So I engaged then NYU graduate student Athena Pierquet to visit the archives to examine portions of the collection in greater detail. She also provided very perceptive commentary.

The NRL was, of course, a critical resource for this effort. Dr. Angelina Callahan, then NRL historian, provided access to Carruthers's colleagues for a series of oral histories, as well as to documents and records. She also co-conducted a number of the oral histories. Holly McIntyre, Gayle Fullerton, and Brian Cadwell provided critical photographic material and clearance permissions, and past NRL historians and archivists, including Callahan, Bruce Hevly, and the late David van Keuran, facilitated interviews with Carruthers and his colleagues from the 1980s through the 2020s.

The oral and video histories I conducted in the late 1980s and early 1990s with NRL scientists were supported by a general grant from the Sloan Foundation with institutional support from the Smithsonian Institution Archives. More recent interviews through the early 2020s were supported by NASM and the American Institute of Physics Center for History of Physics, where Gregory Good, Melanie Mueller, Jon Phillips, and Will Thomas provided critical assistance and advice. More recently, the Lemelson Center for the Study of Invention and Innovation at the Smithsonian's National Museum of American History supported video sessions with Carruthers.

The recent interviews were deftly transcribed and annotated by Kate Bulson, and others were coordinated by the American Institute of Physics. Among the dozens of oral history interviews, those with George and Gerald Carruthers provided key insights. From NRL, most helpful providing both formal and informal information through interviews and correspondence included Robert R. Meier, Harry Heckathorn, Tim Seeley, Howard Russell, Frank Giovane, Charles Brown, Christoph Englert, R. Howard, Vijay Kowtha, Kenneth Dymond, Warren Moos, Henry Pickard, and Robert R. Conway.

Staff at the many archival centers I relied upon were extremely helpful. Notable among them were David Gary and Joseph DeLillo at the American Philosophical Society; Stacey Christen and Lauren Amundson of the Putnam Collection Center (Lowell Observatory Archives); and Cate Brennan, Marie Maxwell, Wanda Williams, Eugene Morris, and Jason Atkinson at the National Archives. Melissa Ann Gauger of the University of Illinois Archives provided rapid access to the Ladislas Goldstein's papers. Amanda Nelson and Jennifer Hadley at Wesleyan University provided access to Thornton Page materials, as did Marc Brodsky, Special Collections, Newman Library, Virginia Tech. Brian Odom, Sarah LeClaire, and Connie Moore of the NASA History Office provided access to the Naugle Papers and access to the NASA images used here.

Marilyn Graskowicz, Jeannie Whited, Kate Igoe, Melissa Keiser, Elizabeth Borja, and Patti Williams of NASM's Archives and Collections departments provided invaluable advice and support in the acquisition and organization of Carruthers's papers and memorabilia. Finally, Lela Sewell-Williams at the Howard University Moorland Spingarn Research Center provided access to Carruthers's records relating to his work for the NTA and S.M.A.R.T., Inc.

The NASM Solution Center more than met my computer needs, and my work was greatly facilitated during the pandemic lockdown by the Smithsonian's participation in the interlibrary loan system (ILLiad), where Wanda West was particularly helpful. The indispensable Astrophysics Data System digital library portal (https://ui.adsabs.harvard.edu/), created and managed by the Smithsonian Astrophysical Observatory and funded by NASA, made researching this book possible during the pandemic. These services kept me connected and sane.

And, of course, my deepest appreciation goes to Joyce Bedi, senior historian, emerita, at the Lemelson Center, National Museum of American History, Smithsonian Institution. She did a magnificent job managing the project, providing editing guidelines and advice, managing links with the publisher and Smithsonian lawyers, and gently but firmly advising me at all the latter stages of the process. She, working with Katie Helke and then Justin Kehoe of MIT Press, also engaged three reviewers, who were most helpful advising me of ways to improve and clarify the messages I was intending to convey. The first two were anonymous, but the third, Barrett Caldwell, corresponded openly and provided deeper insights into the many nuances appropriate for treatment of the subject. In the second round, the Lemelson Center engaged two Smithsonian scholars for comment, Arthur Daemmrich and Eric Hintz, and once again, their suggestions were most enlightening and positive. And in the final stages of editing, indexing, and proofing, I was rescued by Kate Bulson, Audrey McClellan, and Madhulika Jain.

## COLLECTIONS/ARCHIVES

William Baum Papers, Lowell Observatory Archives. WB/LOA.

Bernard F. Burke Papers, National Radio Astronomy Observatory/Associated Universities, Inc., Councils and Committees Series/Naval Studies Board Unit. BB/NRAO.

Benjamin F. Carruthers Papers, Schomburg Center for Research in Black Culture, The New York Public Library. MG433. BC/SA.

George Carruthers Collection, (Acc.2020–0024). Archives Department, National Air and Space Museum, Smithsonian Institution, Washington, DC. GC/NASM.

George Carruthers Papers, Moorland-Spingarn Research Center, Washington, DC. GC/MSRC.

Herbert Friedman Papers, American Philosophical Society, MS Coll 113. HF/APS.

Ladislas Goldstein Papers, University of Illinois Archives. LG/UI.

Karl Gordon Henize Papers, The Dolph Briscoe Center for American History, University of Texas. 2.325-L18. KH/BC.

NASA Records, National Archives and Records Administration at College Park, MD. (Archives II), Record Group 255. RG255/NARA.

John E. Naugle Papers collection, NASA Headquarters Historical Reference Collection. JN/NHO.

Thornton Page materials from both Virginia Tech and Wesleyan University. TP/VT, TP/WU.

Nancy Roman Papers, American Institute of Physics Center for History of Physics. NR/AIP.

Wernher Von Braun Papers, USSRC 807-7. Alabama Space and Rocket Center. WvB/USSRC.

## LIST OF ABBREVIATIONS

| | |
|---|---|
| **AAAS** | American Association for the Advancement of Science |
| **AAS** | American Astronomical Society |
| **AGU** | American Geophysical Union |
| **AIAA** | American Institute of Aeronautics and Astronautics |
| **AIP** | American Institute of Physics |
| **AMPS** | Atmospheric, Magnetospheric, and Plasmas-in-Space (proposed Spacelab payloads) |
| **AOSO** | Advanced Orbiting Solar Observatory |
| **APS** | American Philosophical Society |
| **ARGOS** | Advanced Research and Global Observation Satellite |
| **ASTP** | *Apollo-Soyuz* Test Project |
| **ASTRO** | Autonomous Space Transport Robotic Operations |
| **AURA** | Association of Universities for Research in Astronomy |
| **CCD** | charge-coupled device |
| **CRISTA** | Cryogenic Infrared Spectrometers and Telescopes for the Atmosphere |
| **EMR** | electromagnetic radiation payload |
| **EVA** | extravehicular activity (typically astronauts working in space outside of their spacecraft) |
| **FUVIS** | Far-Ultraviolet Imaging Spectrograph |
| **GSFC** | Goddard Space Flight Center |
| **GIMI** | Global Imaging Monitor of the Ionosphere |
| **HEAO** | High Energy Astrophysics Observatory |
| **HRSGS** | High Resolution Shuttle Glow Spectrograph |
| **HST** | Hubble Space Telescope |

| | |
|---|---|
| **IDEAS** | Initiative to Develop Education through Astronomy and Space Science |
| **ILLIO** | University of Illinois Yearbook |
| **IR** | Infrared |
| **IUE** | International Ultraviolet Explorer |
| **JILA** | Joint Institute for Laboratory Astrophysics |
| **JSC** | Lyndon B. Johnson Space Center (formerly the Manned Spacecraft Center) |
| **LASP** | Laboratory for Atmospheric and Space Physics |
| **LST** | Large Space Telescope |
| **LM** | lunar module, before 1967 called the lunar excursion module (LEM) |
| **MAHRS** | Middle Atmosphere High Resolution Spectrograph |
| **MAHRSI** | Middle Atmosphere High Resolution Spectrograph Investigation |
| **MSC** | Manned Spacecraft Center (later named the Johnson Spaceflight Center) |
| **MSFC** | George C. Marshall Space Flight Center |
| **NASA** | National Aeronautics and Space Administration |
| **NASM** | National Air and Space Museum |
| **NAVSTAR** | Navigation Signal Timing and Ranging |
| **NEOCAM** | Near Earth Object Chemical Analysis Mission |
| **NRC** | National Research Council |
| **NRL** | Naval Research Laboratory |
| **NSF** | National Science Foundation |
| **NTA** | National Technical Association |
| **OAO** | Orbiting Astronomical Observatory |
| **OGO** | Orbiting Geophysical Observatory |
| **OHI** | oral history interview |
| **ONR** | Office of Naval Research |
| **OSO** | Orbiting Solar Observatory |
| **OSSA** | Office of Space Science and Applications |
| **OWS** | Orbital Workshop |
| **S.M.A.R.T. Inc.** | Science, Mathematics, Aerospace, Research and Technology. Originally incorporated as "Project S.M.A.R.T., Inc." |
| **SEAP** | Science and Engineering Apprenticeship Program |

| | |
|---|---|
| **SEC** | secondary electron conduction |
| **SOLRAD** | SOLar RADiation Satellite |
| **SPARTAN** | Shuttle Pointed Autonomous Research Tool for Astronomy, also known as "Station power, articulation, thermal, and analysis." |
| **STEM** | science, technology, engineering, and mathematics |
| **STS** | Shuttle Transportation System |
| **UCLA** | University of California at Los Angeles |
| **USRA** | Universities Space Research Association |
| **UV** | ultraviolet |

# 1
# INTRODUCTION AND OVERVIEW

## ON THE MOON

Houston, the Manned Spacecraft Center, April 20, 1972. There was a problem. *Apollo 16* had launched from Cape Canaveral four days earlier and was now in lunar orbit. Apollo's lunar module (LM) and lander *Orion* had just undocked from the command module. But *Orion*'s propulsion system, critical for a safe landing on the Moon, was not behaving properly; two regulators had failed. Should astronauts John Young and Charles Duke still land? Or should they return to the command module, scrub the landing, and head home?[1]

It took some six hours to make the decision to land near Dolland crater in the Descartes Highlands of the Moon, and during that time, tension in mission control penetrated hearts, minds, and hands. So much was riding on this mission: prestige, the first attempt to land in the lunar highlands, the second lunar rover that would give the astronauts a much wider sampling territory, a myriad of lunar experiments, and the first astronomical observatory to be set up on the lunar surface to view the heavens and the Earth.

The six-hour delay affected not only landing protocols but also the mission's goals, especially the observatory. It was not Apollo's primary mission, as it did not relate directly to the Moon, but it meant everything to the men who proposed it, built it, and planned its mission. There was an immediate problem during the delay: the Moon rotates. That meant that in those six hours, the Sun and stars would rise higher in the lunar sky, and the objects in space they wanted to view, except the Earth, would have moved. So during those six hours, the astronomers madly recalculated where the astronauts would have to point the camera to capture the images of the ultraviolet (UV) universe they had long wanted to examine.

And there was another problem. The observatory, a little golden camera, had to work in the shadow of the LM to keep it in darkness and cool.

But with the Sun higher in the sky, that shadow would be smaller, and the portion of the sky blocked out by the LM would be greater. There was also a greater danger of contamination from being closer to the LM. The astronauts were carefully instructed on how to assemble the camera and to point it at the areas of the sky deemed most important. Would they now make it their top priority, rather than deploying the rover, collecting lunar samples, and setting out the other lunar surface instruments? Not likely.

At hour 106 into the mission since launch, about two hours after touchdown on April 21, 1972, Capsule Communicator Anthony England in Houston advised the astronauts that they were working up new target coordinates for the UV-sensitive camera. After a rest period, Young and Duke acknowledged they would have to put the camera closer to the LM. They deployed the rover about three hours later, and this made the camera accessible on its payload pallet. Young casually commented, "Okay, Houston. I'm about to deploy the old UV here." And characteristically, he quipped for his audience back on Earth, "Look at me carry it! I'm carrying it over my shoulder!!! Ha ha ha! I guess we don't have to worry about dust getting on it. Boy, only one-sixth g is the neatest environment you can find for this kind of work."[2]

In another hour, amidst many other tasks in the first sixteen hours after landing, Young finally set up the camera in the shadow of the LM less than ten feet from the ladder, pointed it almost straight up at the crescent Earth, and started the first of many preprogrammed exposures of selected objects and regions in the sky.

During the delay, Naval Research Laboratory (NRL) space astronomer George R. Carruthers—the man who designed and built the camera and proposed flying it to the Moon—was on hand in Houston, intensely recalculating the positions of the star fields and objects he wanted to observe. One can only imagine how Carruthers managed to keep calm and work effectively during this stressful time. But an eyewitness said that he certainly did. Harry Heckathorn, later to become an NRL colleague, first noticed Carruthers at an astronomical meeting in Chicago in 1967 because he stood out "given the dearth of black astronomers." Heckathorn met Carruthers again in the viewing room at Houston Mission Control during the tense hours of the *Apollo 16* landing. He walked up and excitedly introduced himself and started talking about electronic imaging cameras. But Carruthers was totally absorbed in the moment, "worried about his instrument on the Moon."[3]

Carruthers's deep concentration was understandable. He and his colleagues had worked for years preparing for this moment. They were far from confident that their proposals would be chosen, and then they hoped that the procedures they had developed with engineering staff at the National Aeronautics and Space Administration's (NASA) Manned Spacecraft Center (MSC) in Houston, Texas, would be executed faithfully by the astronauts. It would be a week before the Apollo astronauts would return the camera's film cassette safely to the Earth, and many weeks before anyone would know if the camera had worked to specification, that the right fields were properly exposed, and that the processing would yield useful information about the Earth's uppermost atmosphere and the nature of the UV universe. These were still the days of film recording for the most reliable scientific image data.

But Carruthers—calm, collected, focused, and unflappable—saw it all through. Even as his colleagues met with reporters after the flight to celebrate the achievement, showing off a copy of the camera that was still standing on the Moon, Carruthers stuck to his lab, planning his next projects.

## OVERVIEW

This biography explores how a man like George R. Carruthers, an outwardly unassertive but intensely focused African American, who was raised on a farm and then moved to Chicago in his teens, ended up building the first astronomical observatory that went to the Moon. As a result of this feat and his continuing successes placing his instruments on other space missions to observe everything from comets to galaxies and other celestial wonders, Carruthers became a public figure.

Despite his calm, reserved personality, in the years following *Apollo 16*, he responded positively to countless calls for his presence at conferences, in classrooms, and on local, national, and international publicized tours. His visibility was also a consequence of his passionate desire to inspire and engage children, especially minority children, to be interested in science, to appreciate that science was accessible, and to know that science could be a fruitful career choice, just as it was for him.

As Margaret Weitekamp and other scholars have noted, early histories of spaceflight and biographies of its pioneers did not address "questions of race and ethnicity."[4] More recent efforts have begun to correct that lapse, and there are now collective biographies and social studies in the context of the Civil Rights Movement, most notably *Hidden Figures*, an acclaimed collective biography that explored the lives and contributions of Black women, who were critical but overlooked workers at NASA.

However, Carruthers's employer, NRL, was quick to acclaim him as its "Not-So Hidden Figure," not only for his scientific and technical achievements but for his later efforts making his world accessible to young students.[5] It is in this context that I examine Carruthers's life and work as a scientist, engineer, inventor, and educator, drawing on historical studies of the African American experience.[6] My sense of affinity with Carruthers inspired me to take on this biography not because Carruthers was a Black man, but because he was an extraordinary scientist, inventor, and aerospace engineer who perfected a revolutionary new way to explore the universe.

This biography concentrates on Carruthers's scientific and technical accomplishments and the institutional environment that stimulated and nurtured his inventive genius. The main thrust of the book is the path he took in his life and how that path, especially to NRL, was fostered by teachers in Chicago and Urbana (at the University of Illinois) and then by socially conscious NRL staff like Herbert Friedman. Moreover, it covers in depth how, after he accomplished highly successful work in science, again facilitated by NRL, Carruthers became an icon for activists dedicated to increasing access to and interest in science and technology for minority youth.

To the extent that surviving documentation of his life allows, we explore Carruthers's personality, his view of the world, and his passion not only for science and space but also for making those realms appeal to, and be more accessible to, youth in the Black community. Carruthers did experience racial discrimination, and it will be noted to the extent that the historical record allows. There is no evidence, however, that it ever had an influence on his early life or career. The essence of the book, therefore, is a detailed overview of the science and technology he was a part of and the technical and political hurdles he and his NRL colleagues had to go through while working with NASA on human-tended missions.

I hope to help a broader audience better appreciate the life and work of George Carruthers—not just his successes but also the great challenges he faced as a space scientist who, among others, fervently competed for berths on rockets, satellites, and space probes. To do this, I examine the challenges he had when working with NASA's most visible scientific project in the 1960s, the Apollo program. This helps foster appreciation for the challenges that Carruthers faced, as well as others at the time who were wanting to send their instruments into space.

Among the many challenges in crafting his biography is that, due to Carruthers's reserved nature, his voice is heard here less than the author wished. Even in the several interviews he gave to historians and filmmakers, he maintained his distance in sharing personal feelings about critical moments in his life. Of course, this could also have been due to the interviewers' (including myself) lack of knowledge on what to ask or a lack of persistence seeking tidbits in the style of a *60 Minutes* interviewer. I also never felt as though I was intruding on his privacy; he was aware of my biographical ambitions, and he approved.

Another complexity is that Carruthers's projects, though focused, were not single threaded. He proposed many different missions in parallel, and the gestation periods for them, mainly those to NASA, lasted from months to years, and many overlapped. The decisions depended upon many technical, economic, and political factors. Many of his proposals were complementary, and as each of them progressed, the others could be affected. Thus, the presentation of his work here is episodic and not strictly chronological.

As an NRL scientist, his projects were directed to both military and civilian ends. This book shows how the projects were related, and it illustrates the broad technological capabilities of his instruments. Thus, while his professional career path was linear at NRL, the projects he engaged in were multifaceted, their shapes evolved over time, and many were executed in parallel. He was constantly multitasking toward a single goal, and the complexity of his life grew as his devotion to his first wife, his research, and then his student outreach efforts gained momentum later in his career.

Although he was a loner, Carruthers did not work alone at NRL. At various points in the biography, we provide background on the scientific and technical infrastructure that stimulated Carruthers's work and also made it possible. To do so, we introduce the mentors and team members in Carruthers's personal and professional life, especially members of his family, his NRL colleagues, and his students. Building on that basis, we describe the institutions that were critical in his life to better appreciate the worlds he lived in, benefited from, and gave back to. This is especially important because, as more than one observer has stated, science is a community: "Discovery is buoyed by associations of peers; research is supported by materials and equipment."[7] In their anthropological exploration *Laboratory Life: The Construction of Scientific Facts*, Bruno Latour and Steve Woolgar argue that it is hopeless to try to separate out individuals from teams and communities. They emphasize that "each of our informants was part of a laboratory."[8] This was certainly the case for Carruthers.

Thus, my intention here is to explore the worlds Carruthers became part of and was defined by. These include his childhood on a farm near Milford, Ohio, high school years in Chicago, his college years in Champaign-Urbana, his extracurricular interests in rocketry and telescope-building in Chicago, his graduate school training in a highly productive plasma physics laboratory at the University of Illinois, and his move to NRL, first as a postdoctoral scholar and then as a full-time staff member, where he remained for the rest of his professional life.

Carruthers's career at NRL helps one appreciate what it means to be a space scientist. In an earlier history, I explored this question generally for various institutions like NRL and the Applied Physics Laboratory. There I explored, in the words of a reviewer, "the complex organizational structure" of these early centers for space science, including the "technical and managerial problems that each group of researchers faced in using rockets as instrument carriers."[9] Carruthers worked within this structure, but unlike his predecessors, he had to secure outside grants for his work and so was subject to the strictures of the funding sources, like NASA. In the first years, a series of National Science Foundation (NSF) grants provided the means for his instrument development efforts: successfully creating a fast and efficient electronic imaging camera that could operate autonomously, survive the ferocity of a sounding rocket launch, and somehow provide a stable programmable platform for observations of celestial objects.[10] But for most of his career, although he had access to support from the military, he was subject to gaining NASA's favor—a hugely competitive challenge.

Sounding rockets were relativity cheap by military standards, and NRL scientists commonly used them through the 1950s and early 1960s even though throughout much of the earlier years, observing the heavens and the Earth from a spinning and tumbling rocket was anything but straightforward. They were the first vehicles Carruthers had access to, and his early flights were well supported. By the late 1960s, however, the stakes and complexities grew exponentially when Carruthers and his colleagues began proposing to send his instruments on NASA orbital missions during later phases of Apollo and manned spaceflight after the first Apollo landing, called "Post-Apollo" (we will refer to Post-Apollo flights and programs as "Apollo") programs.[11] He and his NRL supervisors faced daunting challenges interfacing NRL expertise with NASA bureaucracy for high-budget projects, and questions arose regarding whether Carruthers would be capable of managing these administratively heavy tasks. As a result, Carruthers was teamed up with others who were more adept at political maneuvering and multi-institutional coordination. Given their critical roles fostering and assisting in Carruthers's work, we frequently digress to identify these players and their backgrounds, and how they both inspired and complemented his interests.

There were many twists and turns along the road, but with the aid of his NRL colleagues, Carruthers never deflected from his primary passion for instrument design, construction, testing, and execution. The camera that Carruthers designed, built, and tested, for which he applied for a patent in 1966 (issued in 1969), combined all the properties necessary for an Apollo experiment.[12] It was an electronically amplified photographic camera that was small, lightweight, powerful, and, most of all, could be operated by a human in a clumsy spacesuit. We examine the development of the technology that led to his camera in detail.

Decades before the advent of powerful solid-state sensors common today, photochemical photography was the main means of recording images. But photography was highly inefficient and required calibration to produce useful scientific data. To view faint objects in the heavens, you needed to collect more light, and that needed a big telescope. But big telescopes would not fit on small sounding rockets, satellites, or even Apollo. So the most competitive solution was to find a way to amplify the incoming light signal so that scientifically useful photographic recording was possible. That is what Carruthers did.

Carruthers knew well that two-dimensional recording was best done with some form of photographic process. In 1967, a NASA Office of Space Science study pertaining to the need for a manned optical telescope, and the technical challenges, concluded that photography was "still an unrivalled data storage medium at the present time and no purely electronic imaging detector currently available compares in resolution and data capacity." The study recommended that "photosensitive solid-state devices of all types should be studied to find the highest quantum efficiency. Development of electronic imaging devices must be accelerated since the high quantum efficiency of the photocathodes promises to make orbiting telescopes even more efficient."[13]

Carruthers thus combined the best of both worlds: electronic amplification and film recording, which at the time meant vacuum tube electronic amplification. He did not invent the concept, but he improved it and made it capable of exploring the UV universe, that part of the optical spectrum not visible from the Earth's surface because it is blocked by the atmosphere. His design proved to be highly efficient, rugged, and reliable. Throughout the 1970s and early 1980s, Carruthers's designs were competitive, giving them the opportunity to fly on many space missions that were human operated since his photographic film required physical retrieval. But over the same time period, the solid-state revolution had produced lightweight and rugged digital electronic image sensors: charge-coupled devices (CCDs) that could store data onboard and then relay them to Earth efficiently and reliably were promising.[14] Even though at first Carruthers's detectors had wider fields, more UV sensitivity, and spatial resolution than early CCDs, he knew that they would soon compete. Accordingly, he adapted his later designs to incorporate solid-state detectors to achieve more powerful and useful ends, and electronic retrievability. But it was evident by the 1990s that the solid-state revolution had replaced the vacuum technologies in which Carruthers had excelled.[15] As the astronomer Joseph S. Miller observed in 1989 when reviewing the status of astronomical instrumentation, although they were still difficult to procure, "CCDs have become the detectors of choice for much optical observing."[16] And so Carruthers's success rate diminished, and following how he reacted to it provides an intimate portrait of the impact of technological change on an astronomical career.[17]

Carruthers never lost hope. He continued to experiment, refine, and propose. But as his competitiveness waned, he increased his efforts to reach out beyond his laboratory to inspire young minds to get involved in his never-ending quest to create new tools to explore the universe. His great notoriety for being the man who sent the first astronomical telescope to the Moon made him attractive to the ardently dedicated groups that were debating how to bring science, technology, and engineering to people of color. He became a symbol and conduit for their efforts, helping them move from debating how to do it, to doing it.[18]

We interrogated Carruthers about his own perspective on the matter of race. As he claimed, issues of race were, for him, invisible. By that, he meant that all races are equally capable of contributing to science. There is no question that discrimination based upon race was, and remains, real in science as well as in society as a whole and that Carruthers was aware of it. I hope to show that as a visible Black man in science, Carruthers did all he could to erase that distinction by actively engaging in a wide range of minority mentoring programs that provided students the opportunity to personally engage in laboratory practice, thereby convincing them that a life in science was a worthwhile and accessible goal.

This then, in a nutshell, introduces George R. Carruthers. I hope that his biography will stand as an example of what it takes to live a competitive scientific life, specifically in the design and use of new tools in pursuit of new knowledge. The full story reveals far more about the trials and travails that people of all colors and genders faced when meeting the challenges of space research, or indeed of any world where highly competitive training is required for entrée and success. It also offers readers insight into Black physicist and social organizer Hattie Carwell's classic statement in her reconnaissance of *Blacks in Science*: "To be great, an individual must rise above the petty inconveniences of surroundings and circumstances to meet the challenge of competition."[19]

The "inconveniences of surroundings and circumstances" that Carruthers faced, sending his experiments into space and especially to the Moon, were, as we will see, anything but "petty." The challenges he faced—technical, institutional, and political—were immense, and yet he overcame them all as he became the first person to send an astronomical camera, telescope, and observatory from his laboratory to the surface of the Moon.[20]

# 2

# FAMILY HISTORY AND EARLY LIFE

George Robert Carruthers was born at the Catherine Booth Hospital in Cincinnati, Ohio, on October 1, 1939, and was the first child of George Archer Carruthers and Sophia Singley Carruthers. His father, born in 1914 in Chicago, was a graduate of the University of Illinois with a degree in civil engineering. In 1936, he met and married Sophia Singley, who was also college educated and active in the Chicago-area Caribbean social circles. She was born in St. Louis, Missouri, in 1916 to Samuel Robert Singley, a mail clerk for a railroad, and Pearl Edwards. After her family moved to Chicago, she attended Englewood High School on the south side and then the Chicago Teachers College, "the first university in Chicago to unconditionally accept African American students."[1] Soon after their marriage, George Archer was hired by the US Army Corps of Engineers, so they moved to Cincinnati for his work on various Ohio River projects. And soon after that, he was reassigned to projects at the Wright Patterson Air Corps Base in Dayton and commuted from Cincinnati.

At the time of George's birth, since less than 2 percent of Black men and women had completed college in the Chicago area—compared to some 6 percent of the white population—one can surmise that George's immediate family was in the socially active professional middle class. They were not part of the some 40 percent of Black families in Chicago on relief at the end of the Depression.[2] And as we will note soon, earlier generations of the family were socially active in St. Louis and Chicago and were active in academic, fraternal, and religious circles. Most of note, as will be explored below, his uncle Benjamin Frederic Carruthers Jr. was already a professor at Howard University.[3]

George Archer's family had lived for several generations in various parts of central Tennessee, near Lebanon in Wilson County. In the nineteenth century, they spelled their name "Caruthers." Great-grandfather John Francis Caruthers was listed in the 1880 US census as a farmer and, with his wife Susannah Combs, had seven children.[4]

In a brief draft family history, George Archer's older brother Benjamin Frederic described the family as being of "Scotch-Irish-Negro" descent, and the mother's side as French-Creole. As was the custom of the day, the 1880 census listed the family as "Mulatto," or of mixed blood, a typical inference made by a census worker making a visual inspection. It was a common description for many Black families in Tennessee, both free and slave, who were the result of "the amalgamation of African and Scotch-Irish race stocks."[5] Although far from consistent in different regions and times, it carried a mark of social status.[6]

After John Francis Caruthers died, his family evidently faced difficult economic times in the 1880s. Some of John Francis and Susannah's children, including one of their younger sons, Benjamin Jackson Caruthers (born in 1867 in Lebanon), left the family with an older sister, hoping for a better life. They ended up in St. Louis, where the older sister worked for a "kind and generous" family of means. She soon died, however, and so the family adopted Ben and gave him chores. As he grew to adulthood and married, he became their salaried chauffeur and "major-domo." In 1910, recently widowered and serving as usher at the Central Baptist Church, he met "Dollie" Francis Archer, a schoolteacher and a bright attractive woman "having taught 'the grades' in some of the toughest Negro schools in St. Louis." Dollie, her son Benjamin Frederic recalls, was light skinned and regarded as "high yaller" by "her social inferiors" in the parlance of the day. She was a teacher and tutor, was active in social circles, and dutifully supported her husband in the Black faction of the fraternal "Knights of Pythias" of Missouri, where he had risen to become Grand Chancellor.[7]

Their oldest son, Benjamin Frederic, was born in September 1911 in his grandmother's home. His father had changed jobs for more income, becoming a chef on a private railroad car that ran between St. Louis and Toledo, and sometimes to Chicago. He moved the family to Frankfort, Indiana, close to one of the train stops and then in early 1914, at Dollie's insistence, to Chicago where her family had migrated and prospered. Dollie Francis's family lived in and around the Englewood district in Cook County Ward 6, on the south side, where an already large Black population was concentrated. As Alvin Winder notes, "this concentration was enforced through a policy of containment supported by restrictive covenants and discrimination through social pressure."[8]

The move to Chicago—or moreover, from Tennessee, to Missouri, and then to Chicago—personifies what has been called "The Great Migration" of southern Black people northward, a movement that began during Reconstruction and continued through the first half of the twentieth century. Benjamin F. Carruthers's family history romantically described the migration as follows: "Chicago was the city of promise and hope. El Dorado to which all the refugees from all the world came . . . within a few years its Colored population, swollen by the advent of thousands from the Southland seeking freedom and wartime employment was to double and triple."[9]

The Great Migration of Black people from the southern to northern states—seeking acceptance, greater opportunity, and, most of all, relief from the rampant repression and violence of the Jim Crow South—has been chronicled by writers and historians. Isabell Wilkerson notes generally that, at the time, there were labor shortages in the North: "The North needed workers and the [Southern] workers needed an escape." She quotes the *Chicago Defender*: "Treatment [in the South] does not warrant staying."[10] Even though it was far from free from the ills of racism and segregation, Chicago was one of the more popular destinations, possibly because the *Chicago Defender* newspaper had wide circulation in the deep South. It has been cited as a stimulus for the Great Migration.[11] But the dream of "promise and hope" turned out to be very elusive.

We know little of George Archer's early life. He was born in 1914 in Chicago and was attended to by Dollie Francis's "nearest and dearest of kin" in the family home. But there were moments: His brother Benjamin poignantly recalls that when they were living with an aunt, "baby brother George wandered out the back way one hot summer evening" and got lost. He was "picked up by a childless love-starved woman" who kept him at her apartment. "Our family was frantic," scouring the neighborhood and contacting the police. But the woman's neighbor saw the woman and child in a window and, knowing that she had no children, called the police. "Little George, hardly a tot, [was] dirty but happy and unharmed. His dimples and charm had won over all the tough policemen."[12] Evidently, the community was alert and mutually protective.

Due to the competition of the continuing influx of migrants and restrictive policies in Chicago, however, it took Benjamin Jackson some time to find stable work in late 1914. He was first a porter and then a "janitorial" utility man, which then meant taking care of utilities in homes, office buildings, and apartments. But these were unsteady jobs. He later became a "shipping clerk and general handyman with a large wallpaper and window shade manufacturer on the West Side" and held that job for years.[13] His job paid enough to allow the family to move to a five-room apartment on Wabash Avenue in Englewood, where the family supplemented their income by taking in boarders. As with his social activism in St. Louis, in Chicago, he eventually became president of the Woodlawn Citizen's Improvement Association.[14]

Benjamin Frederick entered public school, but soon after, his mother homeschooled him; she feared for his safety in the aftermath of the 1919 Chicago riots and because of the racial pressures caused by the continued influx of Black migrants and "newly arriving immigrants from central and eastern Europe."[15] The overcrowded schools, far from immune from the tensions of the racial and foreign influx, were hotbeds for conflict. In the following years, not surprisingly, Dollie Francis also homeschooled George Archer.

When they got older and more resilient, however, and tensions relaxed, she allowed them both to attend public school. Both brothers excelled in school, likely because of their mother's strong guidance, and both attended college. As did his brother Benjamin, George Archer attended the University of Illinois Urbana-Champaign and received his bachelor's degree in engineering in 1936. At the time, Blacks were less than 1 percent of the student population. They were not allowed to live on campus, and they were denied access to local Urbana-Champaign businesses.[16] George Archer faced this challenge by joining a small activist community of Black students and supporters: he became a member of Kappa Alpha Psi, one of the earliest Black service and social fraternities promoting Black determination.[17] And in the same year, as noted above, he married Sophia Singley. She was then active in Black social circles in Chicago, and after graduation, she became a teacher.[18]

By the mid-to-late 1930s, George Archer's older brother, Benjamin Frederic, who, as we noted, wrote the family history we have been using here, was becoming an intellectual of some prominence. He graduated in education from the University of Wisconsin in 1932 and received a master of arts degree the next year from the University of Illinois in modern and romance languages. He married in 1935. When he announced his intention to marry, his father eloquently responded, "Yours to hand and read with much pleasure. I am only dropping you a line as mother will write you in full . . . We will love our daughter as we love you and George. . . . . I am ever yrs, *Dad.*"[19]

As noted, by the time George Robert was born in 1939, Benjamin Frederic was a professor at Howard University in Washington, DC, continuing his graduate work toward a PhD in 1941. He soon moved to New York City and worked for numerous associations such as the Pan American Union and the International Broadcast Division of the US State Department during the war, becoming active at the United Nations and UNESCO after the war's end. He also became a prolific travel writer, touring the world and reporting on his experiences. He knew women's rights activist Dorothy Height and is noted for collaborating with poet Langston Hughes in the 1940s in translating Nicolás Guillén's *Cuba Libre*. His attainments once again demonstrate that the Carruthers family was academically and professionally strong, and secure enough to seek out a new life.

## LIFE ON A FARM IN HAPPY HOLLOW

George Robert, his two younger brothers, and his sister were each born about three years apart at the Salvation Army's Catherine Booth Hospital in Cincinnati. Anthony Carruthers (born October 8, 1942), after an Army career in nuclear weapons work, became a machinist in Chicago. The youngest son, Gerald Carruthers (born February 5, 1945), like George, was fascinated by science, technology, and rockets and also became a career serviceman and engineer in the Army. The youngest, Barbara Ann Carruthers (born September 22, 1947), eventually worked for the Chicago Postal Service, following her mother's path.

When George was about five or six years old, the family moved from a northeast suburb of Cincinnati called Madisonville to a small farm near Milford, Ohio, some ten miles northeast of Cincinnati and a bit closer to Dayton, some thirty miles distant, where his father was then assigned. It had been a working farm and was, for his father, a chance to be away from the city and live his childhood dream, finding it on "Happy Hollow Road" about a mile east of the Milford town center.[20] (See figures 2.1, 2.2, and 2.3.)

When his father and uncle Ben were children, they had spent summers on an uncle's farm, and this became George Archer's passion. Farming had

Figure 2.1
The restored family farm outbuildings in Milford, as seen from the roof of their home at the top of a hill in the late 1940s. *Source:* Courtesy of Gerald Carruthers.

been in their family, and he may well have been driven "by a nostalgic aftertaste of the old ways" as historian Pete Daniel has observed existed among former tenant families.[21] Living on a small family farm in Milford, but otherwise middle-class, the Carruthers children enjoyed working in the vegetable garden, and their parents managed hogs, a cow, and chickens—all for food and fun. Their father worked in Dayton during the week and came home on Fridays, spending his weekends "constantly working" on improving the ten-acre farm.[22] As Gerald Carruthers remembered, at first there was just an old log cabin, two small outbuildings, and a barn, and they needed a lot of work. The farm was barely livable and required much rebuilding and repair, which George Archer relished. "The first thing he had to do [was] build a bridge to get on the property [and] up to the cabin. He had to do a restoration/renovation to the cabin so we would have some place to live. The way our father worked, like all the time!! But [he] showed us kids how to do things [and] showed us how to live our lives!! George got the most and best. . . . The rest of us were just kids, and too young."[23]

Figure 2.2
(l to r) Gerald, George, and Anthony Carruthers. *Source:* Courtesy of Gerald Carruthers

In their first years there, the family moved back and forth between Madisonville and Milford until George Archer (see figure 2.4) got the place into shape, making an old log cabin livable by the late 1940s. Soon after, there was a main clapboard house, the large barn was restored, the outbuildings became chicken coops, and other small buildings on the property were put into usable shape. Gerald relished his life on the farm and felt especially that he and George absorbed their father's strong work ethic.

The kids played together. Gerald recalls playing checkers, working with an Erector set, and enjoying other creative pastimes. More practically, they all enjoyed slopping the hogs and climbing trees to gather fruit. Their mother (see figure 2.5) kept the chickens to herself. As Gerald poignantly

Figure 2.3
Barbara Ann, 1949. *Source:* Courtesy of Gerald Carruthers.

Figure 2.4
George Archer Carruthers (circa 1950). *Source:* Courtesy of Gerald Carruthers.

Figure 2.5
George and his mother (circa 1948). *Source:* Courtesy of Gerald Carruthers.

observed, she was "a very smart lady and adapted to farm life easily. She ruled the house and farm and us children. She insured that we got our schoolwork done and correctly. She also insured we got to the library, and got our chores done! She too was always showing us how to live on the farm and working like little farm hands!"[24]

In addition to the normal farm-type chores around the house and fields, George relished learning how to fix things as he helped his father around the farm. He particularly recalls how his father was a role model. George was particularly fascinated with his father's training and work: "My father's background was in civil engineering and general engineering, so to a large extent he served as a role model in terms of giving me advice on studying math and science, although the particular area of astronomy and space flight was not really his area."[25] And as George recalled in a 1960 Career Conference speech, he first became exposed to engineering "when [his father] came home from work, he would tell me about all the fascinating things at the base such as high-speed airplanes, as well as things he saw at other bases, such as Edwards Air Force Base when he went on trips for the government."[26]

He also enjoyed reading his father's books, particularly a section on astronomy in the *Book of Knowledge*: "My father explained to me the things which I did not understand, which increased my interest further."[27] There was a school library in local Mulberry, and what little he found he devoured. Unlike his parents, who encouraged his interests in astronomy and space, George keenly recalled, "Needless to say, most of the teachers and students in school were not supportive of me in that type of thing."[28]

Gerald remembered that George did not play much with his siblings and that he was single threaded: "George was a dedicated person. He was focused on one thing, and that's the way he is all the way through life."[29] He apparently had no discernable hobbies, he did not play in sport games, especially with his brothers, and he rarely socialized—again, according to Gerald.

George's grades in elementary school, at Miami Rural in Milford some four miles away, were excellent across the board. In sixth grade, he excelled in spelling, English, science, and geography. He was less successful in arithmetic, scoring 77 out of 100. And he rated B- to C- in "attitude," which most likely meant he resisted direction.[30]

Stimulated by his father, George sought out books and magazines in the library about building flying things in the air and space. There was a comics series he especially liked called *Wings*. It featured atomic spaceships that could fly to the Moon and other planets, and even underwater. George also devoured *Buck Rogers* comics to expand his dreams of space travel and even took his turn writing space travel stories and illustrating them with cartoons of his atomic rocket designs. He boldly sent one of them, his "Design for an Atomic Moon Rocket," to the Atomic Energy Commission and got a blunt reaction: "It'll never get off the ground." However, as encouragement, they sent him brochures.[31]

Other stories he wrote were titled "Jack and Joe Go to The Moon" and "Marty Mars Visits the Earth." They were more cartoons than writing and, as he thought back on them, "quite corny." But it was his passion night

Figure 2.6
George looking through a soda bottle as if it were a telescope. *Source:* Courtesy of Gerald Carruthers.

and day, and even in class, to the point where his parents got critical letters from teachers that he was drawing spaceships in class rather than doing his schoolwork.[32]

George could be playful with his passions. He dreamed of looking at the heavens with a telescope. Before he built one, he mimicked the act by peering at the sky while looking through the spout of an empty soda bottle (see figure 2.6). When he was about twelve years old, George was reading magazines like *Popular Science* and *Popular Mechanics* and came upon an Edmund's Salvage advertisement for a small telescope kit containing an objective lens and an eyepiece, with instructions on how to make a telescope with them. His father helped him buy the kit by supplementing the money he earned from a paper route. Still, with his father's help, he had to scrounge for odds and ends around the farm to mount the lens in a tube, adding a smaller sliding tube to hold and focus the eyepiece. With it, he happily scanned the landscape and the night sky: "I went out and looked at the stars, the Moon, and the planets and that sort of thing. Of course, it was a crude telescope. I didn't have a mounting or anything like that, so it was not anything anybody would be interested in these days."[33]

But it was a start. His father supported George's passion for building things and wanted him to be interested in science and technology, especially aeronautics. But George's comic book readings, especially the space-themed ones, kept him glued to the heavens. Gerald also warmly remembered George as a budding artist, drawing cartoons, especially "drawing spaceships going to and landing on the Moon, and drawing space Christmas cards!"[34]

The glue became stronger when *Collier's* magazine issued its hugely imaginative and persuasive series of articles starting in March 1952 and lasting through April 1954 under the affirmative heading "Man Will Conquer Space Soon!" Gripping articles by space pioneers like Wernher von Braun, father of the German V-2 missile; Willy Ley, German-born expatriate popularizer of rocketry and space travel; the Irish journalist Cornelius Ryan; and the Harvard astronomer Fred Whipple made it all seem so real. And as the editor's commentary—titled "What Are We Waiting For?"—made clear, "What you will read here is not science fiction. It is serious fact." It was both a "scientific symposium" and an "urgent warning" that the West must attain "space superiority" over the Eastern bloc, because "the first nation to do this will control the Earth."[35] Von Braun wrote the first article, "Crossing the Last Frontier," and then Whipple teamed up with the noted UCLA geophysicist Joseph Kaplan to speculate on what studying worlds in space might be like, including the possibility of finding life on Mars. Carruthers especially remembered Whipple's description of what it would be like doing astronomy from space platforms. The illustrations by artists such as Chesley Bonestell must have been spellbinding for the young man. As space historian Roger Launius has noted, von Braun "was able to take his concepts and sell them, both to people who could help him [and] to the public." The series of articles in *Collier's* magazine "made him a household name.[36] He certainly sold it to George. As Carruthers recalled, "All of that came out in the early 1950s, and that sort of shifted my interest once again to the space flight area, because up to that point I had only science fiction to go on. There was no factual information on space flight at all up to that point."[37]

Over the next two years, George, now in Chicago, devoured the rest of the series and became so enamored that he wrote to von Braun stating his interest in "space and all that sort of thing, and he actually sent me an autographed photograph, which I had totally unexpected."[38]

## MOVE TO CHICAGO

By the time he wrote to von Braun, George's mother had moved the family to Chicago. Thirteen-year-old George's quiet rural life changed drastically in September 1952 when his father died suddenly, at only thirty-eight years old. Gerald, only six years old at the time, sadly recalls seeing the ambulance race away from home as he came back from school. His father was dead by the time they reached the hospital and was later diagnosed with a cerebral hemorrhage. Chicago family attended the funeral, and evidently there was sad but hopeful discussion of family options for the future.[39] Distraught and in need of support, after months of trying to keep their lives together in Milford, his mother decided to sell the farm and take her children back to Chicago to live with her mother, Pearl Singley, and her aunt Lucille Dunlap, a teacher. Gerald recalls that he did not want to move; he loved the farm but recalls that George was silent on the matter.[40]

Living on the Black south side of Chicago, at first George and his brothers attended McCosh Grammar School. Barbara Ann remained at home with her grandaunt and grandmother while their mother sought work. The school was about ten blocks away in the Woodlawn area. It was the school Emmett Till attended. Till was then in the seventh grade—a year behind George. By the time Till was tragically murdered while visiting family in Mississippi in 1955, George had already graduated in 1953 and entered Englewood High School, his mother's alma mater. Gerald and Anthony were still there. Today, it is still a heartbreaking reminder of the racist horrors of the day.[41]

In 1955, after working part-time for the stockbroker Hornblower & Weeks, Sophia secured a permanent job as a postal clerk for the Chicago postal service. With her income and savings from selling the Milford farm, and in the spirit of "striving" exemplified in the 1959 landmark play *Raisin in the Sun*, she was able to move her family to a modest home at 7627 South Champlain Avenue—not a Black neighborhood, but closer to her job.[42] George stayed with his aunt and grandmother, however, until he finished Englewood High School because they lived close to his school.

Englewood High School at the time had a growing majority Black population, facilitated by the continuing Great Migration from the South and enriched at the time by a blossoming of Black artists and writers in the city.[43] Its alumni included at least one person who would become known for his prowess in aerospace science and technology: Robert Henry Lawrence (a 1952 graduate). After obtaining a doctorate in physical chemistry and a commission in the US Air Force as a pilot, Lawrence became the first Black astronaut in the Air Force's planned Manned Orbiting Laboratory program. He tragically died in a plane crash in 1967.[44]

George was fortunate to have several science teachers who guided him through laboratory experiments and then encouraged him to enter local science fairs. George became so active in the school's laboratories and could be so intense at times that his teachers were worried that he would detonate something. They tried to discourage him from experimenting too freely in the chemistry laboratory, but he was persistent. Although his general grades were fair, he excelled in science and laboratory work, going quite beyond what the curriculum called for. Accordingly, his teachers urged him to prepare more presentations for more and higher-level science fairs. At one fair, he described a "nuclear powered airplane."[45] At another, he presented a reflecting telescope that he had recently built at the Adler Planetarium. And at a third, he demonstrated different methods to generate and transmit electricity. He won some prizes for these efforts.

George built his reflecting telescope with a mirror he had made at the Adler Planetarium about six miles north of his home on the shores of Lake Michigan. The Adler was the first large planetarium in the United States, featuring a Zeiss Model II optical star projector, the most sophisticated instrument of its type at the time. From its earliest years, the Adler housed an optical shop in its basement and by the late 1950s had developed programs providing access to telescope-making materials and techniques for Chicago's youth.[46]

Adler programs were open to everyone, and it was relatively easy for George to reach the Adler. His aunt took him there first to be sure he would not get lost.[47] There was an elevated train—the 63rd Street "L"—about two miles north of his home that took him straight to downtown Chicago. George wasted no time getting active in Adler's astronomy club and telescope-making class. He also attended informal events and met other dedicated and capable students in the Adler classes.[48]

George relished contact with the astronomers at the Adler. "Certainly, the planetarium was an [order of] magnitude jump over anything that I had been exposed to in Milford, so that caused my interest to go up another order of magnitude, you might say."[49] But it was the telescope-making class that was the real draw. There, as a brochure later attested, "a person with average mechanical ability can, for relatively small investment, make a telescope as good as many expensive commercial instruments."[50] In the mid-1950s, mirror-making kits were generally available through sources advertising in *Sky and Telescope*, but the Adler also supplied the kits as part of the registration cost. George ground and polished a 4¼-inch mirror for a Newtonian reflector, where light is collected and concentrated by a concave mirror onto a small flat mirror that reflects it to the side of the tube to make it accessible to the observer. George's mirror, made of Pyrex, was the smallest commonly available. Commercial kits, including a mirror blank, grinding tool, and a range of grits for the grinding process, cost about six dollars in the mid-1950s, not a small sum for the day.[51] Other students were making larger mirrors, but they cost more than what George's family could manage. The class covered just the optics and left the mechanical construction to the students. George hammered together a simple square wooden tube to hold his optics and made a simple mounting out of ordinary plumbing parts.

By then, George was reading *Sky and Telescope* magazine and telescope-making books. The Adler library was a perfect place to bury himself in such joys, and he remembered especially reading Milton Rosen's *The Viking Rocket Story*, published in 1955.[52] Carruthers readily admitted, "I was really interested in space from the engineering point of view,"[53] and it was Rosen's book that converted the dreams of the *Collier's* series into reality. Viking was the Navy's first successful medium range missile, built originally as a tactical weapon to match and surpass the capabilities of the V-2 missile after World War II. Rosen, who led the project at NRL, made it clear that it could also be America's first step into space. By the mid-1950s, it had become the first stage of what the Navy called "Project Vanguard," America's bid to launch the first world-circling spaceship.[54]

By his senior year of high school, George had gained notoriety for his projects and prowess. Encouraged by his teachers, he had entered the sixteenth annual science talent search for a Westinghouse science scholarship, organized by the Science Clubs of America. In February 1957, competing against some twenty thousand students, he was given an honorable mention and was personally congratulated by Watson Davis, the lead organizer and head of the widely respected distributor and popularizer of science news content, *Science Service*. In May, he was designated a "Star Senior" by the Englewood faculty based on his impressive achievements, which included his induction into the National Honor Society and his involvement in the school's science and mathematics clubs.[55] Paul J. Copeland, professor of physics at the Illinois Institute of Technology, searched him out to congratulate him for placing fifth in a competitive examination in physics among over two hundred students from thirty high schools: "This is such a good showing that I think you should make a real effort to continue with your education beyond high school."[56] And finally, his continued efforts in sketching and art won him a summer scholarship to study at the Art Institute of Chicago.[57]

George's focus during his high school years remained his space-related dreams, centered around libraries, school, and the Adler. He relished his laboratory classes and showed a certain playful fascination with volatiles—an early insight into his adolescent sense of humor. Evidently bored during the graduation ceremony at Englewood, he scribbled out his own variation of the school's theme song, trying to produce an explosive reaction (see figure 2.7). Even though he did not get the equation quite right, combinations of phosphine and sodium bicarbonate could produce a stinky gas.[58]

Despite his adolescent humor, his teachers at Englewood and the Adler astronomers strongly encouraged his efforts and industry, but whenever "I talked about space flight, they told me that was nonsense, because that was before any space flight mission had ever taken place."[59] Their attitude would soon change.

Carruthers recalled little "overt discrimination" or physical taunting in the Milford community—and certainly not from the teachers. Some of the kids "would give me a hard time. . . . Every once in a while they would make racial comments and things like that, and pick fights and things like that, but there wasn't a whole lot of that that I recall, at least not specifically racial in nature."[60] There was more discrimination in Chicago, but George remained relatively immune because its presence was less in his grandmother's neighborhood, where Black families dominated. But when his mother moved the family just a few blocks farther south, where they were one of the first of the few Black families to move in, they were definitely

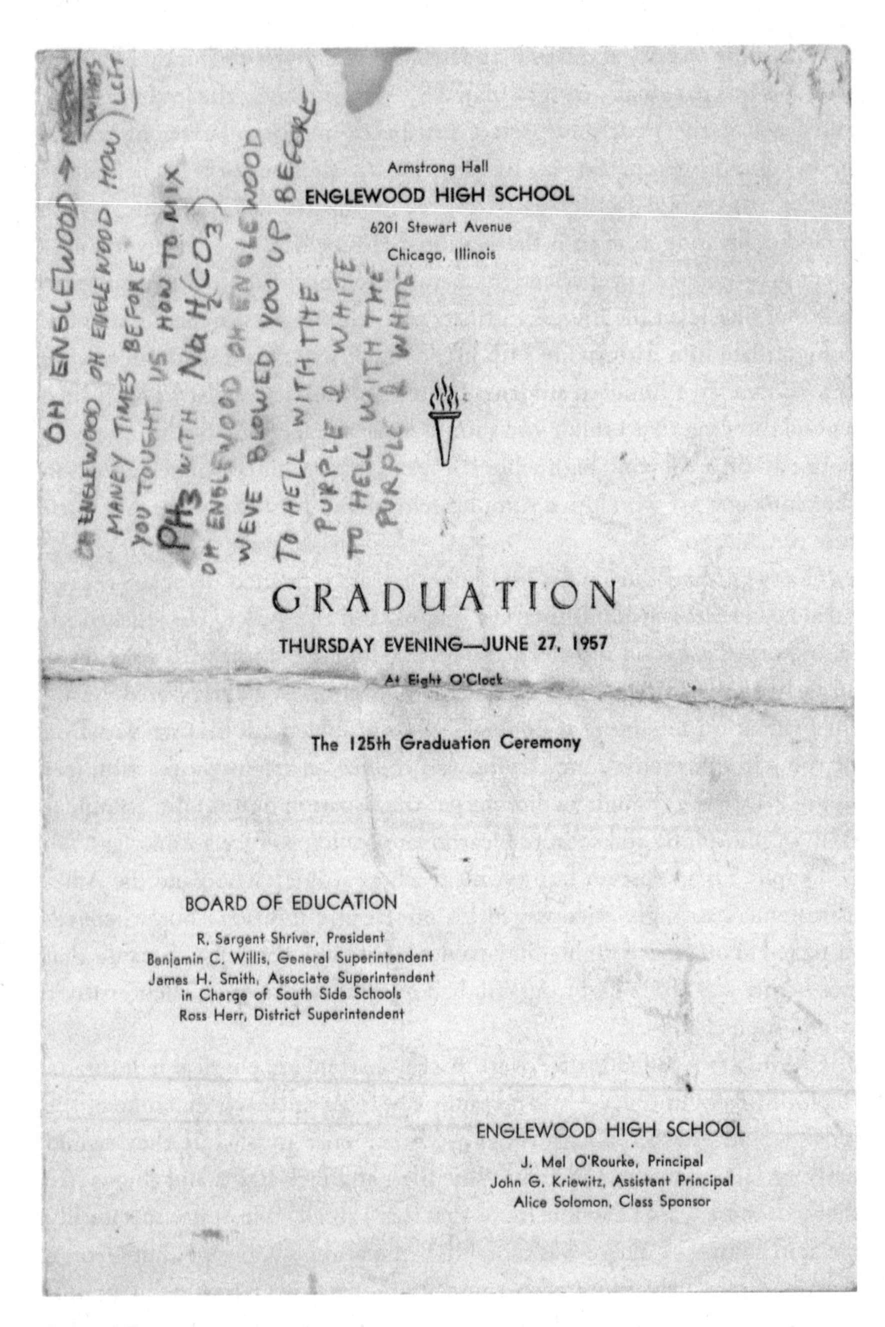

Armstrong Hall

ENGLEWOOD HIGH SCHOOL

6201 Stewart Avenue

Chicago, Illinois

GRADUATION

THURSDAY EVENING—JUNE 27, 1957

At Eight O'Clock

The 125th Graduation Ceremony

BOARD OF EDUCATION

R. Sargent Shriver, President
Benjamin C. Willis, General Superintendent
James H. Smith, Associate Superintendent in Charge of South Side Schools
Ross Herr, District Superintendent

ENGLEWOOD HIGH SCHOOL

J. Mel O'Rourke, Principal
John G. Kriewitz, Assistant Principal
Alice Solomon, Class Sponsor

Figure 2.7
Carruthers's sarcastic version of his high school's theme song. *Source:* George Carruthers Collection, box 1, folder 2, Archives Department, National Air and Space Museum, Smithsonian Institution, Washington, DC.

faced with hate. Gerald believed that they were only the "second family on the block to be of color." He recalls being threatened in the local park, where Black kids were not welcome. If they were caught, "the white guys would beat us up."[61]

George felt that racial reactions in Chicago were different, but it depended upon whether he was with his grandmother, going to high school during the week, or with his mother and siblings on the weekends. He knew that bullying and taunting existed, but it affected him less because, as his brother Gerald recalled, he typically stayed home and did not play in the park with the white kids. But when he did cycle through the neighborhood from his grandmother's home to his mother's home, as Carruthers recalled in an interview with the independent public radio documentary producer, Richard Paul, "I remember when I was a kid in Chicago I used to ride a bicycle in white neighborhoods and a bunch of white guys came out and started shouting out to me so I just gave them the same type of words that they used and kept on—since I was riding a bicycle they couldn't catch up with me. But I've never had anything like that here in DC."[62]

He came to his mother's house only on the weekends until he graduated and went to college, dutifully attending church with his mother. In Milford, they had attended a Baptist church because it was local, but in Chicago, they switched to Christian Science with the rest of the family.

# 3

# COLLEGE YEARS

George graduated from Englewood in June 1957 intent on a college career. Although his parents had always assumed that their children would attend college, George Archer's death changed the landscape. His mother was supportive but not able to meet the cost "because it was all she could do to keep the household going."[1] He considered staying in Chicago and entering the Illinois Institute of Technology. He did not consider the University of Chicago; although they certainly had physics and astronomy, they did not have engineering, "and my goal was to get a degree in engineering because I felt that way I could further support my interest in space astronomy by helping to design the spacecraft and propulsion systems. . . . I had parallel interests in both the rocket engineering and astronomy."[2]

Even more ardent, reflecting that at the time, he was just being "young and naïve," Carruthers wanted to make a point to those who were skeptical about spaceflight. He wanted to make "space astronomy a reality."[3]

Carruthers eventually chose the University of Illinois Urbana-Champaign, his father's and his uncle's alma mater. To be able to leave home, however, he needed support. So he applied for and received both a Cook County Tuition Scholarship and a Cook Foundation Scholarship for his freshman through senior years. He also received student loans and took part-time work as a dishwasher on campus. Away from home, "I figured that I could concentrate on my studies."[4]

According to his brother Gerald, this support was still not enough to afford the cost of a dorm room. But another family connection with Urbana, his paternal aunt Edna, lived just a block or two from the engineering campus and rented rooms to students. She had housed his father when he was there and asked only that George pay what he could, giving him a room with a bath in the basement. Gerald visited him once in his seven years in Champaign to get help on his own student science fair project. However, he chose to stay in a campus dorm. It was "like heaven compared to George's place."[5]

In his first semester, Carruthers's life was changed in October 1957 when Sputnik flew. It "certainly reinforce[d] my goals of going into spaceflight engineering." He also recalled poignantly that "it turned off a lot of the criticism that I received from my peers that space flight is nonsense."[6]

Given the wave of attention and even panic brought about by the first Sputnik, and especially by the larger Sputniks to follow that put a dog in space barely a month later in November 1957—and a month before the United States had failed on its first attempt and did not have a successful launch until February 1958—the national attitude, and support for spaceflight and the critical importance of aerospace training, changed drastically.[7] His friends and teachers no longer mocked his passion, freeing him evermore to focus on his singular goal that would lead his career, or even him, into space. His enthusiasms were also buttressed by activities on campus; as a member of the campus astronomy club, he became involved in a national program to track the first satellites called "Moonwatch."[8] But even with the excitement of the moment, he also found that moving from high school to college was far from easy: "I guess one of the things that came as a shock to me when I went to University of Illinois, which I also warn students about nowadays, is that the transition from high school to college is not a trivial transition. In other words, what gets you an A in high school will get you a C or a D in college simply because the level of competition is an order of magnitude stiffer."[9]

For the first time, Carruthers sensed real competition. He found the professors more distant and the students far more competitive. He took as much physics, chemistry, and mathematics that the curriculum would allow, but he keenly realized that the mathematics he had at Englewood was inadequate for the physics he was now taking, so it took the better part of his first year to adjust. Another major change, possibly related, was that the college population was almost completely white. Englewood was mixed, and he had Black teachers in both physics and general science. And though they were more supportive, their grasp of the sciences was not what he found to be the case at the university. Carruthers poignantly recalled that he "suddenly went to a mostly white university in which I was competing not with inner city kids like myself, but guys from the suburbs."[10]

His grades in his first year were mainly Bs, although he got a C in plane trigonometry and an A in an ROTC course called "Freshman Drill and Theory." He spent his freshman summer in Chicago at home, and he took summer courses on the university's Chicago campus, doing poorly in calculus and badminton (both Ds) and a bit better in general physics and mechanics, gaining C grades.[11] In addition to taking classes, he worked as a part-time clerk during his freshman summer at his mother's post office branch "slinging mail" to raise money for the family.[12]

Little is known of Carruthers's personal life throughout his college years, but he recalled that he sensed only indirect discrimination on the Urbana campus. By then, the university was trying to discourage any student organization that restricted "membership on the basis of race or religion."[13] But it was not wholly effective. Although other students did not overtly challenge him, he felt unwelcome in the community because most students regarded him with indifference. "They don't want to actually socialize with you, either. So if you have study clubs that all get together and exchange notes, you're left out. Not that they would prevent you from coming, but they just don't invite you."[14] He was typically the only Black man in his classes, but he felt that his teachers were fair minded and supportive, as were the more technically oriented student groups like the staff of the campus newsletter, the *Illinois Technograph*, which he joined in his sophomore through senior years (see figure 3.1), and his membership on the Engineering Council in his senior year.

Otherwise, Carruthers generally kept to himself, remaining intensely focused on his singular goal, taking inspiration from the enhanced library facilities, the faculty and their laboratories, and the competition, even if indifferent, from his college classmates. He especially enjoyed the campus Rocket Club, advised by Dr. Shee-Mang Yen, who was one of Carruthers's professors in aeronautical and astronautical engineering, an expert in studying shock waves, and specialized in the theory of rarefied gas dynamics and computational hypersonics.[15] He was the ideal mentor, fostering Carruthers's focus on rocket technologies on campus and keeping in touch when he was in Chicago during the summers.

In 1970, when Carruthers visited the campus to accept an award and talk about his recent work, Yen stated that "George was the most versatile student we ever had in the department. He pursued all learning in depth. He had the knack of reviewing a field and understanding a problem." To Yen, Carruthers was a "very enthusiastic participant in their rocket experiments," to which the news reporter also quoted him, laughing, "And he was very bold in experimenting. We were always worried George might blow up the lab."[16]

Figure 3.1
The *Illinois Technograph* editorial staff, a student organization, promoted engineering interests. *ILLIO* 68 (1961), 188. Copy in George Carruthers Collection, box 1, folder 43, Archives Department, National Air and Space Museum, Smithsonian Institution, Washington, DC. AncestryLibrary.com—U.S. School Yearbooks, 1900–1999. *Source:* Courtesy of "Illini Media/Illio Yearbook," University of Illinois Urbana-Champaign.

## ROCKETRY

Throughout the 1950s, especially after Sputnik, rocketry was both a key to national security and a rising sport for both adults and children, so much so that the newspapers issued cautions, calling for a "parental brake."[17] At home in the summers, Carruthers remained deeply involved in experimenting with rocket motor designs, neglecting his summer classes and almost everything else. In 1956, he became a founding member of a new junior division of the venerable Chicago Rocket Society, which formed in 1946. Carruthers was featured in local papers for designing and building intricate test stands to evaluate the performance of a wide range of propulsion systems.[18]

The junior division was only open to males, ages eleven through twenty-one, and their mission was to conduct static tests under controlled conditions with the hope of actual firings on a local Army base. The senior division of the society teamed up with the local US Fifth Army, and had access to their facilities in Hyde Park to sponsor a series of lectures for the junior division, held at the Illinois Institute of Technology. Beyond the science and technology of rocketry, the lecturers from the Illinois Institute of Technology and from the Armour Research Foundation stressed safety issues, and the Army held out the goal of actual test firings on Army ranges once they had proven their designs. Carruthers, however, did his own testing at home: "I used to make homemade rockets from chemicals in school laboratories and shoot them off in back alleys, etc. However, I soon found out that this was not only against the law but very dangerous, and you really don't learn anything that way."[19]

By the summer of 1959, Carruthers was the head of the Chicago Rocket Society junior division's "research department," engaging local high school students. His brother Gerald was also active. The junior division was funded by the senior division, and Carruthers also received assistance and advice from Professor Yen at college. With this support, he built, in his mother's basement, a plasma rocket motor that promised to be more efficient than chemical propulsion (see figure 3.2),

Ever since Sir William Crookes defined a fourth state of matter in the 1870s, based upon his years of evacuating glass chambers and filling them with rarefied gases and watching how they behaved in the presence of electric fields, physicists have explored this fourth fundamental state of matter, called a "plasma." In a heated plasma, electrons are ripped from atoms, rendering them charged particles called ions, which are then subject to direction and amplification by electric and magnetic fields.[20]

Carruthers was fascinated with the properties of plasmas as a means to propel a rocket. He had read that the resulting ions could be accelerated by a high-voltage electric grid to high velocities, producing thrust. Thinking long term, he envisioned the low thrust but high specific impulse motor not

Figure 3.2
Carruthers in his mother's basement testing his rocket motors. From "Group's Hobby Is Really Out of This World: Rocket Unit Conducts Research," *Chicago Daily Tribune*, September 13, 1959, S2. *Source:* Courtesy of *Chicago Tribune*.

as a launch system but "enough to send a spaceship anywhere in the solar system once it has been placed into orbit by use of conventional rockets."[21] His challenge was to produce a high vacuum in the chamber, which could not be done in his mother's basement. He also needed technical information, so he reached out to aerospace corporations asking for reports not available in public sources.[22] As he reported at a career conference in 1960, "We have just scratched the surface in ion and plasma propulsion. However, we have learned quite a bit from the tests, the failures more than the successes because we know what does not work as well as what does work."[23]

Carruthers evangelized for rocket development. His Chicago Junior Rocket Society colleagues regarded him as a "salesman," urging him to write up his work and give speeches at meetings, which led to his invitation to speak at the career conference we have been quoting.[24]

Carruthers's enthusiasm for rockets, however, resulted in one sorry mishap. At home in the summer of 1958, he decided to have some fun and fabricate several small solid rockets to fire off on the Fourth of July. His brothers were fascinated too, and, unbeknownst to him, they tried to punch up the effect by placing small sparkler fireworks into one of the white phosphorous motors. It was Gerald's idea, but Anthony grabbed the rocket cluster away from him to finish the process, and somehow, he ignited the whole works in an alley, seriously burning himself and others standing close to him, sending him to the hospital. The newspaper reported that George supplied the fuel, which he confirmed in a 2004 interview for the HistoryMakers: "We made some fireworks for the Fourth of July and my brother got burned with one of them. And the police came and busted me, and it took me a while to get out of that because in Chicago, fireworks were totally forbidden. In fact, they still are totally forbidden except for licensed displays."[25]

This was not the only time that Carruthers's experiments got him in hot water. Back in high school, not all his science fair efforts were about rocket propulsion. In one of them, he wanted to demonstrate how electricity could be conducted over long distances. He set up a demonstration that included two transformers and neon signs. Unfortunately, he inadvertently touched one of the terminals with his elbow, which resulted in a spark that "burnt a hole in my shirt (laughter). And that wasn't very pleasant. But what might have been a problem was the principal came over to see the exhibit, and the rim of his eyeglasses touched one of the wires (laughter), and he wasn't pleased."[26]

In the following years, Carruthers's experimentation led to mishaps, and over time, their frequency increased. It was evidently in his nature. It was also to become a hallmark of his methodology. He was so curious about how things behave when you test them that he took risks. He shared one particularly poignant episode with his NRL colleague Ken Dymond:

> When he was a teen, he and some friends were near some train tracks. The tracks featured the electrified "third rail." George decided to demonstrate to

his friends that you needed to have a conductive path for the third rail to harm someone. He told me that he had hopped onto the third rail and walked, tight-rope style along the rail and then jumped off unhurt. The kids told him there was no way to do that if the rail was powered and that he was conning them. George found an old street sign nearby and proved it . . . by tossing it onto the rails shorting from the third rail to one of the others. A huge shower of sparks shot up and part of the sign post vaporized.[27]

Years later, visitors to his NRL office saw a sign on his wall that attested to the fact

**"If it ain't broke, let's see if we can break it."**[28]

Carruthers's grades improved in his second and third years, both in Urbana and in Chicago during the summers. Although he did poorly in calculus one summer, possibly distracted by his rocketry, his physics grades during the school year were better, in the B and C range. But in specialty classes in the upper division—like aircraft structures, metallurgy, atomic physics, aerodynamics, and other aeronautical subjects—he got As and Bs. He was a winner in anything relating to experimental laboratory work (see figure 3.3).[29]

Upon graduation in 1961 with his degree in aeronautical engineering, Carruthers chose to stay on the Urbana campus and enter the graduate program in nuclear engineering. With his scholarships, loans, and part-time work as both a research and teaching assistant, he was now self-supporting. Also, both of his younger brothers had joined the Army, so he did not have to worry about his working mother, now remarried (see below), and younger sister Barbara Ann. Further, his rocketry experiments reinforced his fascination with alternatives to chemical propulsion. So nuclear engineering "seemed promising at the time as a method for spacecraft propulsion."[30]

In his senior year, Carruthers applied for summer jobs in industry and at government laboratories. In 1961, he was rejected by Argonne and Los Alamos and was hoping that either Rocketdyne or Aerojet might come through.[31] This hardly dampened his enthusiasm for rocket technology. One of his projects in his senior year was to explore the "Control of Missiles and Space Vehicles," wherein he explored technical issues like the second-order effects due to the fact that missiles were not "rigid bodies."[32] Finally, after graduation, he was offered a job for the summer of 1962 at the Aerojet Corporation in Sacramento, California—a major name in the

Figure 3.3
The 1961 issue of the Illinois ILLIO featured Carruthers engaged in his plasma experiments to assess the resistance of ballistic missile nosecones to the physical conditions encountered in high-speed reentry through the atmosphere. *Source: ILLIO* 68 (1961), 78. Copy in George Carruthers Collection, box 1, folder 43, Archives Department, National Air and Space Museum, Smithsonian Institution, Washington, DC. Courtesy of "Illini Media/Illio Yearbook."

aerospace field. As noted recently, "Aerojet, combining roles as space purveyor, employer of last resort, and hired gun," probably agreed to hire Carruthers not only for his promise, but as a counter to the perception that "black employment in most big firms was largely tokenistic."[33] One way or another, it impressed his brother Gerald, who, since he regarded his brother as rather cheeky, blurted out, "So you are going to be the first monkey on the moon!" and then signing off, "See you soon racoon."[34]

That summer at Aerojet, as George recalled, "was my first real exposure to what engineers actually do."[35] He was exposed to methods of assessing reliability and quality assurance on projects relating to the Gemini and Titan II programs, both spacecraft and launch technologies, and he had some exposure to Aerojet's nuclear rocket program.

Carruthers had a mixed experience at Aerojet that summer. It was large-scale industrial engineering, and he learned a great deal. But he also learned that he did not like being restricted to a "little bit of a very large project." What Carruthers longed for was a grasp of the "whole picture."[36] He worried that he would not find it in industry but thought he would possibly find it in a government laboratory, like the Naval Research Laboratory (NRL). He had read about NRL in Milton Rosen's *Viking* book and in journal articles citing NRL, including *Aviation Week*, the *Journal of the British Interplanetary Society*, the *Journal of the American Rocket Society*, and even the *Astrophysical Journal*. From Rosen's description, it seemed that the research atmosphere at NRL was what he wanted: to avoid environments where hundreds of engineers worked on pieces of big projects:

> The guy in one cubicle is designing a valve. The guy in the next cubicle is designing a turbine. The guy in the next cubicle is designing a plumbing system. Nobody has the whole picture, at least not at that level. . . . That was what I missed during that summer experience was seeing [how] what I was doing really fit into the whole, because I didn't get that much exposure to the whole except by reading on my own the literature that the company put out.[37]

It was also during his brief stay in California that he experienced overt racial discrimination beyond the bullying he and his brothers encountered as children. Aerojet had arranged a rental for him in the neighborhood, but when he arrived, he was told, "Well, I hate to tell you but we don't take colored in this place."[38] Carruthers wrote home reporting that he could not find a place to stay near his job, and he was wondering if he had made the "right move." This worried his family, of course, but after he wrote again in a few weeks, he was able to reassure them that the situation had improved. Relieved, his now stepfather, Wellington Martin (see below), wrote back assuring him that "I am sure you will never forget those trying and difficult days, no doubt you will laugh many times as we have now that it is all over." Martin added that having the Aerojet experience would look good on his transcript when he started looking for jobs.[39]

Returning to Illinois in the fall to continue his graduate studies, Carruthers left nuclear engineering for aeronautical and astronautical engineering, and he also elected to minor in physics and astronomy. This allowed him to take some courses in astronomy, since several professors there had half-time appointments in the electrical engineering department. Arnold Willard Guess and Gary R. Swenson, for example, directed a satellite-ionosphere project jointly between the astronomy and electrical engineering departments. Carruthers had found his niche in the combined world shared by the aerospace and astronomy professors.

## THESIS

Given Carruthers's evident promise, senior professor of electrical engineering Ladislas Goldstein, the founder of the Gaseous Electronics Laboratory on the Urbana campus, consented to become Carruthers's thesis advisor.[40] Goldstein was then actively exploring the behavior of plasmas when subjected to various forms of radiation, like microwave beams directed by magnetic fields.[41] He was also interested in the effect of radio waves on the Earth's ionosphere.

Under Goldstein's guidance, Carruthers chose a thesis that combined his two worlds: experimentation and space travel. Carruthers's proposal to Goldstein concentrated on testing the theory behind "the recombination of atomic nitrogen behind shocks in the Mach number range 2 to 10" and a description of his planned experimental procedures, including a rough diagram of his shock tube design. But nowhere did he discuss how his work might help to better understand the ionosphere or its influence on missiles reentering the Earth's atmosphere.[42]

Carruthers knew that experimenting with how plasmas behave when exposed to, or are generated by, different stimuli provides insight into the structure of matter, such as the structure of atoms and molecules, and, to the point covered in his thesis, how that structure can be eroded by those stimuli. Although Carruthers did not say it until his final report, it became an experimental means of probing the nature of the upper regions of the Earth's atmosphere—ionospheric physics—which was of major interest to the military services, especially NRL.[43]

However, Carruthers's completed thesis, "Experimental Investigations of Atomic Nitrogen Recombination," supported by departmental grants from the US Air Force and a scholarship from the Cook Foundation of Chicago, did address applications: some of the most challenging aspects of spaceflight, such as the reentry problem—that is, how to design a spacecraft so that it could reenter the Earth's atmosphere without vaporizing. This included knowing how atoms in the Earth's high atmosphere are torn apart and recombined by the shock waves caused by a metallic body entering the atmosphere at high speed, and how this interaction can affect the vehicle by the heat energy generated that is transported in the process. This study, as Carruthers wrote in the opening for his published thesis, "is of importance in the design of vehicles for flight in the upper atmosphere, where the molecules are partially dissociated by solar ultraviolet radiation and by auroral particles."[44] Carruthers listed other areas where the results of his laboratory experiments could be useful, such as in plasma arc-heated wind tunnels and in electric arc-driven shock tubes.

Carruthers concentrated on studying how nitrogen responds to shock waves at high temperature because it was the prime gas in the Earth's upper atmosphere "and probably also of those of Mars and Venus." In Goldstein's laboratory, he constructed a structure of glass tubes, called in the practice "shock tubes," that he could fill with nitrogen and then subject the gas to various forces. He wanted to rip apart nitrogen molecules and then study "over a wide range of temperatures" how they recombine back into a diatomic molecule [two nitrogen atoms].[45] It was well known that upon recombination, nitrogen emits a yellowish afterglow, but at the time, there were few quantitative studies of this phenomenon. That was Carruthers's goal.

For over a decade, Goldstein and his Gaseous Electronics Laboratory staff had experimented with gas discharge plasmas using a wide array of microwave heaters and detectors. Receiving his PhD in physics from the University of Paris in 1937, concentrating in nuclear physics and a wide range of gas discharge studies in the microwave region of the spectrum, Goldstein was a leading member of the faculty and a formal though generous mentor to his students.[46] Carruthers's thesis fit perfectly with laboratory life there and benefited from the guidance of its faculty and staff.

After months of practice and testing in 1963, Carruthers became adept at glassblowing and other laboratory techniques for building his vacuum chamber equipment—considered a "central practice" in experimental physics.[47] He designed and built at least a half dozen different static and flowing gas systems and shock tube systems to study both the photometric and spectroscopic behavior of nitrogen—mainly, how single atoms of nitrogen recombined into molecules at a range of temperatures and in the presence of other elements, like helium and argon. After much experimentation, he settled on two cylindrical Pyrex chambers for his gas-flow shock tube systems. One was a ½-inch-diameter tube some eight feet long, and the other was a 4-inch-diameter tube over ten feet long with several appendages. He mounted them in parallel on long racks with appendages feeding the gas into the tube, heating and tearing apart the molecules with a radio frequency discharge system, and then exhausting it through two separate arms driven by vacuum pumps. After the gas was heated and the atoms split apart (dissociated), the gas would then enter the main tube where it would cool and recombine, releasing photons that produced an afterglow that Carruthers could analyze using various sensors like a red-sensitive photomultiplier—in this case, a commercial RCA tube that converts light energy, or photons, into an electric current and then amplifies that electric current by directing it to a series of charged plates each at a higher voltage. He also employed a monochromator, a device that emits extremely narrow bands of light, and finally a laboratory spectrograph. The next step was to release another gas from the other end of the tube that had been under high pressure. This produced a shock wave that once again could be analyzed by the sensors.[48]

In the course of his work, Carruthers became familiar with both the theory and practice of photoelectronic detection systems, particularly calculating the number of photons striking the cathode of an electronic sensor from its electron output current. And he also explored how molecules could be split, or ionized, by high-energy electrons as well as intense UV photons, mentioning, in particular, "solar ultraviolet radiation." Evidently throughout his analysis, he was well aware of the similarities of his gaseous vacuum experiments and conditions in the Earth's upper atmosphere.[49]

Carruthers spent weeks and months conducting his experiments, varying flow rates and temperatures as well as different mixtures of inert gases with nitrogen. At first, he was challenged by the fact that his results were not reproducible from day to day, but eventually, through "cut and try" alterations to reduce the surface-to-gas recombination rates and other design glitches, he achieved reproducible results. His best results, however, came from his experiments with systems producing higher temperatures quickly, which could be precisely calibrated and controlled, and where the problem of surface recombination was minimal. The same types of photometric and spectroscopic devices were used to record the changes in the afterglow.

By early 1964, Carruthers could sum up his work and compare his results with those of earlier studies, finding his results matched the most recent studies by other workers. And as his thesis committee expected, Carruthers outlined four proposed experiments that could be conducted to further refine the results, suggesting greater temperature ranges, the use of microwave or laser heating, and studies of recombination in the presence of charged particles, all implicitly geared to reach the conditions encountered for spacecraft reentry, which, one might surmise, were also applicable to the study of the upper atmosphere itself.

When he had finished his experimentation and analysis, he wrote up his results, using the stiff pedantic style expected in an academic thesis: phrases like "it is worthwhile to consider the evidence" and, for the opening sentence of an appendix on "The Theory of the Shock Tube" inviting his examiners to visualize his work, "consider a chamber which is separated into two sections, separated by a nonporous diaphragm."[50] By now, he most certainly was aware of the importance of gaining command of the language as tacit acceptance into the discipline.

## FIRST CONTACT WITH THE NAVAL RESEARCH LABORATORY

Walking down the hall one day near his laboratory, Carruthers came upon a recruiting poster from NRL announcing a training program for PhD candidates interested in the space sciences. Carruthers, as we noted, had known about NRL since he was in high school. As he recalls, it was

> when I was in high school that I was reading about [the] Naval Research Lab. It's certainly true that when I was at college, I knew about what was going on in NASA, and, in fact, I did submit applications to various NASA centers, but it seemed to me that since Naval Research Lab was the father of NASA, both the headquarters and the Goddard Space Center at least were started by NRL

> people, and they had the broadest program. In other words, the NASA center, once they branched out, tend to specialize more than the NRL group did. Having read about the Viking and Vanguard and the other projects that NRL was in charge of, it was certainly a logical place to go.[51]

Carruthers had already applied for summer jobs at NASA when he was an undergraduate but does not recall being accepted. None of those with openings evidently were local to Chicago. He recalls, "If I was accepted, I probably turned it down. I really can't say for sure." Still, he "really wanted to get into hands-on experimentation more than anything else."[52] NRL was where he wanted to be.

In May 1963, he wrote to NRL ("Dear Sirs") asking for more information. He identified himself as working on his PhD, hoping to take the preliminary examination in September, at which time he would become eligible for the new NRL predoctoral program in its E. O. Hulburt Center for Space Research (named for the NRL physicist and geophysicist who pioneered the study of the Earth's uppermost atmospheric layers and became NRL's civilian director of research starting in 1949).[53] Carruthers identified his interests as experimental aspects of astrophysics, upper atmospheric physics, planetary atmospheres and interplanetary plasmas, and particle radiation.[54]

A week later, E. O. Hulburt Center chief scientist Herbert Friedman, who established the center and program and named it after his retired boss, responded positively that Carruthers's background and interests were "very appropriate" and that, if his thesis advisor approved, it is likely that NRL would support his thesis work in their laboratories. Carruthers filled out the application forms in July to be a research associate but stated that university regulations required that the thesis be conducted under the direct supervision of his thesis advisor.[55] Still, he hoped he could assume the Research Associateship when his experimentation was complete and he was writing up his thesis. Responding directly to the NRL description of desirable subject areas, he opted for "stellar ultraviolet spectroscopy, aurorae and airglow, and solar and interplanetary medium spectroscopy."[56]

Carruthers's professor Shee-Mang Yen once again strongly endorsed Carruthers's proposal, marking him as truly exceptional: "I have known very few students who have his enthusiasm, persistence and initiative in doing space research." He was "very competent in doing experimental work," and Yen had total confidence that Carruthers would do well in space science at NRL.[57] Friedman was impressed, and even suggested that NRL might be able to support having Yen, or even Goldstein, in residence while Carruthers completed his thesis there. He also invited Carruthers to visit and give a colloquium.[58]

Carruthers was eager to visit and give a talk, but he also felt it was unrealistic to complete his thesis at NRL, "as considerable time and money has already been expended for the construction of shock tubes and related research." But he also expressed his deep desire to work at NRL because his thesis "has application to upper-atmosphere physics, since the gases are partially dissociated by solar radiation and auroral particles at very high altitude." In addition, he felt that his work would be useful to study "shock phenomena in stellar atmospheres, interplanetary and interstellar gases in which the initial gas is partially dissociated or ionized."[59]

Friedman had established the E. O. Hulburt Center Associate program at NRL with a grant from NSF to encourage young scientists into the space sciences, especially because NRL had recently lost a good fraction of its space scientists to NASA. Indeed, in Sputnik's wake, with a national call to improve science education in the United States, Friedman was concerned that available training in the space sciences was inadequate. Hulburt was still alive and deeply touched by Friedman's decision to name the center after him: "Nothing that I can think of gives me more pleasure."[60]

Friedman's concern for the dearth of trained recruits reflected those of many science and engineering centers in the early 1960s. It applied to everyone, but especially to Black scientists. As historian Steven L. Moss concluded for NASA, which applied equally to NRL, the "greatest problem with equal employment was not the hiring of blacks but finding blacks to hire."[61]

Friedman took this personally. A Jew, he experienced discrimination in his early professional life, so, beyond encouraging students to enter the space sciences, he was sensitive to enabling a diverse workforce.[62] Friedman's concern was shared by many astronomers at the time. The first of astronomy's Decadal Surveys, released in 1964 and limited to the future health of ground-based astronomy, pointed to a shortage of trained astronomers and adequate graduate programs augmented by accessible modern telescopes.[63] Friedman felt the same had to be done for the space sciences, and the answer resided in laboratory training and access to the best expertise, not for telescope time, but for laboratory time preparing and flying experiments aboard sounding rockets.

Carruthers visited NRL in December 1963, and spoke in some detail about how the results of his thesis applied to studies of the upper atmosphere and, specifically, aurorae. His performance more than convinced Friedman to propose Carruthers for a "Hulburt fellowship" in light of his "very well-received colloquium."[64] Carruthers's appeal for Friedman and NRL was no doubt heightened because the time—late 1963—was a period of considerable racial unrest. In 1961, President John F. Kennedy issued Executive Order 10925, establishing the President's Committee on Equal Employment. The order stipulated that employers and contractors within the federal government "take affirmative action to ensure that applicants are employed, and that employees are treated during employment, without regard to their race, creed, color, or national origin."[65]

Passage of the Civil Rights Act was still a year away. In the meantime, the United States witnessed atrocities in Birmingham and Martin Luther King's arrest, fueling a demand for federal workforce compliance with the Equal Employment Opportunity (EEO) standards, as a part of a broader effort to support civil rights.[66] This hot political climate could only have strengthened Friedman's conviction that Carruthers was right for NRL.

Carruthers stayed on campus until he completed and defended his thesis for the PhD in aerospace engineering, and it was approved in September 1964. According to his alma mater, he was "among the earliest African Americans to earn an AE degree." And by then, he was fully accepted by the fraternal engineering honor societies Sigma Tau and Sigma Gamma Tau and had also joined the junior ranks of the American Institute of Aeronautics and Astronautics (AIAA).[67]

# 4

# POSTDOCTORAL YEARS AT THE NAVAL RESEARCH LABORATORY

As noted before, Carruthers was the kind of person Friedman was looking for. What Friedman had established became the framework that would shape and support Carruthers's career. We therefore need to examine the motives and drives that led Friedman to establish this center and define its purpose, the scientific problems the center addressed, and the general atmosphere of the work environment.

Located on the Potomac River's eastern shore within the southwest quadrant of Washington, DC, the Naval Research Laboratory (NRL) was established in 1923 as the US Naval Experimental and Research Laboratory. It was the Navy's reaction to public statements by Thomas Edison in 1915 that the United States had to establish research facilities to meet national security needs.[1] Originally focused on high-frequency radio communications and underwater sound propagation, led by Edward O. Hulburt and A. Hoyt Taylor, by World War II, NRL had greatly expanded to meet the many new technologies that could aid military goals. Divisions were later added for physical optics, chemistry, and other applied fields. During World War II, Hulburt added a Communications Security Section within its Radio Division and a Guided Missiles Subdivision responsible for electronic guidance, control, and radio countermeasures. It was imperative that the fleet be in constant radio communication, which required knowing how the reflectivity of the ionosphere depended upon solar radiation to predict radio fade-outs.[2]

Immediately after the war, with the realization that future wars would be fought with missiles, NRL's Guided Missiles Subdivision responded to a US Army Ordnance offer to fly small scientific experiments on their captured German V-2 missiles, being tested at the White Sands Missile Range in New Mexico. NRL's chief goal was to study the medium through which missiles traveled given how critical this type of research would be to producing ballistic missiles in the Navy.[3]

NRL had long been active in ionospheric research as part of its responsibility for long-range radio communications security and saw the Army offer as a means to enhance basic research at NRL.[4] It quickly reformed subdivisions to complement the rocket program, developing spectrographs to study the Sun and the structure of the Earth's upper atmosphere, electronic sensors for studying atmospheric electrical conditions, sampling devices, and even establishing a section to create a missile equivalent to the V-2.

But just what kind of research was NRL interested in? In a 1994 report by the Navy's Office of Strategic Planning titled *National Security and the U.S. Naval Research Laboratory*, D. J. DeYoung described it as "defense science and technology." Any form of scientific practice that could inform the needs of national security, which included "economic competitiveness, environmental health, energy security, and public health and welfare," was encouraged and supported.[5] The NRL administrative structure kept this purpose in mind. As historian Bruce Hevly has pointed out, in the 1950s, "research topics changed to accommodate new Navy technologies," which dealt with radio communications and reconnaissance, and ballistic missile development. For both, upper air rocket research "emerges as perhaps the best example of post war research carried out under military auspices," which he described as the marriage "between the ivory tower and the missile silo."[6]

By 1954, there were some sixty scientists, engineers, and technicians at NRL working on upper atmospheric research, putting instruments on rockets, starting with captured German V-2 missiles and then NRL's Viking rocket and the Navy's Bureau of Ordnance sounding rocket called the Aerobee. The Aerobee became the launcher of choice for scientific payloads by NRL and hence became Carruthers's first step into space.

It is not known if Carruthers was ever concerned about joining a military facility. By this time, as we describe below, both of his brothers were in the military, and his stepfather was a military man. There is no record of Carruthers ever voicing any regrets. His priorities matched those he found at NRL under Friedman. If there was one thing he wanted to avoid, it was bureaucracy. Again, as we shall see in following chapters, throughout his career, NRL—especially Herbert Friedman—protected him from that world.[7]

## THE SOUNDING ROCKET

Carruthers's challenge was to make things work on rockets. Sounding rockets, named in tune with depth sounders that sensed and measured deep ocean conditions, are small rockets that carry sensors to observe and record conditions in the Earth's high atmosphere and then, for a few moments, glimpse the universe from beyond the atmosphere. After World War II, the military services developed small rockets as antiballistic missiles, and some of these designs, rapidly deployable and efficient, were adapted for scientific studies.

The most popular and long lived was the Aerobee sounding rocket. Conceived by physicist James Van Allen, then at the Applied Physics Laboratory, it was a hybrid of the Applied Physics Laboratory's Project "Bumblebee" antiballistic surface-to-air missile program for the Navy Bureau of Ordnance and the Fifth Army's Corporal ballistic missile program's prototype called the "WAC Corporal." The Aerojet Corporation built the rocket for the Navy—hence, the concatenation "Aerobee."[8]

Despite the common sexist quip that it was named for WAC (Women's Army Corp) because it was a small rocket, the WAC acronym here stood for "without attitude control," which meant that this class of vehicle was intrinsically unstable. Thus, the acronym also describes one of the greatest challenges facing those who wanted to fly scientific instruments on these rockets: as described later, they had to be stabilized to point to objects of interest. It would take many years to overcome this problem, but still the Aerobee (originally a 30-foot liquid-fueled rocket that could send payloads of up to 150 pounds to over 150 miles) dominated the field from the 1950s to the 1980s. Over 1000 were flown until NASA cancelled the program in 1985 to limit small payload to being part of a shuttle payload. It had been the launcher of choice for NRL, and it was the clear choice for Friedman and his group.

"Aerobee" and "WAC" are only two of dozens of terms that we will encounter in Carruthers's life, many linked in the institutional collaborations that personify spaceflight. They demonstrate and provide insight into the complex challenges facing a professional in the space sciences.

By the time Carruthers came on the scene, dozens of Aerobees of various designs (see figure 4.1) were launching from White Sands Proving Grounds in New Mexico, from Wallops Island Flight Facility in Virginia, from Fort Churchill in Manitoba, and from other sites. The Aerobee came in several configurations, but by the early 1960s, managed by NASA, the Aerobee 150

Figure 4.1
Launch of an Aerobee 150 from White Sands. The rocket is being launched by a solid-fueled booster from a tower to gain stability. *Source:* Courtesy of Archives Department, National Air and Space Museum, Smithsonian Institution, Washington, DC. Aerobee 150s were the third generation built by the Aerojet Engineering Corporation and first flown in 1955. Length: thirty feet. It could carry a 150-pound payload to about 170 miles.

was the standard: a liquid-propelled and fin-stabilized rocket with a solid fuel booster. It was typically launched through a guidance tower, achieving aerodynamic stability by the time it left the tower. The Aerobee was modified many times to increase payload lift capacity and altitude capability, starting with the Aerobee-Hi in 1955.

The management and allocation of Aerobees was dispersed prior to the formation of NASA, with NRL being the major user. But after the transfer of the rocket group to the Sounding Rocket Branch of the Goddard Space Flight Center (GSFC), NASA became the main provider of vehicles and launch services to universities and other NASA branches, as well as to the military, including NRL.

Carruthers's proposals for flights had to be directed to NASA's Office of University Relations and then, given the size of the project, most likely to its Office of Space Science and Applications at NASA Headquarters.[9] There were also other options for acquiring sounding rocket berths through NASA as well as the Department of Defense (DOD) if funding could be secured through other branches and offices. Winning these proposals and the overhead funding they brought to the lab became a requirement for success as a space scientist at NRL, as did the scientific results. By 1969, Carruthers and his NRL partners had participated in over twenty flights.[10]

## HERBERT FRIEDMAN

The key figure bringing Carruthers to NRL was one of the founders of X-ray astronomy, Herbert Friedman, so here we pause to establish his profile. As NASA Associate Administrator for Space Science Homer Newell observed in a 1980 oral history, "Friedman was something of a genius. In fact, not only was he good at the physics that he was performing, but he was good at making the instruments to fly it."[11]

As noted before, after obtaining his PhD in experimental solid-state physics at Johns Hopkins, Friedman secured a position at NRL in 1940 and quickly applied his expertise developing electronic gaseous discharge sensors to detect and evaluate X-rays. X-rays are very high-energy light photons, powerful enough to "ionize" or tear electrons away from their atomic nuclei, leaving both with unbalanced electrical charges and hence subject to the presence of electrical currents or magnetic fields. His detectors, a class of the popularly called "Geiger counter," were useful to the

Navy for a wide range of quality control and testing in metallurgy and the remote detection of nuclear detonations, and soon he came to E. O. Hulburt's attention as a desirable prospect for his Physical Optics Division to head a new branch for "electron optics."[12]

Of course, Friedman knew what his NRL colleagues were doing, such as Richard Tousey's team in the Physical Optics Division, who had been flying UV photographic spectrographs on V-2 missiles since the 1940s to try to reach deep into the Sun's UV spectrum to assess the intensity and character of its high-energy radiation. They wanted to reach what is known as the Lyman-alpha spectral line, which lies in the deep UV portion of the solar spectrum. Named for its discoverer, Theodore Lyman of Harvard, it is a strong bright spectral line produced when an electron drops from the first excited state in the hydrogen atom to its ground state. This transition creates high-energy photons, which, upon entering the Earth's upper atmosphere, influence the structure of the ionosphere. Anything that could do this would alter the ionosphere's conductivity and hence its ability to reflect radio waves, which as we have noted, is a key factor enabling long-range radio communications beyond the Earth's horizon. Being able to predict the reflectivity of the ionosphere was therefore of central importance to the Navy to keep its far-flung fleet connected. It became one of the central questions NRL addressed in its Space Science Division.

Friedman knew that this was not only of critical interest to the Navy but was also a major justification for conducting upper atmosphere research with rockets, since one needed to get far beyond the absorbing layers of the Earth's atmosphere to make the UV universe accessible.

Friedman did not start to send instruments on rockets until 1949 because he was heavily involved applying his expertise to nuclear test detection techniques, including Project Rain Barrel: collecting and analyzing rainwater as a means of sensing nuclear detonations around the world.[13] He started flying his Geiger counters on V-2 missiles in September 1949 and not only detected solar X-rays but measured the intensity of the Lyman-alpha line of hydrogen.[14] Friedman also showed that the solar X-ray flux was a strong influence on the ionosphere. His instruments also had the advantage of being able to send their data directly to the ground via telemetry.

During the 1950s, Friedman and his technical staff, led by Edward Taylor Byram and Talbot Chubb, aided by NRL's growing network of branch laboratories, created better detectors and trained teams to fly them in a wide range of rocket experiments, from land and shipboard, where they followed an eclipse path in the South Pacific. He also began to experiment with searching for extrasolar X-ray sources but was distracted by numerous other projects.[15]

Sputnik and the hasty formation of NASA resulted in a deep reorganization of the rocketry-related groups at NRL; this was also the case for many of the early rocketry programs in the Signal Corps and Air Force laboratories. NRL's Rocket-Sonde Research Section was one of the largest groups dedicated to sending scientific instruments into the high atmosphere and near space on small rockets ("Rocket-Sondes"). But with the emergence of NASA and the creation of the Goddard Space Flight Center in Greenbelt, in the Maryland suburbs near Washington, DC, NRL scientists were encouraged to move, and most chose to move. Friedman stayed, soon sensing that "there was some friction between our groups at that point, those who went to NASA and those who stayed behind."[16]

Friedman was now designated head of a new division at NRL eventually called the Space Science Division. His staff built Lyman-alpha instruments for the first Vanguards and developed X-ray sensors for the first satellites devoted to monitoring solar X-ray radiation (as well as radio reconnaissance), called the SOLRAD (SOLar RADiation Satellite) series, first launched in 1960. This was also when Friedman established the E. O. Hulburt Center for Space Research.[17]

Friedman created the center to make space science accessible to young graduate and postgraduate students and to grow the pool of talent that NRL needed. He knew that in the growing satellite era, "the opportunities to enter directly into rocket and satellite programs are very limited."[18] Indeed, the lead times and competition for building payloads for satellites were too high to attract graduate students and even recent postdoctoral PhD scientists who had to make their mark quickly to gain a permanent post. Friedman rectified this by promoting sounding rocket payloads, which had planning and development time frames short enough to be feasible for a PhD thesis student. More important, it gave the student a fuller picture of what was required to conduct a space science project using new and creative techniques—exactly what Carruthers was looking for.[19]

In August 1963, the journal *Science* announced the new program at the Hulburt Center: "Graduate Education: Navy Program in Rocket Astronomy Opens New Horizons to University Scientists." It reported that NRL was now taking a "more active role in the education of astronomers and physicists." Doctoral and postdoctoral students would stimulate a "give and take" atmosphere, quoting Friedman, linking their graduate school atmosphere to the world of a scientific military laboratory. This, he predicted, would benefit both NRL staff and the visiting students, and was another example of efforts to "utilize the resources and staff of specialized laboratories operated or financed by the government to increase the supply of scientific manpower in fields in which the supply is short."[20]

The *Science* essay also noted that, in the wake of NASA's creation, NRL's Hulburt Center was something of "an anomaly at a time when most people equated the national space effort with NASA." Citing the fact that military funding for a space program was huge and most directed toward defense interests, "a substantial amount supports basic research" at NRL. Still, the essay recognized that "interagency frictions and rivalries remain, if only because NASA arrived on the scene late and in such a big way." NRL still retained a considerable budget, compared to university laboratories, for activities such as space research, and an infrastructure familiar with but looser than large-scale project management. NRL's remaining rocket astronomy program had a $2.5-million-dollar annual budget to support staff and programming, "about a tenth of the total NRL budget for basic research."[21]

This was far less than NASA allocated of course, but it was supplemented by various Navy subsidies and by the large technical infrastructure in the laboratory. By the early 1960s, this was sufficient support for NRL to launch on average some fifteen Aerobee sounding rockets yearly and a couple of small dedicated Navy satellites, like SOLRAD, which carried X-ray and UV sensors on its spherical skin. Some SOLRADs also carried a payload called GRAB, for Galactic Radiation and Background, which was a cover for the first satellite radio intelligence monitoring efforts in the Navy's ELINT, or Electronics Intelligence, program.

Friedman and his group had also been flying medium-band to broadband UV photon counters, versions of his Geiger counters, on rockets for about a decade when Carruthers joined them. Their goal was to determine if the energy distributions of hot stars fit theoretical models, an astronomical question with some relevance to military interests. But they needed new more sensitive and reliable ways to detect the ultraviolet.

There was plenty of competition. In the early 1960s, for instance, Theodore Stecher and James Milligan at Goddard had similar goals, using an optically fast spectrophotometer to produce narrowband scans of some sixteen hot stars in the winter sky, mainly in the Orion region.[22] Their Aerobee *150* rocket flight from Wallops Island, Virginia, in November 1960 revealed that the far-ultraviolet intensities exhibited by these stars was much less than predicted. Their preliminary analysis showed that the effect was not constant in their sample and seemed to vary by spectral class and distance. This indicated that much was left to discover.

Friedman's team was interested in this question, but from a different angle, one that was closer to pure astronomical interests than to Navy interests. They hoped to find evidence beyond the radio spectrum, for material that made stars. To do this, they looked at the regions around hot young stars to see if there were remnants of their formation: hydrogen and dust grains. These would create a glow in the ultraviolet around hot stars like Spica. By 1963, they found no evidence for it and suspected the problem was due to aberrations in their optics and sensitivity limits in the detectors.

NRL's results differed from those of other rocketry groups at the Goddard Space Flight Center and in academe. This made Friedman's group worry that they were not competitively equipped to explore the phenomenon because their UV technologies were inadequate to the task. As historian Richard Hirsh observed, "The resolution of the nebular glow problem signaled the end of Friedman's involvement in ultraviolet astronomy."[23] Moreover, this realization made it even more imperative that NRL needed to attract new talent who would be capable of filling that critical spectral gap. The question of the glow persisted, and it had wider implications.

Observational phenomena in this spectral range were, of course, critically "relevant to situational awareness" and therefore to the needs of the fleet, which included everything from detecting launches and overflights of missiles and spacecraft. So no anomalies could be ignored.[24]

Friedman's colleagues Byram and Chubb continued the UV observations, building upon the rocket data in early 1963. They examined the UV brightnesses of some fifty stars collected by 4- and 6-inch telescopes feeding Friedman's gas-gain ionization chambers looking out the side of the spinning rockets. They then compared the observed brightnesses to predicted brightnesses based upon values derived from theoretical stellar black body (ideal radiator) curves. None of them agreed for data in the medium UV range, but their data at bluer wavelengths did fit theoretical predications. In none of the observations did they detect a nebular glow, but overall, they concluded passionately, "All we can really say at this time is that additional experimental data on the nebular glow are badly needed."[25] Indeed, even by 1966, the general view was that "observations of stellar ultraviolet fluxes have not yet firmly established whether there is any major discrepancy with theory."[26]

During the early satellite era, Friedman was convinced that sounding rockets were still critical for this kind of work, and also for attracting new talent: "We believe . . . that rockets offer an invaluable proving ground for many of the experiments planned for future satellite observatories."[27] His sentiments were more than supported in 1965 by the Working Group on Optical Astronomy of the Space Science Board, a National Academy of Sciences committee, then meeting at Woods Hole, Massachusetts. Their first recommendation was that, to attract new talent, "the number of coarse-pointing sounding rockets available each year for optical space astronomy should be increased to twice the present level." Following that, they called for concentrated efforts at developing new, light, and versatile electronic detectors.[28]

Friedman, a member of that elite board, took its recommendations as a blueprint for his plans. He and the board knew that, at the time, NASA Headquarters associate administrator for space science, Homer Newell, a former NRL scientist, was realigning Goddard's priorities to make its "Manned Space Flight Network" first priority, followed by large scientific satellites, then smaller and simpler "Explorer class" missions, named for the early Explorer series of satellites. Explorer ranked twelfth and sounding rockets sixteenth in NASA's priority, above just "biosciences" and "miscellaneous."[29] Friedman and the Working Group tried to fight back, arguing that it was essential to have a constant flow of young researchers join the lab, bringing in fresh talent and ideas, and this could best be done with short-term small-scale projects like those on sounding rockets. His plans for the Hulburt Center reflected this view and gave NRL "the opportunity to discover . . . the best ones in any year's group and would try to hang onto them."[30] Again, this model fit Carruthers's ultimate career path "to a T."

## CARRUTHERS'S APPLICATION TO NRL AND HIS FIRST POSTDOCTORAL YEAR

After Carruthers's visit and colloquium, Friedman told NSF that he was "fully qualified" and that his research plans "fit well" with NRL interests.[31] In letters back and forth, Carruthers had also done his homework, making it clear how his laboratory research was closely related to, and would inform NRL's interests in, areas he well knew were Friedman's interests—particularly, the nature of the Earth's uppermost atmospheric layers called its geocorona and the nature of stellar atmospheres and the interstellar medium.[32]

In May 1964, Carruthers reported that he hoped to complete his thesis by August and that he would like to report to work immediately, maybe the first week of September. He was willing to either join one of the teams in the division or continue his own work. Clearly, he was anxious to get to Washington, but he also wanted to assure Friedman that he would fit in.[33] In early July, in another long letter to Friedman, once again describing his many interests and how they applied to NRL programs, Carruthers admitted that he did not feel ready to take on an independent line of original research. Possibly to assure Friedman that he would be a team player, he felt he needed guidance in developing rocket payloads and needed a broader grasp of the literature. He therefore proposed that he join one of the ongoing teams, engaging in "experimental work, as well as in the development of the instrumentation and rocket vehicles." His first choice was spectroscopy of "the brighter stars and nebulae in the extreme ultraviolet." He also identified studies of the aurora and airglow by other NRL groups.[34]

Friedman responded positively and advised him to prepare his formal proposal immediately to meet NSF deadlines. Evidently feeling that NSF preferred independent proposals aligned with NRL's interests, he suggested that Carruthers identify a specific program and, following Carruthers's stated preference, agreed strongly that he concentrate on low-resolution stellar spectroscopy in the far ultraviolet, which "can provide answers to such basic problems as the interstellar density of molecular hydrogen." Friedman was also supportive of Carruthers's other options, supporting NRL interests, such as the observation of glowing gases in the Earth's upper atmosphere and how they were influenced by solar storms. But what NRL needed most was a successful stellar UV program "with emphasis on improvement in quantitative accuracy."[35] Such improvements would benefit other problem areas. As Chubb remembered the situation, "We sort of felt that the next greater progress would be done with UV imaging, and we sort of turned over the ultraviolet stellar program to Carruthers."[36]

Carruthers's and Friedman's goals synchronized nicely, which propelled Carruthers into action. Even though he was just then putting the final touches on his thesis and preparing to defend it before the faculty, he also prepared his formal proposal to NSF within two weeks. His proposal, closely aligned with Friedman and Chubb's "A Study of the Far Ultraviolet Spectra of Stars and Nebulae Using a Rocket Spectrograph" was for a year of funding to start in late September or October "to further develop the study of stellar spectra in the vacuum ultraviolet." Citing work done to date by NRL astronomers, including Friedman, Byram, and Chubb, and recent work by Goddard Space Flight Center astronomers Theodore Stecher and James Milligan, Carruthers's goal was to improve the quantitative accuracy of the measures by both spreading the spectrum wider to reveal detail and rendering the details of the spectrum more clearly (what is known as higher dispersion and resolving power). He would do this by designing and building a new form of camera to be more sensitive to far-ultraviolet radiation, at least as far as the first line in the hydrogen series, Lyman-alpha.[37] This is just what Friedman was hoping for.

By then, Friedman had assigned Carruthers to Chubb's "upper air physics branch" and, in accordance with Carruthers's proposal, Chubb directed him to design an improved UV-detecting system that would solve the technical problems Friedman and Chubb's group were experiencing with UV photon counters. They all hoped this would resolve the nebular glow mystery. Such a camera could also be useful in aeronomy, the core of Chubb's efforts to better understand the Earth's upper atmosphere and its dependence on solar activity, a part of the field called "heliophysics."

Chubb encouraged Carruthers to center his efforts "on an experiment that they were already developing."[38] In support of the nebular glow problem, it was a search for the existence and abundance of molecular hydrogen in interstellar space, as Friedman suggested. Thus, Carruthers chose to reexamine a few of the stars Friedman's colleagues had examined—but now with higher dispersion and more resolving power, reaching shorter wavelengths with greater sensitivity that could explore the Lyman-alpha region—using a new form of camera that he would develop at NRL.

## CARRUTHERS'S CAMERA PROPOSAL TO NRL

Right off the bat, Carruthers proposed that, instead of sending photon counters or a scanning spectrophotometer on a spinning rocket into space, as NRL had done in the past, "it is proposed to investigate the use of the Lallemand electronic camera."[39] French astronomer André Lallemand developed his hybrid camera in the 1930s at the Strasbourg Observatory. It converted optical images into electron patterns that could then be electronically amplified and magnetically focused. Carruthers was familiar with this technology based upon his laboratory experience with gaseous discharge phenomena at Illinois.[40]

Carruthers had already read a persuasive 1956 article by William Baum in *Scientific American* titled "Electronic Photography of Stars."[41] Baum described both the promise and the challenges of the hybrid technology that were created by Lallemand's application of photoelectronic vacuum tube technologies, and his continuing efforts and frustrations attracted other experimenters in England, France, and the United States, including Baum in the 1950s. As Baum excitedly concluded his essay, "If an ideally efficient electronic image tube were combined with the 200-inch telescope at Palomar, it could see as far into space as an unaided telescope with a 2,000-inch reflector!"[42]

Stimulated by Baum's appeal, Carruthers described in detail how he would adapt the technique to make it sensitive to the far ultraviolet. He promised orders of magnitude greater sensitivity but admitted that the feisty technology was far from stabilized and required his close attention in preparation and operation.

## ELECTRONOGRAPHY

The method of photographing or imaging a focused electron beam is variously called "electronography" or "electrography," and it has a complex history.[43] Carruthers made his proposal at a time when the technique was far from stable, which was undoubtedly attractive to him. In 1961, NASA's Ray Hembree summarized the situation at an international image tube symposium in Britain, with the following conclusion:

> It is obvious that each new advancement is the product of much intense effort. Some expressed the belief that the state of the art is advancing slower than was anticipated a few years ago; however, the general feeling was one of optimism. Most of the speakers spoke enthusiastically about their future plans for research and development in their respective areas. It is apparent that this is a business for serious professionals who have chosen a difficult field.[44]

Carruthers fit that description. His plan was to create a photosensitive "electronographic" camera that not only could detect far-ultraviolet light, like Friedman's detectors could do, but also could produce two-dimensional images, or pictures, in that light. First, light collected by a reflecting telescope would be focused onto a photocathode that was made of a chemical compound that could be disrupted by that light, releasing a stream of electrons. Then electrically charged wire coils called "solenoids" would focus the electrons into a beam and also amplify that beam, directing it onto nuclear emulsion film, photographic film sensitive to electrons. The combination would be far more sensitive than photography alone.

Carruthers, repeating Baum's arguments, stated in his proposal that the typical quantum efficiency of regular photographic emulsions was barely 1/10 percent, meaning only one in a thousand photons would be recorded. But his proposed camera could, he predicted, yield as much as 10 to 35 percent efficiency—hundreds of times more sensitive than photography alone—and the response would be linear: "virtually every electron that strikes the plate will cause blackening." Moreover, since this device was an imager, it "combines the high photon efficiency and linearity of the photomultiplier with the integrating capacity of the photographic plate." He proposed laboratory tests to find the best chemical compounds for the photon-to-electron converter, or the photocathode, looking for one that was insensitive to visible light but was primarily sensitive to the extreme ultraviolet.[45]

At the time, there were three general families of image intensifiers based upon vacuum tube technologies in use: signal-generating image tubes, electronographic (or electrographic) tubes, and phosphor output tubes.[46] The first produces a raster scanning electronic signal (a flying spot) that can be read out like on a television screen. The second creates an amplified electronic image that is recorded by photography using an emulsion sensitive to electrons (a nuclear emulsion). And the third projects an electronic image onto a thin and semitransparent phosphor screen that can be photographed or scanned. So-called image orthicons and vidicons form the first and third classes; variations ranged from classical television tubes to modified UV-transmissive systems such as the Uvicon image tube detectors (UV-sensitive vidicons) employed by the Smithsonian's "Project Celescope" that flew on *OAO II* in the late 1960s and provided a first rough map of the UV stellar universe.[47]

Hembree's concerns also have a long history and help one appreciate Carruthers's contributions and the technical hurdles he faced. Telescopic astronomy has always been about gathering more light and concentrating it to make it brighter and therefore recordable. By the 1950s, some leading astronomers, like Baum, echoed concerns that were first expressed in 1944 by Joel Stebbins: there was a "law of diminishing returns" implicit in any effort to build bigger telescopes.[48] But there was another route: as Carruthers pointed out, astronomers were limited because photographic emulsions were highly inefficient. If a more efficient way to record light could be found, then small (40- to 60-inch aperture) telescopes could become as powerful as the largest (200-inch) telescopes.

Baum's *Scientific American* article came out about the time astronomers gathered at the University of Pennsylvania to consider how to make medium-sized telescopes more competitive. They invited those at the forefront of electronic amplification techniques to report and get feedback in search of the best designs. They reviewed different designs for photocells (metallic surfaces that could convert light into electricity). Some were better than others in reducing the many "spurious effects" caused by accelerating electrons in a vacuum tube with high voltages. There were many formulations to consider, as well as designs for vacuum systems that had to be stable and reliable at the focus of a slowly moving telescope, or more to the point, in the nose of a spinning rocket.[49] Between the mid-1950s and the early 1960s, there was definite progress, mainly with one-dimensional or point-source recording. Based upon the Lallemand design, there were a few efforts. A few astronomers and instrumentation specialists at universities and in industry tried to improve Lallemand's camera, making it more efficient and, most important, usable at a telescope where there was typically no climate control. None of the efforts were fully accepted by the astronomical community. Carruthers knew what he was up against.[50]

Carruthers's main contribution was to change the cathode design. Cathodes were typically transmissive, meaning that when the photons hit the cathode, they were converted to electrons that passed through the cathode, emerging on the backside. Given this, cathode formulations were critical: the various formulations of photosensitive electron-emitting alkalis caused different absorption problems. Carruthers had encountered these issues during his graduate studies with vacuum tube plasmas, and so he was aware of the specific formulation options and their relative resistance to environmental changes in the tube. Once at NRL, faced with problems that called for far-ultraviolet sensitivity, he explored options until he hit upon the idea of making the cathode reflective, minimizing absorption and cathode oxidation. It made sense: extreme UV light could not be transmitted through glass, but glass with the proper coatings could reflect it. More on this later.

Carruthers proposed several stages of work at NRL. First, there was the design and development of the camera alone, which could focus light into the cathode, either directly or spectroscopically with a dispersive element like a plane optical grating. Next, he had to design a means to amplify and focus the electron beam created by the cathode, as well as a cassette holding a roll of electron-sensitive film to record the electron beam. He also needed new laboratory equipment, like a suitable testing chamber using simulated stars in the laboratory. And then, when the tests were passed, Carruthers had to construct the flight instrument and again test it before integration into the nosecone of the rocket, monitoring its flight and retrieval. In his proposal to NSF, he included costs for parts, labor, and travel, but not the cost of processing the observational data, possibly because this was a proposal just for the first year.

Friedman agreed completely with Carruthers's objectives. He did caution that the scope of the proposal was "ambitious and we shall have to make compromises." Among the problems were making the design compatible with the payload section of the rocket and the availability of a pointing and stabilizing system for the rocket, which "has not progressed very well and it is uncertain whether it will be feasible soon." By then, NRL flight payloads had achieved a pointing accuracy of one degree (twice the apparent size of the full Moon) on a spinning and gyrating rocket, but Friedman held out hope that this would improve soon, for a variety of applications, a small part of which was Carruthers's goal of homing in on stars.[51]

## MOVE TO WASHINGTON AND NRL

After completing and defending his PhD thesis about a month later than expected, on September 24, 1964, Carruthers was ready to move to Washington. Apparently without fanfare, he touched base at home with family in Chicago and then with NRL funding flew to Washington. He was well aware that he was moving into a new world. He felt generally welcomed at NRL but also sensed a "little bit of skepticism because my degree was in engineering, so I had to convince people that I really wanted to do science as well." He also sensed that

> there was still a dichotomy between engineers and scientists in that when I talked to engineers, especially when I wanted to get parts made in the machine shop, they had sort of a negative attitude towards scientists because they felt that scientists didn't know how to design things, they weren't skilled at putting things together, they were all thumbs. They sort of found it strange that I was claiming to be a scientist, yet I was doing all my own drawings and doing a lot of my own assembly of parts. Of course, I had learned to do that when I was in graduate school, because we didn't have the resources to get other people to do design work and even, to some extent, machine shop work.[52]

Carruthers had now arrived at a place where, as Harry Heckathorn recalled, "Anything that he envisioned, and for which he could muster support, could be built on-site. There were facilities to machine and coat metal parts, fabricate delicate optical and electronic components, and assemble and test instruments in simulated space environments."[53] All true, but the question was, Who would perform these tasks? Carruthers left no doubt that territorial lines between scientific and technical staff were something he resisted:

> I had been used to doing everything myself, so to speak, [and] when I came here I got the impression that people were not used to doing everything themselves. If a scientist wanted a rocket payload designed, he hired an engineer to do it. If he wanted parts made, he went to the shop and had them made. There were very few people who used the lathe or drill press. Certainly glassblowing was something that not very many people did themselves, whereas that was something I had to do myself to a large extent at the university. So I noticed right away that there was a difference in style. I think that the fact that I did have a background in both engineering and science really paid off to a large extent because I was able to get things done quicker and less expensively than people who were used to just sticking with the science end, having someone else do the engineering, because one of the things that people seem to overlook is that there is the problem of communication between the scientists and engineers that makes it less efficient on a manpower basis to have an engineer and a scientist working together, at least on a small project than having one person do the job himself.[54]

From the start, however, as we saw in his reaction to Aerojet, Carruthers wanted to do everything himself, and his striving for independence created "outright hostilities" from the NRL technicians who, he believed, did not "like the fact that I was doing things myself." They felt it was their job and their identity and that he should just stick to the science. When some became overtly confrontational, he just told them to "buzz off."

Some of this territoriality could have stemmed from recent labor management issues at NRL that were only partly ameliorated in 1964 by a labor management contract with the Washington Area Metal Trades Council (AFL-CIO) establishing boundaries, but there were lingering hostilities. Carruthers became aware that there was also overt hostility from some of the "older established engineers and technicians" who scoffed at a young PhD "with no experience [who] really didn't know what they were doing when it came to hardware, which was true to some extent. I have to admit that certainly experience helps a lot."[55]

One can only speculate about the role race played in the reaction of the older scientists and technicians to Carruthers invading their worlds. Despite these early tensions, Carruthers did not recollect any overt racial discrimination at NRL—in his direction, at least. But there were very few Black people who were not in service positions. He was not the first Black scientist at NRL, and there was a younger Black chemist in the division at the time, but they were few and far between.[56] He felt he was well supported by scientists sympathetic to doing things themselves, such as Chubb, Byram, Charles Johnson, Harry Merchant, and especially Julian Holmes: "[Holmes] actually helped me quite a bit in the early years, because after I did become permanent, I didn't really have a staff of my own."[57]

Carruthers soon proved himself and gained acceptance designing, assembling, and testing his instruments in his own way. He had imported the Illinois do-it-yourself–style framework and architecture in constructing his laboratory, which Talbot Chubb noted years later as quite distinctive.[58] He well knew that his "Illinois" laboratory style stood out at NRL, reflecting the fact that he learned his trade "at the university where resources were rather limited and certainly manpower was limited."[59]

The laboratory atmosphere Carruthers encountered can best be described by introducing those Carruthers worked with most closely. Holmes, trained in physics at Bowdoin College, arrived at NRL in 1951 working first under Friedman and then with Charles Johnson on a broad range of related projects from electronic countermeasures to ionospheric physics and aeronomy. He was a liberal citizen activist and was especially open to mentoring Hulburt fellows.[60]

Talbot Chubb was an amiable boss. A 1944 graduate in physics from Princeton with Army service in the Manhattan Project at Oak Ridge, Chubb obtained a PhD from the University of North Carolina in 1951. NRL hired him in the fall of 1950, and because of his background in gas discharge physics, Friedman assigned him to work on Geiger counters for the UV and X-ray regions of the spectrum. Through the 1950s, as he recalled in an oral history, "So I naturally was, I guess, more or less assigned the responsibility for all the sensors that we put on the rockets."[61]

Byram was Chubb's key colleague. Byram arrived at NRL in 1947 from the University of Toledo and was assigned to Friedman's branch. An electrical engineer assigned to the science section (a fact that was noted from time to time in territorial disputes), Byram's first task was to build a payload for Friedman's first V-2 flight of UV and X-ray Geiger counters, the instrument that flew in September 1949 demonstrating that "there was tremendously more radiation in that part of the spectrum than anybody had expected."[62] Together, they formed an amiable team; Byram, always forcefully focused on the experimental problem at hand, complemented Chubb's leadership, making Friedman's Geiger counters work on V-2, Viking, and Aerobee rockets and continuing with SOLRAD and Skylab in the 60s and 70s, and the huge High Energy Astronomical Observatory (HEAO) through the 1980s. During an interview with Chubb, Friedman, and Robert Kreplin, Byram was asked if he was ever frustrated by trying to find the best Geiger tube to perform the mission, to which he gruffly replied, "I was never frustrated, I enjoyed fighting them. It wasn't a frustration; it was a challenge: It was mind over Geiger tube."[63]

Then when asked how he did that, he mildly added, "By keeping at them. Just looking at them more or less constantly, checking them." In the same spirit, Charles Y. Johnson in Chubb's aeronomy section passionately testified that "you constantly keep testing, testing, testing, just to be sure everything is the way you want it. The last test is when you fire it." Byram, Chubb, and Johnson personify the world that Carruthers fit into so well.[64] This was the world Carruthers entered.[65]

## A NEW WAY TO DIVE DEEP INTO THE ULTRAVIOLET

By August 1965, building upon his laboratory work at Illinois, Carruthers had evaluated various commercial electrostatically focused tubes and focused on designing and building a new and hopefully more stable magnetically focused system that used front surface reflecting photocathodes. He experimented with different cathode formulations, eventually preferring potassium bromide, and tested the quantum efficiencies of all these variations against commercial products. By this time, he had moved away from the highly complex Lallemand and Kron designs, preferring to develop a smaller and more efficient linear design similar to those developed by the Carnegie Image Tube Project and developed at the Farnsworth Laboratories in Fort Wayne, Indiana (now ITT).[66] The Carnegie design employed accelerated electrons to bombard a static phosphor that then could be photographed. Carruthers, however, felt that phosphor screens and multiple amplifying stages in these designs introduced "additional statistical processes" that would complicate the processing of the data.[67]

Carruthers wanted to design, build, test, and prove that he could produce stable and scientifically useful two-dimensional images both directly and spectroscopically. Direct imaging in spectroscopy, catching the whole spectrum at once rather than scanning, would greatly improve the efficiency of the process.[68] Carruthers also knew that his instrument had to be designed to work in the vacuum of space. This created challenges, but also simplifications. The Lallemand-type instrument, designed for ground-based work, had to be encased in glass and had detector elements sensitive to the visible range of the spectrum. In space, "front-surface photo-cathode devices," Carruthers argued, "do not require the use of a window at all, provided that the necessary high vacuum is provided by operating the device within a vacuum chamber, or in outer space."[69] Further, as noted, most image intensifiers of that day focused electrons onto a phosphorous screen that was then photographed in the visible range. Carruthers's design recorded the electronic signal directly using fine-grained nuclear emulsions with linear response.

In his first year, working largely on his own, Carruthers was relatively free to move forward at a pace unhindered by the typical bureaucratic requirement of filling out requests and then waiting in line for technical assistance for parts and labor. Even more than in college, Carruthers kept to himself at NRL as much as possible. He got help when he needed it but rarely if ever engaged in social activities outside the laboratory. His background, talents, and interests fostered his isolation.

Carruthers knew that his colleagues were well aware that "it was a little risky in having young PhDs go off and do their own rocket experiments."[70]

So in his first year, Carruthers was guided by Julian Holmes, who was adept with hands-on laboratory work, as well as staff in NRL's Electronic Imaging Section. Holmes, as well as others like Harry Merchant, mentored the young postdoctoral workers.

Toward the end of his first year, Carruthers knew that his instrument would not be ready for flight and that the needed pointing controls were still being refined and tested. Pointing an instrument at a celestial source from a flying rocket required some form of stabilization—of the rocket and the instrument. The challenge, since the late 1940s, was to stabilize narrow-field instruments in all three degrees of motion on spinning and gyrating sounding rockets. It was achieved for solar spectroscopic investigations in the early 1950s. The Sun as a target was relatively easy to home in on because, due to its brightness, a cluster of small photosensitive sensors driving a servo-feedback gimballed system could be used to focus on the image and hold it steady for the few seconds required for exposures. Stars were another matter. In 1956, reviewing the state of the problem, Richard Tousey knew that "many experiments can be imagined if an accurate pointing control were developed."[71]

In the early 1960s, Friedman and others had been hoping to reach a pointing accuracy of one minute of arc. By the time Carruthers arrived, they were closer to their goal—close enough to support Carruthers's plans with his relatively wide-field cameras. So Friedman reported positively to NSF in August 1965 that Carruthers's work had progressed well but his fellowship had to be extended for another year so he could fly his completed instrument on a suitably stabilized Aerobee sometime in 1966. NSF quickly approved "pro forma," telling Friedman that it was "based upon favorable impressions from your staff."[72] Indeed, Friedman had praised him to the skies (see figure 4.2).

## SECOND POSTDOCTORAL YEAR AND FIRST AEROBEE FLIGHT

In his yearly report to the American Astronomical Society in December 1965, Friedman hailed the accomplishments of the E. O. Hulburt Center for Research—namely, the work of the fellows in its first three years. About half a dozen young scientists had prepared payloads for Aerobee and Black Brant rocket flights. Two came from UCLA and one each from the Catholic University of America, Yale, and the University of Illinois, as well

Figure 4.2
(r to l) Carruthers and Friedman inspecting an early Aerobee payload circa 1967. The payload camera is in spectroscopic mode, which at first utilized a mosaic of four plane gratings to feed the camera. NRL photograph P-1051(6). *Source:* Courtesy of Naval Research Laboratory.

as a young professor from Cornell, Martin Harwit, who, encouraged by Friedman, was pioneering infrared observations from rockets.[73]

Friedman's report on Carruthers's work came last, was the most detailed, and evidently was intended to climax his presentation. He described Carruthers's test prototype for a camera that could produce both images and spectra. Friedman made much of the detector design he called "electronography" and highlighted that it would be directed to find interstellar molecular hydrogen in the far-ultraviolet region of the spectra of early-type stars.[74]

A flight was scheduled for spring 1966. But Carruthers was already developing other potential payloads (such as more traditional UV image orthicons), assisting on other projects, and beginning to prepare preliminary reports and publications.[75] He well knew that he was not the only game in town trying to explore the Lyman-alpha spectral region, though most were looking for different aspects of it. By then, NASA's Orbiting Geophysical Observatories (OGO) were flying and carried UV-sensitive ion chambers built by NRL staff working in the aeronomy branch headed by the theorist Phillip W. Mange.[76] There was also continuing collaborative sounding rocket activity at both NRL and Goddard, where Yoji Kondo and James Kupperian were flying payloads to try to determine what was causing the diffuse nighttime radiation near Lyman-alpha, speculating that it might have multiple origins.[77] In academe, Princeton astrophysicist Lyman Spitzer led a team exploring these questions with sounding rockets and was actively planning for satellites. So Carruthers was not alone searching for discrete sources. But Carruthers's camera was unique, having the dynamic range and field of view to sense the faint immediate environment of a source.

There were many wrinkles in the refining process. Carruthers had now modified his first designs to reduce weight by switching to high-voltage electrical solenoids to produce the focusing fields. But these also proved to be hard to shield, and they needed a lot of power. So he switched back to heavy circular magnets to focus the electron beam. Initially in solo rocket flights, this would be fine, but in a multi-instrumented flight, as NRL colleague Charles Brown recalled, the magnets also needed an outside covering to nullify the field and avoid confusing the rocket's control systems and other experiments, and this added more weight.[78]

Carruthers benefited from the expertise of others in the Upper Air Physics Branch. Grady T. Hicks, who directed the Electronic Imaging Section, and George G. Barton spent months testing and evaluating the payload, and by July 1965, after inserting both a direct camera and a spectrograph into the instrument section of an Aerobee, they hoped that it would be ready by next spring to fly in an Aerobee 150 from the Army's White Sands Proving Grounds—then the largest overland launching facility—some thirty-five miles northeast of Las Cruces, New Mexico.

NASA's Sounding Rocket Program managed launch operations in cooperation with the Naval Ordnance Missile Test Station there. In contrast to NRL's collaborative team approach preparing the payload, White Sands had, understandably, a highly formal, multifaceted "Operations Requirement" for the launch. Carruthers sent a copy of the sixteen-page directive to his former thesis advisor, Ladislas Goldstein, evidently to show him the challenges he faced in his new career.

A handbook for NASA's Wallops Island facility, another launching site, described the overall challenge and the responsibilities of the two main groups: "The launch site is a place where fluctuation is almost inevitable, and flexibility is a constant requirement. The Mission Team is focused on operations, the PI's [Principal Investigator] team is focused on science requirements and everyone is adjusting to the safety and operational rules governing the location."[79]

The April 30, 1965, "Operations Requirement" directive that Carruthers sent to Goldstein spelled out the many steps and the workforce requirements for what was planned to be a launch in late June 1966. The NRL team consisted of eight people. Friedman's most experienced technical man, Edward T. Byram, was the project manager and Carruthers the project scientist.

The NRL contingent was expected to be on site some twelve days before the expected night launch. Housing was arranged, and they were allotted a large room in the "upper air laboratory" equipped with tanks of dry liquid nitrogen. After inserting the payload into the rocket, they were expected to complete "horizontal checkout" for weight and balance at least four days before launch and "vertical checkout" a day before, with constant checking up to the launch (see figure 5.5). NRL required access to the experiment doors containing the payload during the vertical check, so they requested a specific launch tower, "Navy Aerobee Tower 'A,'" which made the payload reachable. It was also tiltable to account for variations in wind patterns. The operations requirements also identified the fuel mixture, weights of each of the sections, overall dimensions, and expected general performance.[80]

In addition to assembly and testing, there were constant safety procedures. Among the larger components to oversee included a "Command Control/Destruct System" to terminate propulsion if necessary, flight safety circuits controlling detonators to deploy the payload recovery parachute, battery-powered communications and control systems, and two 20-kV power supplies in the experiment section that "may still be active after payload impact."[81] NRL would be responsible for handling them. The launch preparations also brought in several contractor teams, including Space General Corporation's Field Service Group for the attitude control system, and the Physical Science Laboratory at New Mexico State University for the telemetry.

In early July 1966, Carruthers and his team air shipped their payload to White Sands for integration, and as NRL scientists said repeatedly, the rest was "testing, testing, testing." Integration included "dynamic balance," assessing the moment of inertia (the body's resistance to rotation) and center of gravity of the overall ensemble as well as vibration testing, magnetic calibration, and finally, most critically, optical alignment.

The Army closed Highway 70 on the afternoon of July 18, after all-day preparations for a night flight by Naval Ordnance Test Facility personnel. Balloon soundings by the US Army Atmospheric Sciences Laboratory assessed wind patterns in the upper atmosphere to predict where the rocket would land (hopefully in Army territory). At 20:36 hours MST, the Aerobee cleared its launch tower.

The rocket reached a peak altitude of only 105.7 miles in some 222 seconds and landed, after 480 seconds, about 44 miles from the launch tower. The nosecone and payload had separated from the rocket beyond peak altitude, an essential action that preserved the payload and data. But it took longer to reach the ground by parachute, and the prevailing winds carried it farther into the desert. It was a less-than-perfect flight for this normally reliable rocket, giving Carruthers's payload little if any time at the altitudes his experiment required to completely evacuate the chamber, and to observe the target areas.[82] Further, the critical pointing system that kept the cameras on the target balked, and some electric arcing occurred in the high-voltage systems that Carruthers's payload carried.[83]

Prompt recovery was a final critical step and "absolutely essential to the success of the mission, as the data is recoded on film, and this film may deteriorate if exposed to desert heat. It is very important that the payload not be moved or disturbed in any manner by other than qualified personal." Accordingly, the morning after the launch, Carruthers and other NRL staff raced to the landing site via helicopter.[84]

This first flight was unsuccessful, but Carruthers did not skip a beat. In September 1966, he updated his former thesis advisor, Ladislas Goldstein, in frank terms, specifically blaming the faulty attitude control system and the Aerobee's failure to reach peak altitude. And in addition to planning another flight of the Aerobee payload, he was also adapting it for "a similar experiment on an Apollo Earth-orbital mission scheduled for late 1968."[85]

Carruthers also kept his family informed. In another September letter, to his uncle Ben, he apologized for being silent "for so long." He explained the technical reasons for the failure but confidently assured his uncle: "However, we hope to try again in January. We are making changes in the experiment; hence it will not be exactly the same as the one described in the reprint."[86] And as he did for Goldstein, he enclosed a reprint of his first publication in the "Report of NRL Progress" that described his instruments before the launch attempt.[87]

## SOME FAMILY CONTACT

Carruthers's report to his uncle was mainly about his work. Only at the end did he share personal family information—that brother Gerald ("Jerry") had returned from military duty in Korea and was now stationed at the Redstone Arsenal in Huntsville and that Anthony ("Tony") was returning from Vietnam in a month: "So we all hope to be home for Christmas. . . . Love, George."[88] Carruthers's hope to connect with his family in 1966 sheds a bit of light on family history, which is worth a short diversion to catch up on his personal life.

Carruthers occasionally visited his family in Chicago. His mother, Sophia, had remarried in 1960 to Wellington S. Martin, a retired Army officer, and in 1963, nurtured by his brother's efforts, Gerald had made a name for himself, enjoying local notoriety by winning science contests at the city and state levels in high school with a project on ionic rocket propulsion.[89]

Sophia Carruthers's marriage to Martin, whom she met while working at the post office, brought the Army into the family. Martin was soon called to active duty as a warrant officer in 1961 during the Berlin missile crisis and moved the family (but not George, then in college, or Anthony, who had enlisted in the Army) to Fort Lee, Virginia, where they stayed for about a year.

Anthony was then training as a nuclear weapons specialist at Los Alamos, becoming a skilled machinist and serving at posts in the United States, Germany, Vietnam, and Korea. In the 1970s, he was an instructor at the Aberdeen Proving Grounds in Maryland, providing expertise in machining.

After he graduated high school in the early 1960s, Gerald also enlisted and was posted at the Redstone Arsenal in Huntsville, Alabama, which was then being transformed from a "cotton town" into a "space research mecca" by NASA.[90] He served a tour of duty in Korea, providing Sergeant missile technical support, and then returned to Redstone. Supported by the Army and sharing his brother's passion for rocketry, Gerald worked with all types of rockets and missiles. He received advanced training at Redstone's Missile and Munitions Center and School to develop, maintain, and deploy the guided missile systems being developed there.[91] The Army also supported his education at Alabama A&M, then a vocational school and the "main historically Black college in the area."[92] Gerald graduated with a bachelor of science degree in business management in 1977. He remained at the Redstone Arsenal where he worked on the development and deployment of the Sergeant missile system, taking short-term assignments to other missile facilities in Italy, in California, and at the Aberdeen Proving Grounds in Maryland. After retirement from the Army in 1985, he worked in Saudi Arabia, then returned to Huntsville, where he took the "Technical Training Master" course at A&M for a master's degree in 1996, and then worked as a technical trainer and field service engineer in energy management.[93]

George had also been active as an R.O.T.C. cadet, rising to the rank of brevet second lieutenant in April 1957, his senior year of high school.[94] He continued in R.O.T.C. at the University of Illinois because it was still required at public universities at the time.[95] At NRL, he was, of course, not exempt from the draft, but he sought and received deferments through testimony from the NRL administration.

For instance, when his number came up in 1969, Friedman and NRL's administrative chief, Captain S. N. Ross, speaking by "direction of the Director," stated bluntly, "We do not have a person of his skill and ability who would be able to carry out this work, if we were to lose Dr. Carruthers' services." Carruthers had by then more than demonstrated his relevance to the "needs of the fleet" as we note throughout this biography. As Ross made clear, Carruthers's value went beyond the science, and he stated, "In addition to his unique expertise in "special TV sensor[s] . . . he is in charge of a night vision development program of considerable interest to the Navy."[96]

# 5

# IMPROVING HIS CAMERA

Just as he would later support Carruthers in his appeal for deferments, Friedman provided sterling endorsements to NSF for renewing Carruthers's first Hulburt appointment: "Dr. Carruthers is not only an outstanding designer and experimenter, he is also a penetrating scholar who has a broad understanding of the literature of his field, as exemplified by characteristic articulate expression of great clarity in oral presentation of his work."[1]

Near the end of his second postdoctoral year, Friedman brought Carruthers onboard as a full-time staff member at NRL, remaining in Chubb's group. Unlike other Hulburt fellows, who like Martin Harwit, Richard Conn Henry and Paul Feldman maintained ties with academe during their fellowships, there is no record that Carruthers sought, or was sought out by, academic or commercial firms in aerospace.[2] NRL had become his home.

In his continuing efforts to make the instruments he designed and built work on rockets in the late 1960s, as a staging ground for satellite berths, Carruthers made the first detection of molecular hydrogen in the space between the stars, a major discovery confirming how stars are born and what the universe is made of. In 1966, as he told his uncle, he was hoping for success; by 1969, he was much closer to success; and by 1971, he had done it. And he promised to do much more. But it was far from a simple path to success. Here we detail that process and describe the instruments and the history of the technology in some detail.

Carruthers and his colleagues quickly evaluated the arcing problem after the first flight of his payload in July 1966, but it took many months to make suitable alterations and refinements to the cameras and to the pointing control systems. Typically, NRL payloads were not evacuated prior to launch and depended upon the payload outgassing in the vacuum of space. The flight's low peak altitude and short flight duration did not provide enough time for exposure to a true vacuum. NRL staff knew of this problem before the flight since the voltages as high as those he required "had never been proven in rocket flights."[3]

Carruthers appealed to NRL's infrared group for advice because they worked with techniques for evacuating their payloads, especially after Martin Harwit from Cornell had joined the group and incorporated cryogenic cooling into their designs. Carruthers's first modification, therefore, was to create a payload that could be evacuated just before flight within the sealed nosecone of the Aerobee. Once the rocket was at altitude, a cover would open, exposing the instruments to space. So Carruthers "decided to borrow the infrared group's technology and use an evacuated rocket payload, which was, at least, in the UV field, the first time that anyone had done that. That was probably part of the reason that I felt that they encouraged me to pursue that, even though everyone was skeptical about the high voltage problem."[4]

Carruthers's first formal presentation of his camera design, one of those praised by Friedman and the one he sent to his uncle Ben, had appeared in the internal publication "Report on NRL Progress" in July 1966, just a few days before his first flight. He described his system as capable of being flown on sounding rockets and eventually piloted spacecraft to probe the universe electronographically, for imaging and spectroscopy. Calling his design a variant of the Lallemand design, he emphasized its high quantum efficiency, possibly higher than photography alone. Unlike the Lallemand, however, or other variations of image tubes that were designed for the visible regions of the spectrum from ground-based observatories, his were for the ultraviolet and used reflective calcium bromide photocathode formulations that were less sensitive to the environmental factors affecting those working in the visual region. "The elaborate precautions associated with electronography when using visible sensitive photocathodes," he wrote, "are not necessary with the alkali-halide photocathodes."[5]

Carruthers's patent application also went into great detail showing how it was distinguished from all earlier electronographic systems.[6] He emphasized that it was the first modern electronographic system that did not require transmission through any substance because his optics (save for the thin correcting plate in the Schmidt design option; see below) and his photocathodes were all reflective. This provided higher quantum efficiency, critical for detectors in rockets and satellites. As one reviewer stated, "Many aspects of this camera are novel, but of particular interest is the use of a reflective photocathode."[7]

The patent was approved some three years later, and during that time, Carruthers continually modified his designs for different applications. As a patent issued to the Navy, it was available for any government application. Although patents were not rare among Black inventors by then, Carruthers's patent has been noted frequently by those who chronicle the history of Black inventors.[8]

Throughout the decade, Carruthers continued to craft proposals for a variety of missions and to build an assortment of electronographic cameras and spectrographs as well as photon counters for a wide range of far-ultraviolet applications. He also started publishing general and highly informative reviews of the technology. In the March 1969 issue of *Applied Optics*, he described the history of the technique and the various designs that he and others had created. He then detailed his own efforts: their development, testing, and use. Emphasizing the promise of the electronographic technique due to its "very high quantum efficiencies," he described that it could, like classic photography, "integrate, simultaneously, the photon flux at each point of a two-dimensional image." In other words, it could create a "picture."[9]

His review covered the history of the technique, analyzing Lallemand's design in detail as well as designs by other pioneers like Vladimir Zworykin, James Dwyer McGee, and Gerald Kron. His theory section reviewed extant literature in succinct and accessible terms, and then he went on to explain why the use of modern nuclear track emulsions, consisting of a silver halide formulation that could record ionizing particles and therefore be sensitive to an electron flux, were superior in speed and resolution to the most sensitive high-speed photographic emulsions.[10] Beyond sensitivity, nuclear track emulsions also record linearly, recording every incident electron, and so the image density is directly proportional to the initial photon flux on the photocathode. He emphasized this point, adding again, as we have seen, that his UV-sensitive alkali halide formulations were reflective, or front surface photocathodes, unlike his predecessors.[11] In this arrangement, photons striking the photocathode were converted into electrons that bounced back in the direction from which the photons had come. His cathodes were also not sensitive to visible light, so they were not affected by scattered light or by overlapping orders of visible portions of higher-order spectra from the spectrograph's dispersive grating.

Carruthers emphasized that his detectors were relatively insensitive to environmental factors. Unlike the visual range of the spectrum—Lallemand,

McGee, and Kron detectors—they would not be seriously affected by exposure to dry air, by the violent outgassing of the system during the initial rocket flight, or even by the outgassing of the nuclear emulsions once in total vacuum. But he was also aware of the fact that he was introducing a technology that had not been fully accepted; astronomers were generally reluctant to deal with the complexities of the Lallemand and its successors.[12] So he once again emphasized that his design differed from, and was much simpler than, earlier electronographic designs as well as commercial electrostatic image tubes then available. In an electrostatic tube, the electron beam fell on a transmissive phosphor that then converted it into visible light for recording, and, as with the commercial RCA Vidicon, the image on the phosphor was electronically scanned and amplified to produce an electronic image. Variations on the latter design, the basis of television, were critical for transmitting data from a satellite to a ground station. But Carruthers was thinking only of situations where the film could be physically retrieved, like on a sounding rocket or a human-tended spacecraft. In sum, he argued, "Electronography gives the highest resolution, as the electronics are recorded directly without the additional statistical processes introduced by the use of phosphor screens, multiplier stages, etc."[13]

Overall, Carruthers argued that his detector design was the most efficient available at the time. Indeed, in an internal NRL report in 1966, he described the advantages in detail: transmissive cathodes blocked the far ultraviolet, so "much higher quantum efficiencies may be obtained" due to the "elimination of transmission losses." He also argued that all-reflecting optics and cathodes made "the system most simple and compact."[14] And in sum, Carruthers reported that his detectors had demonstrated gains "of as much as a factor of 20 over conventional photographic techniques."[15]

He could make his reflecting photocathodes as thick as needed to convert virtually all the incoming photons into electrons, which would then be reflectively emitted and accelerated and focused by the magnetic field. And unlike ground-based electronographic designs, no entrance window was needed in the vacuum of space, so there would be no absorptive losses. Even the optical focusing element could be part of the camera's electronics, acting as the ground for the electrode.

Carruthers used two basic telescopic designs to collect and focus light onto his photocathodes. The first was an all-reflecting compound design created by the German astronomer Karl Schwarzschild in 1905 that had a flat focal plane free from image errors and distortions called astigmatism and coma. The Schwarzschild reflector produced high magnifying powers and a narrow field of view, so Carruthers also applied the Schmidt camera design, created by Bernhard Schmidt in the early 1930s, for wide-field spectroscopy and imaging. Light enters the Schmidt camera through a thin glass optical "corrector" that adjusts the beam so that it can be focused properly by an optically fast, short-focus spherical primary mirror to produce a wide field that images stars precisely on a slightly curved focal plane (see figure 5.1). Carruthers made the corrector plate out of a UV-transmissive glass, typically lithium fluoride, which reduced UV light loss.[16]

Like the Schwarzschild, the Schmidt could look directly at the sky to image objects or, by placing a large flat optical grating in front of the camera, could produce spectra of objects in the field. Carruthers's early Schmidt cameras used either 3- or 4-inch primary mirrors with the optically fast focal ratio f/1.5, which focused light directly onto the photocathode. Again, the electrons emitted by the front surface photocathode would be magnetically focused through a hole in the primary mirror and into the film transport.[17] The magnetic focusing system surrounded the optics (see figure 5.2).

The Schwarzschild and Schmidt designs provided versatility, ranging from moderate- to wide-field capabilities, but Carruthers went even further, designing, building, and testing a third camera for wider field imaging, with a 3-inch primary mirror, operating at an extremely fast f/1.0 focal ratio that provided a 20-degree circular field of view. Exhibiting great range and depth in exploring instrumental design, he was out to cover a wide range of opportunities, from examining selected objects and fields to mapping the entire UV sky. Typical for an inventor, just like Herbert Friedman, and in words variously attributed to Mark Twain, Abraham Maslow, and others, "to he who invents a hammer, everything becomes a nail."

## A SECOND LAUNCH

For his second Aerobee flight in March 1967, Carruthers tried to eliminate the voltage problems and erratic magnetic contamination his solenoid-powered electromagnetic focuser produced by switching back to a series of shielded permanent magnets. This also eliminated the power supply for the solenoids and reduced the problem of heat dissipation, but it increased

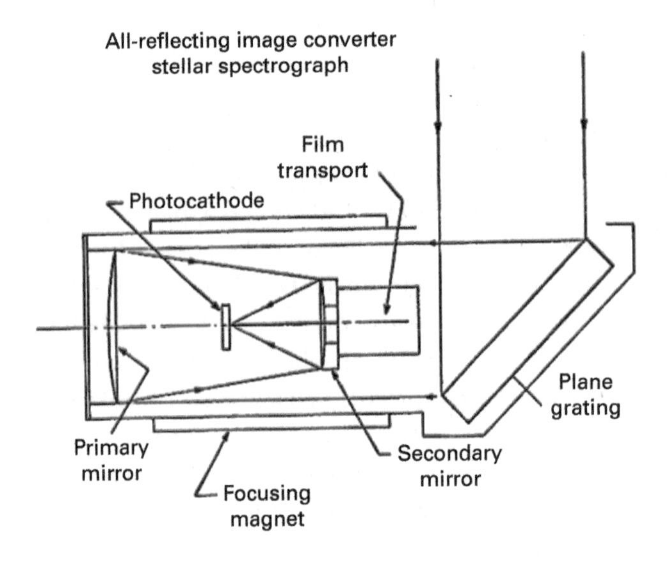


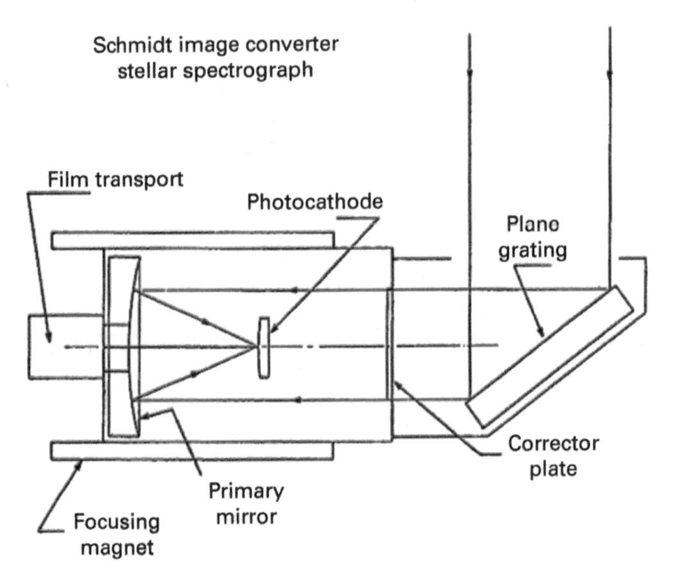


Figure 5.1
Schematics of Carruthers's 1968 versions of the Schwarzschild and Schmidt telescopic cameras in spectroscopic mode for Aerobee flights. In both, once at altitude, the payload door was ejected, emitting light onto a reflecting plane grating that sent spectral light to the left, where it was collected and reflected by the concave primary mirror. The Schwarzschild design secondary mirror is concave, which further condensed the light onto a photocathode. The detector then converted the light into a beam of electrons. That beam was then amplified and imaged by a focusing coil back through a hole in the optical secondary mirror and onto a strip of electron-sensitive nuclear emulsion in the film transport. In the wide-field Schmidt configuration setup

Figure 5.2
Carruthers's Schmidt 3-inch camera with external optical grating. The large diffraction grating is on the right. The lithium/fluoride entrance window is visible as are the magnets that amplify and focus the electronic image onto electron-sensitive film in the small cassette on the back of the cylinder. *Source:* Figure 2 from Carruthers (manuscript), "Further Developments of Magnetically Focused, Internal-Optic Image Converters," paper presented at the Fifth Symposium on Photoelectronic Image Devices, Imperial College, London, September 16, 1971, folder "NRL, Geo. Carruthers Mat., 1972," box 19, HF/APS. NRL photograph 75144(3). *Source:* Courtesy of Naval Research Laboratory.

Figure 5.1 (continued)
for spectroscopy, once the light is dispersed by the plane grating, it then passes through the lithium fluoride Schmidt corrector plate. After reflecting and being focused by the large spherical concave mirror on the left onto the convex photocathode, the resulting electronic beam is focused through a hole in the primary mirror and into the film transport. *Source:* George R. Carruthers, "Magnetically Focused Electronographic Image Converters for Space Astronomy Applications," manuscript draft, October 18, 1968, figure 3: "Objective-Grating Image Converter Spectrographs for Far-Ultraviolet Stellar Spectroscopy," folder 1966–1969, "NASA, Carruthers Proposal, 1966–1968," box 41, HF/APS. Published in *Applied Optics* 8, no.3 (March 1969), figure 3. Image reprinted with permission from © The Optical Society.

weight from the heavy shielding. After rounds of preparations, testing, and calibration in his laboratory, which included the integration of the payload into the interior skeletal framework of an Aerobee instrument section, Carruthers and Holmes decided they were ready to try a second launch for spectroscopic observations. At NRL, and again at White Sands, the entire instrument and structural framework were fitted into a specially constructed chamber that could maintain a high vacuum before launch. These were typical stages in preparing for and retrievals at White Sands, Wallops Island, and other sites in the late 1960s through the 1970s (see figures 5.3 through 5.6).

Once at White Sands and after final integration and more testing, the Aerobee was launched on the night of March 16–17, 1967, and this time reached some 120 miles. The payload included his Schwarzschild system as well as two collimated UV photon counters to sense the intensity of the Lyman-alpha line and to act as fine guidance sensors. By then, on a NASA contract managed by the Marshall Space Flight Center, Space General Corporation in Southern California, a merger of portions of Aerojet and the Applied Physics Laboratory, had improved the all-important pointing control system by adding a "cold gas jet-gyro referenced control system" that could roughly stabilize and orient the rocket after burnout, and then employ "inertial-reference gyros" calibrated by optical sensors to home in on the desired stellar objects.[18] They were programmed to direct the spectrograph to sequentially fix on some eight blue stars in the southern constellations of Vela and Puppis during the flight. After the pointing controls automatically found their targets, exposures commenced, and the spectrograph reached deep into the far ultraviolet, far enough to record wavelengths as short as 1,030 Angstroms, which included Lyman-alpha at 1,216 Angstroms.

After retrieval at White Sands and post-flight laboratory testing at NRL, Carruthers analyzed the photographic record of all stellar spectral features in the region, deriving their physical intensities and line widths. His spectra revealed strong bands at 1,110 to 1,140 Angstroms, probably produced by dissociated water outgassing from the Aerobee. But there was no evidence of the known Lyman-alpha line. The flight was deemed technically successful, but the scientific return less so.

The measurements of continuum intensity in the region, however, confirmed that there were deviations from what was expected from theoretical

Figure 5.3
Carruthers's NRL laboratory circa December 1967 showing a modified Aerobee payload with an improved single full aperture grating. The outer skin of the Aerobee payload section containing the aperture door is visible on the right. In back are vacuum test and calibration equipment with an Aerobee payload frame. George Carruthers photo. *Source:* Courtesy of Gerald Carruthers.

calculations of the spectral energy curves of the hot stars. So in his report, Carruthers went into great detail describing all possible instrument errors, as well as unknowns in the astronomical realm that could alter the observed intensities, like absorptive reddening by dust in the interstellar medium or by what is called "line blanketing," where an increase in metal abundance in the star blocks the ultraviolet and reddens the spectrum.[19] He also concluded that there was less atomic hydrogen than expected, and no direct

Figure 5.4
Calibration and testing. The camera is enclosed in the Aerobee payload section at the bottom of the skeleton frame of the nosecone. The calibration lamp is the dark cylinder on the left, which is the direction the camera will look at the sky. *Source:* George R. Carruthers, “Magnetically Focused Electronographic Image Converters for Space Astronomy Applications,” *Applied Optics* 8, no.3 (March 1969), figure 7. Reprinted with permission from © The Optical Society.

Figure 5.5
Final assembly and testing of an Aerobee 150 at White Sands, May 5, 1972. Carruthers in white shirt with back to camera. NASA Aerobee B.019 DG. *Source:* Carruthers Collection, box 5, Archives Department, Smithsonian National Air and Space Museum (NASM 9A19881).

evidence of molecular hydrogen. Most of all, he made clear what improvements were needed in future flights to improve the results.

## CARRUTHERS'S GOALS

Carruthers was led to these problems from many directions—from his extensive reading as well as from his direct interests at NRL and elsewhere—and here we briefly cover them. There is no doubt he wanted to demonstrate the superiority of his design. But, following Friedman's and Chubb's advice, he also wanted to answer important questions in aeronomy, astronomy and solar physics. Case in point: following Friedman and Chubb's earlier efforts, he decided to focus on cool dusty regions surrounding hot stars to determine if molecular hydrogen is there, and how much there is. Certainly, it was a question far from the interests of the Navy, but as with Friedman and Chubb's efforts as well as the efforts of many researchers in

Figure 5.6
Retrieval of a nosecone at White Sands on August 11, 1967, by (r/l) Carruthers, Julian Holmes, an Army soldier, and N. Paul Patterson. *Source:* Courtesy of NRL. Copy in the C.Y. Johnson Collection, box 5, folder 10, Archives Department, Smithsonian National Air and Space Museum (NASM 9A19890).

military laboratories, these were the types of questions that could lead to new and more useful technologies.

But why would this be an important finding in astronomy? A significant portion of the gases between the stars—the interstellar medium—was known to be composed of atomic hydrogen ($H_1$) from radio observations, but what about its neutral molecular ($H_2$) form that one would expect to find in the cool vastness of space? Stars evidently formed from clouds of hydrogen, so it was important to know how much was available to make stars. Since the 1920s, atomic hydrogen was well known to exist throughout the universe as the primary component of the Sun and stars, and to exist between the stars. But astronomers wondered how much molecular hydrogen existed in the depths of space, possibly on cold interstellar dust grains and far from the ionizing radiation of the stars. Theorists had predicted it since the 1930s, and finding it was a primary goal of Lyman Spitzer's proposed orbiting UV telescope, the third planned Orbiting Astronomical Observatory 32-inch telescope called *Copernicus*, which he hoped in vain would launch in the 1960s.[20]

An even broader aspect of this problem lay in the perennial question astronomers ask: Is that all there is? This question had been brewing since the 1930s, ever since astronomers started noticing that matter in the large-scale universe, like clusters of galaxies, held together and did not disperse even though the galaxies themselves were moving fast, too fast to be part of a stable system. The issue gained wider attention in late 1960 when the famed Princeton astronomer Martin Schwarzschild, son of Karl, addressed a combined meeting of the American Astronomical Society, the American Association for the Advancement of Science (AAAS), and the American Geophysical Union (AGU) in New York City. Promoting *Project Stratoscope*, a 36-inch balloon-borne reflecting telescope program he headed, Schwarzschild was quoted by Walter Sullivan in the *New York Times* that its infrared capabilities at the top of the atmosphere might reveal this missing matter in the form of "billions of 'cool' stars whose light is too deep in the infrared spectrum to penetrate the atmosphere." Schwarzschild also speculated that the missing mass may "be in the form of hydrogen molecules spread as a thin gas throughout space," which would have a UV signature. Even though his telescope was tuned to the infrared, Schwarzschild admitted that, of the portions of the spectrum blocked by the Earth's atmosphere, "paramount among these are portions of the ultraviolet spectrum."[21] This was not a one-time-only pronouncement. At meetings, Schwarzschild would sometimes quip in his warm lilting German accent, "Ninety percent of the universe, we don't know what it is!"[22]

Spitzer and his group at Princeton and others at Goddard Space Flight Center and NRL were all trying to catch a glimpse of the UV universe, focusing on blue stars hot enough to render gases in their vicinity visible in the ultraviolet. Similar questions were also being posed for unseen matter between the galaxies. Friedman knew that his negative results for nebular glow around hot stars was not conclusive, and this stimulated others to continue the search.

As he was advising Carruthers, Friedman knew that another Hulburt fellow, Richard Conn Henry, now at nearby Johns Hopkins University, had teamed up with Byram, Chubb, and John F. Meekins, to get a first estimate from an Aerobee flight of X-ray proportional counters. It would eventually fly in September 1967 and hopefully find some evidence for a "hot, dense intergalactic medium."[23] It was promising but also provisional.

These were important questions in astronomy, and they still are. In his "Annual Report" from the Hulburt Center for 1968, Friedman stated that they were still trying to assess the density of hot intergalactic gas that might "provide enough material to eventually stop the expansion of the Universe by the action of gravity."[24]

## CONTINUED TINKERING

Carruthers stuck to what he knew: how to improve his instruments. Throughout the late 1960s, he tried all sorts of variations of the optical systems and electron amplifier/focuser field mechanisms. First, he replaced the mosaic of four gratings with a single flat grating. Second, he reduced both the weight and the energy requirements for the magnetic focuser. And third, he searched for ways to simplify the optical system and the camera to achieve "higher sensitivity and resolution."[25] By 1969, after many interim steps, Carruthers had designed, built, calibrated, tested, and constantly modified several variations of both his Schwarzschild and his Schmidt optical systems, with the help of Chubb's group, including Julian Holmes, Harry Merchant, and others. He was ready to fly again.

Carruthers found no definite evidence of molecular hydrogen at first in 1967 and 1968. But his precise calibration and improved instrumentation gave him a sense of the faintest features he should have seen, which in turn gave him an upper limit for molecular hydrogen between the stars and the Earth. Comparing his work with similar rocketry efforts in 1966 by Princeton astronomers Donald Morton, Ed Jenkins, and others of hot stars in the heart of the Orion Nebula, and in 1967 by Henry, Carruthers was able to provide an upper limit for molecular hydrogen in the interstellar medium, which still was far less than predicted by theory.[26]

Although Carruthers could state an upper limit, he could not yet assign an actual value for the abundance of molecular hydrogen in the interstellar medium. Nevertheless, his exhaustive error analysis and discussion in a second paper more than set the stage and direction for further work. His initial analysis in 1967 was cited some forty times by 1971, and as recently as 2017. It would take another two years and more flights before Carruthers unambiguously detected molecular hydrogen. But most of all, keeping him going, his payloads worked beautifully.[27]

As we noted earlier, Friedman was deeply impressed with Carruthers's first analysis in 1967, even if his observations were not successful. In the set of notes he clipped to an NSF report on Carruthers's activities, Friedman added that the instruments aboard the rocket provided "the first successful stellar glow observations at wavelengths shorter than 1000A." Further, Friedman highlighted the astrophysical importance of the work. Going a bit beyond Carruthers's own stated conclusions, Friedman described his major findings: interstellar molecular hydrogen was less than one-tenth the concentration of atomic hydrogen, and the brightnesses of stars in the extreme ultraviolet were less than theory predicted, a result that always raised more questions in science. Once again, in conclusion, Friedman cheered, "Dr. Carruthers is an outstanding scientist who brings great credit to the Washington DC community."[28]

But getting to that point was not straightforward. As Carruthers continued to refine his detectors with the continuing advice and assistance of Hicks, Holmes, and Merchant and to plan for his next Aerobee flights to nail down molecular hydrogen, he was also becoming deeply engaged in proposing payloads for flights in the Apollo program—both orbitals and landers. These and the next years would be filled with complex negotiations within NRL, and between NRL and NASA, about who and what would be flown to the Moon, or in orbit around the Earth or Moon. In his 1966 letter to his uncle, Carruthers had hinted that he was planning to propose for the Apollo program. It was likely his "A Program of Astrophysical Studies in the Far Ultraviolet Using Image Converters," which was his response to a NASA "Announcements of Opportunity" inviting experiments for various Apollo missions including the Apollo Telescope Mount and other Apollo Applications Program missions. His first proposal had been rejected by NASA, but given the complexities of the process and the lengthening response times of a growing bureaucracy, Carruthers had probably not received the decision by the time he wrote to his uncle. These were turbulent and uncertain times. But, come what may, Carruthers dutifully pursued molecular hydrogen.

An NRL press release in May 1967 only increased the pressure, strongly asserting that Carruthers's methods were the most promising to answer the question about molecular hydrogen. It repeated the continuing conundrum that only 50 percent of the matter in the galaxy was contained in stars. It also repeated deeper questions that had been brought to the fore by Schwarzschild and others (and would become critical after Vera Rubin's observations in the late 1970s).[29] Could this "hidden matter" account for the understood mass of the galaxy based upon its dynamical behavior? Where was the rest? And as before, the apparent lack of hydrogen between the stars made astronomers wonder if our galaxy had exhausted its star-making abilities and was in the process of dying. The press release asked these questions and then described Carruthers's detectors, concluding that they were the best hope of answering these central questions in astronomy: "Ultraviolet Lyman-Alpha spectroscopy as a means of probing the hydrogen gas clouds of interstellar space is much more sensitive than radio astronomy [and promises to be an] ultra-fine probe for future exploration."[30]

Carruthers said much the same thing in a highly detailed and comprehensive review in *Space Science Reviews* in spring 1969, covering the "present state of knowledge" and how to rationalize it both observationally and theoretically. He discussed the theoretical mechanisms that might be used to rectify why radio observations and theoretical predictions for atomic hydrogen differed by a factor of ten over his and others' estimates of molecular hydrogen's Lyman-alpha absorption. He cited various unknowns, like where in space the two widely differing techniques were sensing or the possibility that there was a form of microwave amplification by maser action (microwave amplification by stimulated emission of radiation) taking place in the interstellar medium to amplify the 21-cm radio observations, proposed by Russian theorists in the recent past. None of these, Carruthers argued, were definitive. He supported his conclusions by focusing on molecular hydrogen, describing its spectrum and how it can be formed and destroyed, giving specific attention to chemical exchange reactions that might take place in the interstellar medium. Reviewing all the arguments and evidence, he noted that "there are still many questions to be answered and details to be resolved" concerning the marked discrepancy between the radio and UV observations, concluding that "atomic and molecular hydrogen in interstellar space will remain an important and exciting field of investigation for some time to come."[31]

Carruthers defined the overall problem, describing both the techniques of observation and theory in accessible terms. As in his unpublished proposals to NASA and elsewhere, Carruthers may have been primarily preoccupied with building better tools to answer important questions, but he was sensitive to, and also deeply engaged in, exploring the whole of the science and making it accessible to a wider readership.

A year after he submitted his review, Carruthers reported on his latest Aerobee flight from White Sands on March 13, 1970. This time, he flew an ambitious array of all-reflecting Schwarzschild cameras for spectroscopy and photometry with an improved pointing system that was programmed to home in on hot stars in the constellations Perseus, named after the Greek hero, and Monocerotis, the unicorn.

Once again, the rocket did not reach peak altitude, but its longer flight duration allowed the payload enough time to evacuate after the nosecone door ejected. So with his improved designs and the faster and more accurate pointing controls available by 1970,[32] his payload acquired excellent data on two stars, Epsilon (ε) Persei and especially Xi (ξ) Persei. Multiple exposures ranging up to twelve seconds produced clear, high-resolution spectra. The observations did not meet all his goals, like determining the temperature of the molecular hydrogen. But, most significantly, he finally found direct evidence for its existence and from that estimate the abundance of molecular hydrogen in space, concluding that "nearly half of the total hydrogen in the line of sight to ξ Persei may be in molecular form."[33]

This definite observation, finally agreeing with theoretical predictions, garnered more than sixty citations in the first few years. This attention established Carruthers as a highly promising space scientist. In 1971, the influential *Annual Review of Astronomy and Astrophysics* cited it as an "important piece of evidence, less indirect . . . for the existence of large amounts of $H_2$ in dust clouds," adding most critically that Carruthers's observations "quantitatively confirmed" theoretical estimates.[34] Among the accolades, he was invited back to the University of Illinois to lecture on his work. The significance of his work, as interpreted by the campus paper *Aerospace Alumni News*, stated simply that we know what makes up a star—mostly hydrogen, helium, and a smattering of the heavy elements. But the question is, where does the hydrogen come from to make new stars? "Scientists figured $H_2$ had to exist in the dust and has among the stars, but the assumption had to be proved, and this he has done."[35] Indeed, he scooped one of Spitzer's delayed *Copernicus* satellite goals, which would finally launch in 1972, but it did not come easily.[36]

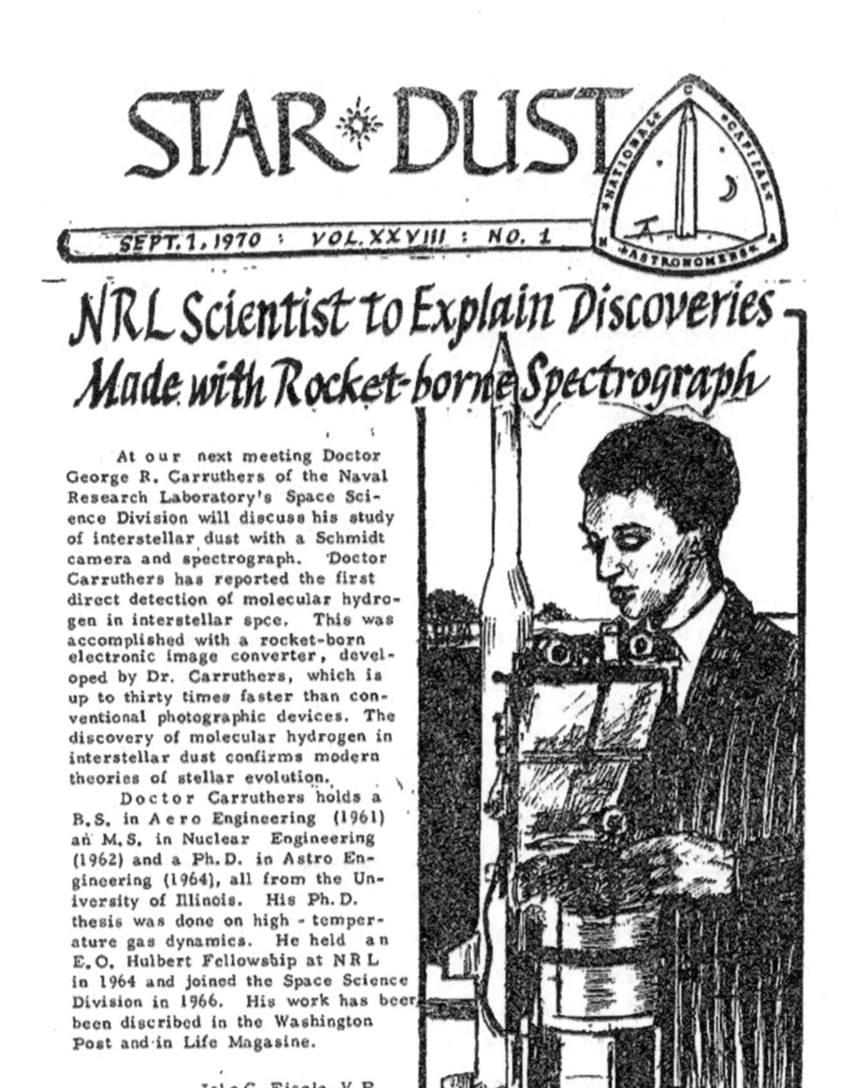

STAR DUST

SEPT. 1, 1970 : VOL. XXVIII : NO. 1

NATIONAL CAPITAL ASTRONOMERS

NRL Scientist to Explain Discoveries Made with Rocket-borne Spectrograph

At our next meeting Doctor George R. Carruthers of the Naval Research Laboratory's Space Science Division will discuss his study of interstellar dust with a Schmidt camera and spectrograph. Doctor Carruthers has reported the first direct detection of molecular hydrogen in interstellar spce. This was accomplished with a rocket-born electronic image converter, developed by Dr. Carruthers, which is up to thirty times faster than conventional photographic devices. The discovery of molecular hydrogen in interstellar dust confirms modern theories of stellar evolution.

Doctor Carruthers holds a B.S. in Aero Engineering (1961) an M.S. in Nuclear Engineering (1962) and a Ph.D. in Astro Engineering (1964), all from the University of Illinois. His Ph.D. thesis was done on high - temperature gas dynamics. He held an E.O. Hulbert Fellowship at NRL in 1964 and joined the Space Science Division in 1966. His work has beer been discribed in the Washington Post and in Life Magasine.

John C. Eisele, V.P.

DR. CARRUTHERS DEMONSTRATES SPECTROGRAPH BEFORE LAUNCH

DN.

Figure 5.7
Front page of the National Capital Astronomers newsletter, "Star Dust," September 1, 1970. *Source:* Courtesy of the NCA and Jay Miller.

Carruthers gained wide as well as local notoriety for his discovery. The venerable amateur association, the National Capital Astronomers (NCA), invited him to speak in October 1970. The NCA advertised the event and illustrated the advertisement with a cover sketch depicting Carruthers with his 1967 payload. It highlighted that "the discovery of molecular hydrogen in interstellar dust confirms modern theories of stellar evolution" (see figure 5.7). His invitation was significant in NCA's history because before that year, the NCA did not allow Black people to join their ranks.[37]

Over time, his paper gained over 200 citations and some 1,455 reads in the Astrophysics Data System.[38] It was cited as late as 2022 as the primary evidence for molecular hydrogen, considered by then "by a few orders of magnitude, the most abundant molecule in the Universe."[39]

# 6

# PROPOSALS TO NASA POST-APOLLO

It was a fact of life that, in Homer Newell's words, "until after the success of *Apollo 11*, science was the least of Apollo engineers' concerns."[1] This attitude was shared throughout much of NASA. But there were proponents both inside and outside NASA looking ahead to a time after the Moon had been reached, what was called the post-Apollo period. Wernher von Braun, head of the Marshall Space Flight Center, was worried about the future, beyond the first landing.[2] As *Saturn* was well along, NASA Headquarters started to shift some personnel from Marshall to the MSC. This prompted von Braun and others there to start asking, "What do we do after Apollo?"[3]

Von Braun's question reflected President Lyndon Johnson's 1964 directive to NASA. In the aftermath of the Kennedy assassination but building on Kennedy's pledge to land a man on the Moon before the decade was out, Johnson, concerned for the future of the MSC in Houston, an important economic driver in his political territory, called on NASA to make bold plans. So even though, as historians William D. Compton and Charles D. Benson have argued, NASA's decision for a lunar orbit rendezvous effectively suppressed open thinking about an Earth-orbiting space station, those dreams were never dead. In fact, they were revived in the mid-1960s as NASA responded to Johnson's directive and began to echo von Braun's query. Marshall and MSC competed with plans for an Earth-orbiting laboratory to face the challenge of living and working in space over extended time intervals. A sympathetic Congress provided generous support to explore all possible applications, and, as Compton and Benson assert, "a manned laboratory had obvious uses in the conduct of scientific research in astronomy, physics, and biology."[4]

Contrasting with Compton's and Benson's observations, *Science* magazine painted a darker picture for space science in the second half of the decade. By 1967, "NASA had lost much of its innocence" in both Congress and the public, and the space race with the Russians was now something akin to a "celestial stock-car race."[5] It predicted that although the Apollo budget was stable, the overall NASA budget would probably be cut and would be "felt most by NASA's scientific and post-Apollo programs." However, the science most closely connected with Apollo would probably be hurt the least.[6]

NASA issued calls for long-range proposals in the mid-1960s, and NRL was quick to respond. Byram, Chubb, Friedman, and others crafted lengthy proposals for upcoming NASA missions, and Carruthers eagerly joined them, even though he still did not have a successful sounding rocket flight. They proposed a series of missions in the Apollo program, robotic missions to the planets, and *Aerobee* payloads in collaboration with newer Hulburt fellows. Here we step back to examine how Carruthers' proposals were received in the years before his spectacular molecular hydrogen confirmation in 1970—in other words, before he had convinced NASA of the promise and powers of his camera.

His first major proposal for Apollo was based directly upon his rocket work: "A Program of Astrophysical Studies in the Far Ultraviolet Using Image-Converters." Carruthers prepared the proposal in response to the formal "Announcement of Opportunity" releases that NASA issued on January 1, 1966, and March 11, 1966. NASA was looking for scientific payloads for what would eventually become, after what would be a seemingly endless series of modifications and negotiations between competing parties, the Apollo Telescope Mount. The Apollo Telescope Mount would be a part of a proposed human-tended Earth-orbiting space station called Skylab, planned for launch in 1973. Carruthers proposed a refinement of his instrument that could be handled by an astronaut in a scientific airlock in Skylab's Orbital Workshop or in a spacewalk. His designs would go through many versions and would be directed toward different goals as the nature of this program changed.

NASA originally envisioned multiple missions, for stellar and solar research, and they were called by various names. One of the first large proposals was for a manned space station carrying a cluster of instruments called the "electromagnetic radiation payload" (EMR). Carruthers was excited by this prospect because, other than sounding rockets, it made possible far longer exposure times and was the only way to physically retrieve his data.

The challenges and frustrations Carruthers experienced with his sounding rocket work during this time were minor compared to what he faced now. His proposals were highly ambitious, as we shall see, and all the more remarkable because he never let up on his sounding rocket work but deftly incorporated it as a step toward Apollo.

The key point here, as we noted, is that Carruthers was preparing these Apollo proposals well before he had a fully successful Aerobee flight that returned the first evidence of molecular hydrogen ($H_2$) in space from a twelve-second exposure of the Lyman-alpha line in the spectrum of the hot star ξ Persei in March 1970.[7] That he pursued these parallel paths so energetically provides insight into his acceptance of the complex and competitive world he had entered. To appreciate this world, we take a quick look at its origins and the major factors contributing to the competition, both scientific and intra-institutional, experienced by scientists proposing for the Apollo program.

## COMPETITION IN PLANNING AFTER THE MOON LANDING

EMR had its origins in post-Apollo planning—what Apollo missions would do after the first landings. Planning and goals morphed many times over the next few years, so to appreciate the complexity of this world, we review some of the steps that brought Carruthers to EMR.

In 1965, NASA was flying its Gemini missions as prelude to Apollo and a Moon landing. The huge multistage Saturn launch system was under development at the George C. Marshall Space Flight Center (MSFC) near Huntsville, Alabama. What would become the Johnson Space Center in Houston in 1973, then called the Manned Spacecraft Center, was developing the payloads. Both centers had their hands full, but as we noted at the outset of this chapter, leaders like von Braun expressed worry about the future.

To make matters worse, when NASA canceled what was called the Advanced Orbiting Solar Observatory (AOSO) in 1967, a much-hoped-for expansion of NASA's robotic solar missions (a series of small missions called OSO, or Orbiting Solar Observatory), the scientific community saw this as evidence that NASA downrated the importance of science.[8]

NASA's growing interest intensified the competition between Marshall and MSC, as well as the other centers, and relations with the scientific community. But Apollo seemed to be the best game in town for anyone proposing experiments that had to be retrieved, like Carruthers's cameras.

Although there were free-flyer military reconnaissance systems capable of film return, their cost and complexity made them impracticable for civilian applications. For the same reasons, high-resolution electronic imaging of faint objects requiring long exposures on stable robotic platforms was still problematic. AOSO would have needed only short exposure times to scrutinize the Sun, of course, but the stabilization requirements for high-resolution imaging and raster scanning were still a challenge for longer exposure times. There were other technical issues with the project, such as a suitable and reliable Sun tracker. OSSA (Office of Space Science and Applications, a merger of two existing units in 1963) "budget cuts" were stated as the cause of the cancellation. Although the precise reasons for canceling AOSO are not known, its cancellation worked to Carruthers's favor.[9]

Friedman knew that Carruthers's dependence on film for data recording required physical return, and so he responded positively when von Braun and his associate Ernst Stuhlinger at MSFC, reacting to the criticisms of the scientific community, announced plans for a large Saturn-launched manned system for science called the Orbital Workshop (OWS), which would house the telescopes and spectrographs from the canceled AOSO that were modified for film recording, plus at least one more from NRL that also required film return: a suite of Carruthers's UV cameras. Likely for that reason, NRL retained good relations with MSFC, receiving positive support in return.

Within a few days after receiving the January 1, 1966, announcement from NASA, a member of Talbot Chubb's technical staff, Donald D. Brousseau, contacted the Marshall Space Flight Center asking for more information on the OWS, which in varied forms eventually became the Apollo Telescope Mount mission on Skylab. He was told that it was now planned to use a spent Saturn IVB stage already in Earth orbit as a "manned orbital laboratory" sometime in 1969. The mission would last a year, and there would be "as many as four separate, long term spaced, dockings of astronauts largely for maintenance work." A Marshall liaison was expected to visit NRL later that month for further briefings. Brousseau made one thing clear: the project was much in flux and "is in the process of being completely changed." Still, they had to meet the deadline set by NASA Headquarters one year later for review of all "final experiment description[s]."[10]

As vague as the mission requirements were—"no prime interfaces can be described or anticipated at this time"—Friedman pushed his staff to vigorously respond to the Announcement of Opportunity. First, in early March, Friedman, Byram, and Chubb submitted an "X-Ray Astronomy Proposal for Apollo-Earth Orbital Missions." Soon after, Carruthers made his pitch, cosigned by the same trio as we noted above, along with another NRL specialist in electronic imaging. Carruthers, after all, was still a postdoctoral research associate and needed institutional affirmation.

In March 1966, Carruthers prepared a long and detailed proposal calling for a careful step-by-step process of designing and building payloads first for sounding rockets and then for spacecraft, both robotic and piloted. Written even before his first Aerobee flight, Carruthers boldly outlined the power and promise of electronic image converters and amplifiers. Thinking of the upcoming series of Orbiting Astronomical Observatories (OAO), explicitly the third in line, *OAO III* led by Princeton's Lyman Spitzer, Carruthers predicted that if its 32-inch aperture telescope had one of his image converters, it would have "a detection capability in the far UV equivalent to that of the 200" Palomar telescope in the visible."[11] He envisioned both wide-field surveys and narrow-angle detailed studies, making indirect and direct references to planned astronomical missions.

Carruthers envisioned an exceedingly enthusiastic four-to-five-year program of instrument development starting with small Schmidt and Schwarzschild cameras and spectrographs for wide-field sky surveys of hot young ("early-type") stars, interstellar and intergalactic space, the Magellanic Clouds, stellar chromospheres, as well as photometric studies of emission from the Earth's airglow and aurorae. He audaciously envisioned larger devices, ranging up to a 16-inch f/10 Cassegrain spectrograph for narrow-field spectroscopic studies "of specific objects which appear interesting."[12]

He laid out five phases of work in two parts: exploratory sky surveys and detailed studies of specific objects. These five phases are as follows:

Phase I: Preliminary sky surveys with Aerobee rockets, both for spectroscopic and wide-field photometric mapping of selected portions of the sky. He claimed that his instruments, though small, would be as sensitive as the larger OAO-class telescopes because of his image converters.

Phase II: Broader sky surveys with similar instruments on a common platform aboard Apollo orbital missions, to be operated by astronauts. One of his cameras would employ a UV-sensitive TV monitor to allow the astronauts to center on desired fields. He suggested several alternatives for the TV system. All the data would be photographic, requiring return to Earth by the astronauts. He assumed that since the pallet of the service module was designed to be stable, the drift rate would be so small that long exposures could be conducted without tracking.

Phase III: In parallel with work in Phase II, Carruthers proposed additional Aerobee 150 (or larger) flights to test out 12- to 16-inch narrow-field Cassegrain systems with image converters for direct imaging and spectroscopy. These required fine stabilization and attitude control.

Phase IV: Build a single platform for long term missions in Earth orbit to carry the Cassegrain spectrograph designs in Phase III as well as two UV image orthicon instruments, one an active real-time guide telescope for long-term exposures and the other a star tracker. The guide telescope would be used by the astronaut to acquire fields and to monitor the exposures. Carruthers felt that if the Apollo Telescope Mount platform was available by then, it would make a perfect podium.

Phase V: Use of his image converters with an "observatory class" astronaut-tended telescope (30- to 40-inch aperture to start) capable of matching the depths and faintness range of the 200-inch Palomar telescope. Carruthers admitted that a telescope of this class "would probably be a joint effort of several institutions."[13]

To achieve the Phase V goal, Chubb had advised Carruthers to contact Perkin-Elmer Corporation in Danbury, Connecticut, a leading producer of precision optics for astronomy, and ask about the reflecting optics that were being planned for Lyman Spitzer's *OAO III* mission that boasted a 32-inch mirror. Indeed, laboratory demonstrations of Carruthers's detector had already convinced Friedman to change course in instrument choice, and so he also encouraged Carruthers to think larger. As Friedman reported to Spitzer in February 1966, considering Spitzer's own recent successful sounding rocket flights observing Lyman-alpha bands, NRL had dropped their own efforts with a scanning Geiger counter spectrometer and "decided instead to try to develop image tube instrumentation in the hope of achieving a very substantial increase in sensitivity."[14]

As Carruthers moved deeper and deeper into the technical details of his proposal, it is clear that the early stages were well thought out, incorporating both his existing designs for Schmidt and Schwarzschild systems with his image converters and his experience developing Aerobee payloads. But even though none of these designs were successful at the time of his proposal, he described every step in detail as if it were a known procedure. Visualizing the process in real time, Carruthers described the procedures the astronauts would take in the Apollo orbiting missions:

> The astronaut turns on the television equipment and, on the monitor, sees the starfield toward which the instrument is pointing. He approximately orients the spacecraft to the desired right ascension and declination. He makes the appropriate changes in the spacecraft orientation to bring the desired starfield into view. Then, the stable platform is uncaged so as to stabilize the instrumentation in inertial space against small drifts and vibrations in the primary spacecraft.[15]

Even when the platform took over, the astronaut would remain involved, using an image orthicon to monitor the field. Carruthers wanted the astronaut to be directly involved and trained to make judgments about exposure times depending upon what they saw on the TV monitor.

Astronauts were essential to the plan, but there were serious tradeoffs. Humans are not inert; they move around and cause the system to vibrate, which would blur out any observations requiring precision pointing. Carruthers here reflected concerns scientists had about the errors caused by a human presence.[16] Now, in his scenario, Carruthers suggested ways to make the combination work with humans onboard. After all, they were essential to bring the photographic data back, which the astronaut would retrieve by taking a spacewalk.[17]

Throughout this proposal, and many subsequent ones, Carruthers demonstrated considerable knowledge of the Apollo spacecraft's stabilization capabilities, which were then under development at MSC in Houston. He also frequently referred to the known capabilities of hypersonic rocket-powered X-15 aircraft carrying stabilized platforms. Only when it came to the technical details of the instruments themselves, the area of his greatest intimacy, do we see alternatives expressed and choices voiced. Here he provided designs, and alternatives to those designs, which would be debated as his proposed program progressed. Here is where his heart and mind resided.

He described his all-reflective wide-angle Schmidt and Schwarzschild designs, expressing full confidence in them, but, as always, he offered several alternatives. These were partly scale-ups of those he had in hand, but they were also refinements based upon instruments NRL was now testing in ground-based wide-field meteor and comet studies, and stabilization systems proven in X-15 flights.

Beyond the phased instrumentation stages, he outlined in detail how the data gathered from an Apollo mission would be analyzed. Here he had both experimental and theoretical arguments for the limits his instruments would reach, which were far beyond what NASA had managed to achieve to date. To help his reviewers appreciate his proposal, he compared the capabilities of Nikon 35-mm cameras using fast (ASA 1600) film, what astronauts typically used, to what his proposed instruments could do. He stated that with larger apertures (16-inch and up), his instruments could gather vastly greater amounts of useful data on the nature of the outer chromospheres of late-type (cooler, redder) stars and on the nature of hot young stars and their surroundings, and could discern the composition of planetary atmospheres by how the far-ultraviolet spectra from their reflected light varied from a pure solar spectrum.[18]

Even though he was well aware of the problems caused by humans in spacecraft, which beyond shifting mass was also pollution from outgassing by the life-supporting systems of the spacecraft, Carruthers made strong arguments for why his instruments had to be human tended. Beyond those already noted, he called for "on-the-spot changes in the experimental procedure as the situation demands," including adjustments and repairs in the event of "nonoptimum performance."[19] The astronaut had to be trained in all of this, of course, and NRL had the facilities to provide that training with their laboratories and a new 16-inch telescope near their Washington laboratory.

All the observations could be performed in low-Earth orbit but needed to be at least two hundred miles altitude or higher. A complete mapping of the UV sky by the astronauts would require observing time over at least fifty-five orbits, but only a few orbits would be needed to provide information vastly beyond what would be available from sounding rocket flights. Finally, he hoped at some future time to be able to observe from high orbits, beyond the Van Allen radiation belts, because they interfered by their "strong Lyman-Alpha background glow" from scattered sunlight, which would limit the all-reflective systems.[20]

He capped his gargantuan proposal with extensive appendices, repeatedly emphasizing the advantages of applying his designs. He also included detailed illustrations and shop scenes at NRL showing the instruments being tested. These scenes gave his proposal a certain reality, and indeed, his prototyping, which he relished, made it even more real.

## ALIGNMENT WITH THE MARSHALL SPACE FLIGHT CENTER

Carruthers's ambitious proposals have to be set in the context of enabling priorities at both NRL and Marshall Space Flight Center. Friedman, who was also proposing a huge one-hundred-square-foot array of X-ray detectors to survey the sky from the Moon, submitted both his and Carruthers's proposals to Ernst Stuhlinger at MSFC who, as director of its Space Sciences Laboratory, was leading the campaign for OWS.

In mid-May, Conrad F. Swanson, contracting officer at the newly formed "Scientific Payloads Branch" of MSFC, issued a NASA-Defense Purchase Request of $140,000 to initially support Friedman's X-ray packages, emphasizing that NRL was in complete control of the scientific objectives and that MSFC "looks to NRL for 'leadership in the engineering design.'" It was Marshall's part to "supply the willing hands and engineering competence to put the show on the road" even if this meant "putting fires under the appropriate pants" of both NRL and MSFC staff to get the job done.[21]

Swanson and others at Marshall were eager to team up with NRL. As we've noted, the roles and responsibilities of the various NASA centers after the first landing, including Goddard, Marshall, and MSC, were in flux regarding who would build and fly scientific payloads and who would provide spacecraft infrastructure and launch capabilities. Marshall was responsible for the launch vehicles and MSC for the habitats for sending humans into space. But the "turf division" between the NASA centers caused constant bickering, as historian Michael Neufeld has covered in some detail. In 1966, an agreement between NASA centers partially settled the debate, with Marshall responsible for the mission module and MSC the command post. So-called experiment modules, however, were the province of both centers, not fully resolving the debate. But roughly, MSC would take on the life sciences and Earth resources, or human-related activities, while Marshall would handle astronomy. This related mostly to Apollo orbiting missions, like the Apollo Telescope Mount on what would become Skylab. But when it came to lunar surface programs, MSC got exclusive control—for the moment at least.[22]

Swanson listed the payloads they were then preparing to include in a "Saturn-Apollo-LEM orbital mission," identifying the launcher, the spacecraft, and the "Lunar Excursion Module" landing craft, and detailing what additional information they required to "sell the Apollo astronomy project up the line." That project included Friedman's X-ray array, a gamma ray experiment from Oak Ridge National Laboratory, three UV-imaging instruments from the University of Arizona for "direct photography," a microwave experiment from MSFC, and, last and "hopefully," Carruthers's three UV image-intensified instruments.[23]

Swanson was especially impressed by Carruthers's proposal and stated that it would be taken up for consideration soon, suggesting an informal meeting between NRL and MSFC teams in Huntsville in mid-June to work out all details—mechanical, electrical, and procedural. In sum, and in a friendly tone, Swanson suggested that in their meeting in June they consider how Marshall might take over "much more" of the mechanical fabrication of the prototypes and integration responsibilities for the payloads themselves. "The more MSFC accomplishes, the more your funds can be conserved," Swanson promised—leaving unsaid how this would also improve Marshall's visibility and its centrality in the payload arena versus the MSC.[24]

Reacting to Swanson's friendly letter, Carruthers sweetened the pie, providing, as Swanson suggested, more detailed drawings for his Apollo payloads. He was also more than willing to collaborate with other proposers like William Tifft of the Steward Observatory in Arizona to combine their platform instruments and remove others to non-platform berths. And responding to Swanson's strong hints, he also outlined what responsibilities Marshall might assume, including the purchase of the large and expensive reflection gratings, the "design and fabrication of front-end assemblies" to mount the gratings, and the design and construction of the film transports, providing for one-hundred-foot length reels rather than the ten-foot length reels designed for his sounding rockets.[25]

Carruthers was more than willing, evidently with Friedman's approval, not only to respond positively to politically advantageous cooperation but also to work collaboratively with scientists from other institutions. As a result, through continuing contact with Swanson and others, Carruthers felt confident that his camera would fly on an Apollo mission.

In his September 1966 letter to his uncle Ben, reporting on his recent rocket flights, Carruthers also proudly announced that "we have been accepted in a proposal to fly image converter instrumentation for astronomical work, similar to those being used now in rockets, on the manned Apollo orbital missions, scheduled for late 1968 or early 1969." As a result of this exciting news, he admitted the following:

> I have been running around like a chicken with its head cut off trying to get the project underway. I have been making fairly regular trips to the Marshall Space Flight Center, in Huntsville, Ala. . . . and made a trip to Tucson, Ariz. a few weeks ago to discuss the experiments with another of the project scientists, who has an experiment on the same flight. Between this project, and the trips to White Sands for the Aerobee shots, I have really been doing a lot of traveling.[26]

Carruthers was apparently unaware of the problems Friedman and Chubb were facing. Even with the clear support NRL was getting from Marshall, the selling campaign up the line to NASA Headquarters was met by a warning of a flat turn-down from NASA Associate Administrator for Space Science Homer Newell. In a terse note to Friedman in August 1966, Newell privately but bluntly stated that NRL's proposals, both for X-ray and UV astronomy, probably "could not be included in the planned AAP [Apollo Applications Program] mission" and referred him to James M. Walden, NASA Apollo Applications Program scientist, if he had any questions about the decision.[27] Newell's letter was not an official rejection, but a warning that their package was not assured of a flight unless it was approved by the appropriate NASA Headquarters committees.

Newell's rejection had a history that illustrates the complexity of the situation, especially the decision process, and requires deeper review to appreciate the world Carruthers was entering as a newly minted and highly idealistic PhD.

The "turf division" between NASA centers that Michael Neufeld has described only became more heated and complex when other organizations external to NASA were involved, like those proposing projects and missions and payloads to NASA.[28] Who was responsible for what? And who decides what is flown? The NASA Office of Space Science and Applications, headed by Newell, had various disciplinary subcommittees, and one of them was the Astronomy Subcommittee, headed by NASA's program chief for astronomy Nancy Roman. She and her committee knew about Marshall's plans for OWS but felt they still had cognizance over the experiments that would be flown on it in the EMR payload. Apparently, von

Braun and Stuhlinger did not agree, and so in October, possibly unaware of Newell's blunt turndown, Stuhlinger wrote to the secretary of the subcommittee saying that Marshall was considering four payloads, including Friedman's and Carruthers's, which were part of EMR. The subcommittee was enraged, demanding that "astronomical payloads for Saturn 1B missions should be selected from experiments selected by the Astronomy Subcommittee." As Karl Henize, a scientist-astronaut and member of the subcommittee, wrote in his notes of the session, there was a "long discussion of the EMR payload. Strong feeling that a mistake has been made in the selection of the payload . . . directly expressed by Nancy, etc." In a meeting the next day, they remained angry about MSFC's perceived obstinance. Henize also noted that there was "debate as to how hard we should slap at MSFC. We slapped hard."[29] Naturally, Friedman and Carruthers were implicated in the fight. To appreciate the state of affairs, we need to look at the history of the subcommittee and Roman's regard for NRL.

## THE ASTRONOMY SUBCOMMITTEE OF THE OFFICE OF SPACE SCIENCE AND APPLICATIONS

In the early 1960s, Roman formed the Astronomy Subcommittee of the Space Science Steering Committee with in-house NASA staff from Headquarters and the Jet Propulsion Laboratory. Although she later included members from the academic community, she recalled with some wistfulness how she enjoyed "incredible freedom" in these early years handling her program. Apparently, she felt that there was little peer review at first, and "I decided on the acceptance or rejection of proposals largely on the basis of my own knowledge."[30]

Roman's poignant recollection oversimplifies the process and so is worth examining to appreciate how Carruthers's proposals were treated at first. It is correct that she played a strong role in selecting proposals, and her committee definitely worried that Carruthers's instruments, mainly their magnetic fields, might interfere with other experiments in the tight quarters of the spacecraft. Carruthers was also flooding her office with proposals, so in 1969, after an only partially successful flight, she responded in frustration that "both your results and the difficulties you encountered during your recent rocket flight" made her feel that his never-ending proposals and the schedule he wanted for more flights "does not sound reasonable." She predicted that present levels of support for NRL flights would be less in following years and that she wanted to "leave it to NRL" to determine its own priorities.[31] Up to that time, NASA was allotting NRL a set amount, some $200,000 per year for support, adequate for sounding rockets but not much more. But NRL staff were still proposing well beyond that budget limit. In the future, this amount would not be assured, she warned.

It has to be appreciated that Nancy Roman had a most challenging role at NASA ever since she was invited to join the new agency, arriving in February 1959 to head the nascent astronomy program. Born in 1925 and trained as an observational astronomer at the Yerkes Observatory where she made significant contributions that were only appreciated later, and aware of the lack of academic advancement opportunities for a woman in academe, Roman joined NRL, working in their Radio Division. She soon found "that the research atmosphere at NRL was freer than the research atmosphere at Yerkes, in terms of freedom to do what interested me."[32] She performed geodetic studies bouncing radio signals off the Moon but did not engage the Rocket Division at NRL in her work. She liked NRL but was not interested in becoming an electronic engineer or "hardware person." So while attending a colloquium at NASA by Harold Urey and meeting Jack Clark, who had left NRL for NASA, Clark asked her if she knew "anyone who would like to come to work for NASA and set up a program in space astronomy."[33] It was her exit clearance.

Moving to NASA, however, "she faced daunting challenges, not least of which was to help two very different groups of people learn to talk to one another." First, she had to convince astronomers that there were opportunities in space, all the while trying to rein in those who "were so gung-ho that they were campaigning for swift work on the largest possible space observatory." However, she was responsible for increasing attention and support within NASA to the space sciences—astronomy, in particular. This was a most delicate balancing act.[34]

However, in a 1982 oral history interview with Frank Edmondson, she expressed "very little sympathy for the space program at NRL" during her years in the Radio Division. She had felt out of place in an environment that, she perceived, concentrated on tool building rather than scientific problem solving. She was particularly annoyed with Friedman for a number of reasons, and overall she felt that NRL staff, although excellent at building instruments, even by the 1960s had "no one on [Friedman's] staff at the time who had an astronomical background and really understood enough about astronomical sources." She also knew that "NASA did not fully respect the quality of the astronomy coming out of NRL."[35] Her advice to Carruthers in 1969 reflected doubts she had harbored for some time, but it also reflected a time of budget shortfalls.[36]

After Newell's first rejection in 1966, regrouping in the fall, NRL and Marshall trimmed their cooperative plan for EMR, developing a combined X-Ray and UV program at a cost of $225,000. Marshall would take the lead, arguing that the proposal is now a "logical continuation of the program of rocket borne UV astronomy now underway at NRL." Friedman's payloads, moreover, fit exactly Marshall's plans for a stabilized platform "comparable to the Apollo payload cross-section." For these reasons, Marshall urged that "a continuation of the NRL x-ray astronomy experiment be developed as quickly as possible." Both arrays of instrumentation, Marshall claimed, were necessary to provide needed observational data to better evaluate theories of solar and stellar evolution.[37]

To make the connection even stronger, both NRL and Marshall now explicitly identified a Marshall engineer, William C. Snoddy, as the lead to preserve the "experimental integrity throughout integration of payload and during prelaunch, launch and flight activities." Marshall would now be responsible for the preliminary mock-up of all the flight components, ensuring they were compatible with NRL's payloads. They would then prototype and construct the flight unit in a proposed time frame starting immediately and lasting well into 1968.[38]

Reflecting these changes, Carruthers simplified his phased study. It preserved his primacy for the detector systems but eliminated his initial proposal for a sounding rocket phase. By now, Carruthers was a permanent NRL employee, so he had more options for gaining support. He separated his sounding rocket work from his larger Apollo proposals by applying separately to NASA's Office of University Affairs to supplement those available from various Navy sources.[39] The office did have some discretion for smaller projects but was really a clearinghouse for proposals from universities and military laboratories like NRL.

The Office of University Affairs also engaged advisory committees to perform its duties. Like Roman's Astronomy Subcommittee within the OSSA, NASA advisory subcommittees were divided into broad disciplinary classes. They would determine which NASA program was most appropriate for the goals of the proposal, and then a subcommittee within that program would evaluate it on four criteria. As Homer Newell later described the questions: "One, was it appropriate to space? Two, was the experimenter proposing something that was feasible and timely? Three, was the experimenter a competent individual? Four, did the experimenter have the proper backing in the way of resources of his organization?"[40]

Subcommittees rated proposals by these four questions, and they were then carried up the line depending upon the disciplinary nature and structure of the program, ranking them in one of four categories. Category I was the winner and IV the loser, regarded as incompetent on all counts. Category II was "good," and III was mixed—deemed a good goal but not yet ready to be implemented, for any one of a wide range of reasons. And given the competition, not all Category I proposals could be flown, and the others had to be recast in ways to make them acceptable for later flights. In at least two cases, as we shall see, Carruthers's proposals for berths, first on the planned EMR championed by Marshall (noted above) but now to be reevaluated by the OSSA, and then a proposal for a Pioneer mission, were rated only "III," which incensed Friedman (see below). His sounding rocket proposals continued separately with some success from the Office of University Affairs.

Carruthers's new and more restrained three-phased study for Apollo flights would now carry 4-inch and 6-inch Schmidt and Schwarzschild cameras and spectrographs with his image converters to make wide-field surveys first, and then at least one flight for a detailed study of selected faint objects of special interest with exposures ranging up to forty-five minutes. Most of the proposal repeated arguments that he made before, but now, after continued ground-based experimentation with exposure times and sensitivity limits, he was able to predict more precisely what the capabilities should be.[41]

The instrumentation in his new proposal was similar to that proposed previously, though more limited and with simplified designs. The complex mosaic of gratings, for instance, was replaced by a single large grating. The astronaut was still an essential part of the process. But this time, Carruthers was more explicit about what type of person would be required: someone with both basic and advanced knowledge of astronomy, who had participated in training sessions with the prototypes and working models, and who was able to recognize targets of opportunity during the mission. He was, very likely, thinking of himself.

## RAMIFICATIONS OF CONTINUED AEROBEE FLIGHTS

While he proposed for Apollo mission berths on the proposed OWS, as we noted before, Carruthers never let up proposing and executing a series of *Aerobee* flights. What needs to be appreciated here, however, is how his projects and plans worked in parallel, especially how he used his growing success with his first Aerobee flights, and especially his highly acclaimed success in 1970, to demonstrate the promise of his Apollo proposals.

After his partly successful flight in March 1967, he celebrated the fact by issuing a detailed status report, sending it to the Astronomy Subcommittee and now adding arguments for why his designs were reliable and effective. Carruthers had no qualms claiming this flight gave his program and his NASA proposals "a great boost."[42] In it, he stated that this flight demonstrated that the exposed high-voltage system required for the image converter worked safely on a rocket "and hence, certainly in a satellite vehicle." There were continuing concerns about this design, however, because of the problems associated with electronographic detectors in the visible range of the spectrum and because his first efforts did produce magnetic interference and thermal problems that interfered with other instruments in the payload. Carruthers thus felt obliged once again to explain that ground-based visible detectors used cathodes with a cesium-antimony formulation that were "extremely sensitive to contamination by water vapor and other substances which outgas from the nuclear track emulsion in vacuum." To counter this, these systems required "liquid-nitrogen cooling of the emulsion and cathode, the use of barrier membranes [as in the Lallemand design], or the use of a vacuum gate valve [as in the Kron design] between the emulsion and cathode." None of this was the case for his UV-sensitive alkali-halide cathodes, which were also insensitive to outgassing from common payload components such as O-rings and epoxies, "which one would not dare use near a cesium-antimony or similar photocathode."[43]

In his May 1967 status report to Nancy Roman for EMR, Carruthers addressed other questions that were raised about the performance of his instrumentation by the reviewers. He could now account for the anomalous "strong bands" of emission in the Lyman-alpha spectral region he had noted in his published findings that were submitted to the *Astrophysical Journal* a week before. Subsequent analysis revealed that it was due to residual water outgassing from the rocket. But there was still a nagging nonuniform "ionic background" glow observed on his film strip, he reported privately to Roman, which had to be addressed.[44]

Carruthers tried to account for this faint glow as being due to residual outgassing of the *Aerobee* instrument chamber. It was harder to explain, he admitted, because although the ionic background decreased with time, there was no correlation found with altitude or movement of the pointing control. Thus, it could not be due to lingering gas pressure in his payload, but only to outgassing from the vented nosecone above the payload. He also sent Roman photographic evidence from the successful March flight—four frames from the image converter spectrograph—which convinced him that the "observed ionic background was due to outgassing of parts of the rocket external to the experiment housing, and hence was in no way the fault of the image converter."[45]

None of this convinced Roman or the deliberative subcommittee she headed, and so they once again gave Carruthers's proposal the low rating of "III," which recognized the value of its goals but still questioned its readiness for consideration for a berth on an Apollo payload. Now more incensed, Friedman wrote to Henry J. Smith, deputy director for the physics and astronomy programs, and Roman's direct boss that he was "very unhappy about the category 3 rating which was assigned by Dr. Roman's Committee to Dr. Carruthers' proposal for the EMR payload." Rebutting the committee's conclusions, "it seems clear to me, beyond any reasonable doubt," Friedman continued, that the glow was residual outgassing from the short Aerobee flight and that the problem would not arise in a satellite flight. "Even if it existed in the last flight, it had a minimal effect on the performance." Friedman went on to hail Carruthers's last flight as "a major achievement in ultraviolet astronomy," adding, "every knowledgeable astronomer who has heard him present these results has recognized the validity and importance of the observations."[46] He also went so far as to assert that the committee could only have come up with the low rating because "they were unfamiliar with the experiment and were inadequately and incorrectly briefed." In sum, he called for a full reevaluation and a "fair review and a proper rating."

The committee had deliberated over several days, comparing Carruthers's proposal to others for EMR.[47] Carruthers's proposal, to cover the 1,000–2,000 Angstrom range, was meant to complement a similar one from William Tifft for the 2,000–3,000 Angstrom range. Tifft was a highly respected Steward Observatory astronomer at the University of Arizona, a specialist in photoelectric photometry, and an advocate for "manned space astronomy," who would later develop a sophisticated "electron optical focusing system" for Steward's 90-inch reflector at Kitt Peak.[48] In 1965, he was a finalist for the first round of selecting a "Scientist-Astronaut" and so, like Carruthers, harbored strong motives to fly on missions. Tifft's instrument designs were also more traditional, since access to the 2,000–3,000 Angstrom range was easier.

Henize presented both Tifft's and Carruthers's proposals at the subcommittee meeting, expressing some reservations about Carruthers's detector. Al Boggess, one of the NRL scientists who left for Goddard, agreed but felt that the "NRL detector has great potential, long term value apart from this particular experiment" adding, thinking about Carruthers's first Aerobee flights, that "the past experience is bad, however, this experiment has worked in the laboratory and that may be all we can ask for at this time." Tifft had also secured some top-level astronomers as allies, which the committee felt was a "good point." He got high ratings, even though some members felt that the 2,000–3,000 Angstrom range, the "near-ultraviolet," was less important than Carruthers's range of 1,000–2,000 Angstroms, the "far ultraviolet." Still, as Henize reported in his notes, "Carruthers took a beating but [Tifft] held out." On the bright side, the committee voted unanimously that Carruthers's design was "very promising" and warranted "further development."[49] This was a frequent reaction to his early proposals.

It is true that Carruthers's designs required high voltages and strong shielding from the necessary magnetic fields, and this rightly concerned members of Roman's committee because it added weight and could interfere with other instruments in the payload. At one point, as we noted, he switched to solenoid-produced magnetic focusing, but that required high voltages, so he soon reverted back to heavy permanent magnets, which required heavy shielding and hence more payload weight. In 1969, his proposal for sending his cameras to the Moon admitted that they would create a magnetic field that "may be appreciable at distances up to two to three feet from the instrument." It was a requirement that could be met on the Moon but not so easily in a cramped payload.[50]

This problem would plague his designs for years, yet with successful Aerobee flights in the period 1968–1970, his cameras became more competitive because of their sensitivity and linearity. Charles Brown, one of his younger colleagues, reported in an interview, "If you fly a strong magnet in space, they're heavy. And you have a magnetic field to deal with. So we had to put all these magnets all over the payload to cancel out the detector's magnetic field," emphasizing, "but that was our only way to get a detector. So we did it."[51]

Given these problems, not only political but technical, we can appreciate why Friedman's plea had no noticeable effect on NASA's decisions in 1967. And as Roman warned, space science funding would not increase significantly in the future and could, in fact, drop. In the late 1960s and through the 1970s, authorizations for space science were substantially less than requests. This was, as we noted, partly a result of overall drops in NASA funding in the 1970s: as a *NASA Historical Data Book* explains, "space science was not high on Presidents Richard M. Nixon's and Jimmy Carter's list of space objectives." A task group recommended a greater emphasis on "space applications and national defense."[52]

Roman's concerns and advice to Carruthers reflect her overall opinion of NRL's space science work, and of Friedman in particular, but they also reflect larger issues. As we have noted, NASA's creation resulted in territorial friction over the management of space science, both internally with NASA centers and NASA Headquarters, and externally with NSF, and, as we know, with NRL.[53] This episode in Carruthers's career reflects the situation as much as it does the technical hurdles he faced sending his cameras into space. In the end, EMR and the hope of an Apollo Telescope Mount devoted to stellar and galactic astronomy never materialized, although one for solar observations did.

## CONTINUING AEROBEE MISSIONS

Fortunately, Carruthers had multiple sources to support his smaller projects. He continued to prepare flights of ever-more refined cameras and spectrographs aboard Aerobees funded by NASA's Office of University Affairs and by the Office of Naval Research, as well as through joint programs with the Air Force. His third flight from White Sands in November 1967 carried instruments with higher spatial resolution, and even though the pointing controls still did not perform adequately for long exposures, he felt that he could eventually reach "stars as faint as 5th magnitude in 10-second exposures, given better pointing stability."[54]

Carruthers also collaborated with several Hulburt fellows, including Richard C. Henry and N. Paul Patterson, with support lasting through June 1969. Their proposals called for flights of payloads on the Aerobee *150* and its new larger sibling, the Aerobee 350. The first flight was set for March 1968, and two more were proposed to last through to the end of the year. For those, they proposed altering the rocket to have a fully ejectable nosecone and, instead of the balky pointing controls and delays in their improvement at NASA's Goddard Space Flight Center, to utilize a gyroscopic system to orient the entire rocket.[55]

A simplified version of this payload was finally flown on September 21, 1969, on an Aerobee 150 that reached 100 miles. After the nosecone was ejected, the rocket was oriented to record hot stars, gas, and dust in the Orion Nebula, observing for some three and a half minutes and exposing thirty seven-second frames, covering most of the constellation as well as the circular emission nebula known as Barnard's Loop. Although three cameras were requested in their January 1968 proposal, only one was flown. Still, it was enough to extend knowledge of the UV characteristics of the gas and dust in the region, specifically that its intensity in the far ultraviolet was less than in the near-ultraviolet range of the spectrum, providing new insights into the structure and composition of star-forming clouds, especially the "nature of the newly discovered dust component."[56]

Between the flights, there were more proposals for X-ray telescopes enhanced by image converters, in conjunction with Friedman's team. And, by late 1968, something of a diversion: a detailed proposal to the Office of University Affairs for a planned Pioneer mission that would pass by Jupiter using it as a gravitational "whip" to accelerate it into a "Grand Tour" of the outer Solar System.[57] As with Carruthers's latest Aerobee proposal, Chubb served as consultant and Friedman as administrator.

## A DIVERSION: *PIONEER F & G*

Carruthers's constant focus in the 1960s was proposing payloads for his electronographic cameras on missions that could physically return his instruments and data. But on one occasion, he strayed from his focus, proposing for a planetary mission that required traditional detector technologies that could telemeter data back to Earth. Just why he took this turn remains unknown but demonstrates that, on occasion, he could be stimulated by a unique occasion.

In an interview with Glenn Swanson years later, when asked generally about taking advantage of an opportunity, he responded that it was worth the effort because "you only get one chance to go to Jupiter."[58]

Pioneer, NASA's first space probe program dating from 1958, became its first series of solar orbit probes in the 1960s and planetary space probes in the 1970s. The solar missions lasted through *Pioneer 9*, provided valuable experience in operating remote spacecraft, and brought back useful data on solar activity like the solar wind as well as the nature of the interplanetary medium.[59] In 1968, a National Academy of Sciences committee recommended a "grand tour" of the outer planets based upon a 1965 prediction that a rare planetary alignment would greatly reduce travel time. NASA responded by announcing that the next two *Pioneer* missions (F and G, or 10 and 11 once they launched) would visit Jupiter and then use Jupiter's intense gravitational field as a "whip" that would send the probes to the outer planets.[60] In the late 1960s, *Pioneer 10* was planned to be the first mission to the outer planets, gathering important data on the solar wind and interplanetary medium along the way.

Carruthers, possibly urged on by Friedman and Chubb, proposed using *Pioneer's* close approach to Jupiter to study its atmospheric structure and composition, especially its hydrogen content, from observations of the progressive selective absorption of sunlight by the Jovian atmosphere as the spacecraft rounded the planet's night side. In so doing, his instruments would observe "the attenuation of solar far-ultraviolet radiation as a function of time, using narrow band ultraviolet photometers, as the spacecraft passes through the shadow of Jupiter."[61] Film return was impossible, of course, so Carruthers, in a rare departure from his focus on electronography, proposed photoelectric sensors provided by the NRL Engineering Services Division and the Electron Tubes Section of the Electronics Division. No major new facilities would be needed for this project since most of the components had flown on NRL sounding rocket flights. The total cost was substantial, in the range of $223,000 for a prototype, three flight units, travel, and data reduction and analysis. In his proposal, Carruthers would be involved one-quarter time and would be "responsible for the over-all concept," and for the "reduction and interpretation of the data." A full-time draftsman/project engineer and a machinist would round out the effort, with Talbot Chubb acting as consultant, drawing on the expertise of NRL's Electron Tubes Section and Engineering Services Division.[62]

His payload would be a nested set of seven narrowband UV photometers optimized to monitor specific spectral signatures of elements in the Jovian atmosphere. The photometers would be linked to similarly filtered photocells and ion chambers in a combined package that would weigh less than five pounds. It would have to sit on the craft in such a way that it would be continuously directed toward the Sun. Overall, he argued that this experiment would greatly improve knowledge of the Jovian atmosphere, complementing ground-based infrared and visual stellar occultation studies. All of this, he concluded, would be of great aid in improving knowledge of Jupiter's upper atmosphere, which would assist in the "design of more advanced probes which will descend into the atmosphere of Jupiter."[63]

Even though this proposal deviated from NRL's focus on the Earth's upper atmosphere, and certain was a departure for Carruthers since it did not involve instrument development, it was still potentially useful for understanding the overall nature of planetary atmospheres as well as the abundance of hydrogen in the universe. Its great popular appeal may have been a factor, and both Friedman and Chubb enthusiastically endorsed it because it was employing their proven technologies.

But once again, as with his initial proposals for Apollo payloads, NASA eventually turned it down. The rejection took a year to be communicated and tersely stated in boilerplate that "while the proposal was of interest to NASA," it was not assigned sufficient priority to be included because of "other work considered more urgent to our program requirements."[64] There were more than two hundred proposals for payloads on the two Pioneers, which by 1970 were designed to carry some eleven science experiments with a combined weight of only sixty pounds.[65] The one that was selected with similar goals to Carruthers's proposal came from a successful collaboration on previous Pioneer missions. It proposed a broadband extreme-ultraviolet photometer that also collected data on the solar wind through the cruise phase, and then in the encounter phase, searched for structure in the Jovian upper atmosphere, including composition and auroral phenomena.[66]

This episode illustrates the fact that there was extreme competition everywhere, among scientists, scientific and technical teams, and NASA centers, especially in the Apollo missions and the planetary programs.[67] There was also some concern among members of the scientific advisory committees that NASA Headquarters was placing too much "emphasis on the 'team' approach, as contrasted with the previous Principal Investigator approach," which was preferred by scientists, especially Carruthers, who epitomized the solitary investigator.[68]

Carruthers, undeterred, prepared more proposals with detailed descriptions of refined prototype cameras constructed and tested in his labs. His proposing allowed him to do what he most loved to do: constantly experiment by prototyping variations of his basic design, each optimized for a specific problem. One indication that both he and his NRL colleagues cherished these prototypes is that during my interviews with them in their offices, they would proudly point to samples on their shelves. So, indeed, it is not surprising that in the midst of these rejections, Carruthers was always thinking about new possible applications, including the Moon.

Despite these rejections, approvals continued for sounding rocket flights. So by the end of 1968 and well through 1970, Carruthers was becoming known, publishing his Aerobee results and being well cited for his efforts, especially after his successful detection of molecular hydrogen in the interstellar medium. Through 1971, his papers were cited some five hundred times in primary journals, his image converter design was patented, and he had written extensive review papers in *Applied Optics*, *Science*, and the *Astrophysical Journal*.[69] In December 1968, NRL's Director, Capt. James C. Matheson, nominated him for the Annual Award for Scientific Achievement in Physical Sciences from the Washington Academy of Sciences. With rhetoric likely provided by Friedman, Carruthers was described as having "indefatigable zeal and imagination" and showing "ingenious adaptation" in developing electronographic technologies and proving them on rockets.[70] And this was only the beginning.

# 7

# LIVING IN WASHINGTON, DC, AND AIMING FOR THE MOON

## LIFE IN THE 1960S

The mid-1960s were a terrible time for racial relations in the United States. The *Washington Post* reported that "everyone is frightened" in the District of Columbia from sporadic riots, arson, and constant looting. It was a dangerous time for everyone; even though this lawlessness was a reaction to the ills of the world in that day (the Vietnam War, Civil Rights, poverty, and, to be blunt, excessive government spending on programs perceived as "white"), Howard University psychiatrists concluded specifically that youth harbored "nonspecific anger, impulsive license and above all, a lack of concern for consequences."[1] Martin Luther King's assassination in 1968 "triggered immediate and intense reactions" in cities across the country, especially Washington, DC.[2]

In the mid-1960s, Carruthers lived less than two miles from NRL, in a two-story brick row house on 739 Congress Street in Washington's southeast quadrant in an area known later as Congress Heights. By the time Carruthers arrived, the area was racially mixed and was experiencing rising tensions. One commentator recalled that if you crossed into the wrong territory, "you might get a beer bottle thrown at you, and the next thing you know, you're getting your ass whipped."[3] After King's assassination the city suffered one of the worst riots in the nation lasting four days, over a dozen deaths, 1000 injuries, and 6000 arrests.

Carruthers was a familiar sight as he biked daily to work and around the NRL campus, staying mostly aloof from anyone on the street and keeping hours well into and through the night—a habit that, according to a *Washington Post* article by Thomas O'Toole in late 1971, made NRL colleagues worry for his safety as, being a Black man, he was likely to be "shot as an intruder by security guards" or harassed on the streets. O'Toole, who covered space activities for the *Post*, described him as "one of a few black scientists in the space program" who was "also painfully shy, so much so that he is the butt of 'bashful' jokes at the Naval Research Lab."[4]

O'Toole also interviewed Carruthers's NRL colleagues and came away with a profile and imaginative quotes about Carruthers's early life and character. He made racial assumptions, claiming that in Carruthers's early years, Chicago's south side was "a black ghetto that is scarcely a spot for budding young astronomers." He implied that Carruthers had to overcome many obstacles, from the death of the three *Apollo 1* astronauts inside their capsule during a test, to the recently announced cancellation of later Apollo flights, to the fact that observing celestial objects from the Moon was a low NASA priority. Only the last applied directly to Carruthers.

O'Toole also reported that Carruthers had no hobbies and never took vacations. He regarded his work as better than hobbies, and his visits to White Sands, as O'Toole recounted, quoting Carruthers, were "better than vacations" because they provided momentary relief from Washington bureaucracy. No matter where he was, O'Toole added, Carruthers never wavered and remained wholly absorbed in his work, up to fourteen hours a day, seven days a week, preparing proposals and building and testing prototypes.[5] And public radio documentary producer and writer Richard Paul, interviewing Carruthers more recently, reported that he seemed unaffected by the rampant segregation at the time, when Black people were not welcomed or even allowed to patronize downtown businesses, and "police stopped them randomly if they walked in white neighborhoods at night." A vigorous urban renewal effort in the southwest quarter of DC added more fuel to the flame as it was directed to displacing the Black population.[6]

Carruthers's neighborhood, near NRL and the Bolling Air Force Base, still racially mixed but largely Black and middle class, was a good distance away from the centers of violence, protest, and strife across the Anacostia River in northeast and northwest Washington, though it could not have been untouched. "By 1967, whites were 37 percent of the population in Anacostia. After the '68 riots, that number soon dwindled to a handful." Carruthers, however, claimed, "Of course I may not have been reading the newspaper every day," bringing Paul to conclude that Carruthers "had his nose to the grindstone all those years."[7] And given how focused he was on his work, Carruthers may well have rarely ventured downtown in the 1960s, but he could not have been oblivious to the rapid changes going on.[8]

Indeed, Carruthers was an extreme example of the focused scientist, described by historians Howard McCurdy, Ann Roe, Susan Cain, and others, who could apparently shut out the world around them if it did not directly impact their work.[9] And work he did, pouring out proposal after proposal with, in Herbert Friedman's words, "indefatigable zeal and imagination" for instruments to be carried aboard low- and high-orbit Apollo missions, missions to the planets, and missions to the surface of the Moon. With his mind and body focused on such distant shores, one can imagine that he transcended even very real daily fears and frustrations.[10] One of his colleagues, Richard Conn Henry, recalls a story about a secretary at White Sands who checked in on Carruthers at the usual quitting time, saying, "George, it's five o'clock." And, as the story goes, "George looked up, bleary eyed, and said 'AM or PM?'"[11]

## PRIORITIES FOR SCIENCE ON THE MOON

Taking science to the Moon was anything but a simple matter, as Donald Beattie's titled history affirms.[12] Underneath the politics, territoriality, and institutional priorities, there were many ideas for what kinds of science could be done on the Moon in the early to mid-1960s. Most, logically, were geological, geophysical, and geodetical, building transport vehicles that could broaden the area to collect samples and take measurements. Even more ambitious, reflecting the enthusiasms of the early Apollo years, there were also plans for automated LEMs that would land near piloted LEMs to vastly increase the amount of payload equipment and to provide pressurized habitats for longer stays.

There was little question about scientific priorities. As Homer Newell, who spoke for the Office of Space Science and Applications, assured the MSC director, "Within time available to the astronaut for scientific experiments, priority is given first to the collection of lunar samples to answer basic questions in the fields of geochemistry, petrology, geology and bioscience." Second priority was "emplacing and activating [the] new Apollo Lunar Experiment Package (ALSEP)," which contained mainly lunar-related experiments.[13]

Planning contracts in the millions of dollars were given over to commercial firms to develop tentative designs for these ambitious payloads. One of the studies was for an astronomical observatory on the lunar surface—an idea, according to Donald Beattie, that was "supported by some, but not all, in the astronomical community." In 1965, MSFC awarded a $144,000 contract to the Kohlsman Instrument Corporation to design a telescope for an automated lander. Kohlsman knew that Goddard scientists were then proposing an ambitious 36-inch telescope for one of the OAOs and teamed up with them to design a version for Apollo.[14]

Staff in the Research Projects Laboratory and the Advanced Systems Office at MSFC thought that the so-called Goddard Experiment Package for the robotic orbital missions, although quite large, would also be desirable as a blueprint for a lunar-based observatory, if it could be modified to have astronauts manage it as part of the Apollo Applications Package.[15] It was proposed as an automated digital high-resolution spectroscopic camera fed by a telescope with a 38-inch aperture mirror. Kohlsman would determine the types of equipment that would be capable of addressing questions like the spectroscopic nature of the interstellar medium and, oddly, the "investigation of galactic periods of near-by stars." There would also be photoelectric observations, equally oddly expressed, "to test the brightness-color-distance relationships leading to further classification."[16] Other rather vague thoughts included using photon counters to study the solar outer atmosphere and the corona just before sunrise from the lunar surface, to investigate the Magellanic Clouds during the lunar day, and to acquire and track any "typical star (constant or variable) of magnitude 10 or brighter."[17]

The Kohlsman telescope would be managed by astronauts, but it would also be controllable remotely from Earth. In the remote mode, only photon counters could be used. But in the tended mode, they could use both photon counters and photography since the astronauts could bring the film home. The telescope itself, a reflector, would be over eight feet long. The optical system would be an improved wide-field version of the classic Cassegrain called a Ritchey-Chrétien after its designers, George Willis Ritchey and Henri Chrétien. The mounting would be a box-shaped structure that could be adjusted for altitude and azimuth. And for long exposures, there would be a motorized servo tracking system on both axes to follow the stars, planets, and other celestial objects. It would be deployed on the top of the docking hatch of the LEM/shelter and weigh some 2,500 pounds.[18]

This wildly audacious plan was proposed for follow-on landings sometime in the early 1970s. Future missions were envisioned in two categories: evolutionary, adding a high-resolution TV link to Earth for imaging, and smaller diffraction-limited systems for simultaneous observations. And to be sure, there were even more fanciful post-Apollo "Applications Package" program concepts for telescopes in the 60- to 120-inch range, equatorially mounted.[19]

Very few proposals have been found during this time for reducing the sizes of the telescopes by increasing the quantum efficiency of the detecting system. One suggestion, encouraged by Martin Harwit of Cornell, was made in 1966 by E. J. Sternglass of the Westinghouse Research Laboratories describing a form of UV-sensitive cathode that could vastly increase sensitivity so that a relatively small telescope (12-inch aperture) observing from the lunar surface would be equivalent to a 170-inch telescope "without the increase in weight proportional to the cube of the aperture diameter."[20]

The Kohlsman report was distributed widely within and beyond NASA, but no identifiable astronomers or astronomical institutions appeared on the distribution list. Such visions were a product of rather wild Apollo-era enthusiasms, and as the program continued through the 1960s, the telescope and other experiments were drastically scaled back and the enthusiasms stifled when the overall Apollo Applications Package program was reduced by NASA cuts mandated by the Office of Management and Budget due to rising costs of the Vietnam War and the Great Society programs.[21] But a vestige of this proposal reappeared in one application from MSC several years later, simultaneously and in competition with Carruthers's proposal from NRL.

## A FLURRY OF PROPOSALS FOR A LUNAR-BASED OBSERVATORY

Between April and October 1969, Carruthers prepared three distinct proposals for a lunar surface mission as well as two more for related instrument development. On April 25, 1969, Carruthers submitted a hybrid proposal: a payload that could be operated from Earth orbit or lunar orbit (the Apollo command and service module), or from the lunar surface (now called the "lunar module," formerly the "lunar excursion module"). Its purpose was to record time-lapse images of the full-disk geocorona in high-resolution UV light. This was the region of the Earth's outermost atmospheric layers where aurorae occurred. This work, he claimed, would "provide a basis for study of the day and night morphology of the global auroral systems, their development with time, and their conjugate point relations."[22]

Auroral activity had been observed and described for centuries and, by the late nineteenth century, was linked to explosions, or "flares" on the

Sun. A great store of historical information was collected on their structure and dynamics. During the International Geophysical Year in the late 1950s, there were many efforts, based at NRL and elsewhere, to broaden the picture by taking simultaneous observations from widely separated stations.[23] But there was not yet "a true global picture" that included both day and nighttime activity.[24] By then, intensive observational studies by NRL and the University of Colorado, from sounding rockets and now aboard the Orbiting Geophysical Observatory-IV (OGO-IV), launched in July 1967, were able to distinguish auroral activity in the daytime from normal dayglow in the tropical latitudes, but no one could explain the source of the emission.[25] OGO's highly inclined eccentric Earth orbit, however, took it barely beyond 550 miles altitude—not enough for a fully global perspective, nor enough to be beyond the extremes of the Earth's outermost atmospheric regions.[26] A camera in orbit around the Moon, or on the Moon, would hopefully have that perspective.

Carruthers identified several questions that could be addressed by global observations from a distance. How do auroral arcs form and migrate? How do they split and combine? And, most important, what are the possible causes for these migrations? Do northern and southern auroral systems show systematic behavior, and how is this behavior correlated with other geophysical parameters like the interaction of the Van Allen radiation belts and the particle stream from the Sun, the solar wind? Extended observations over days might also help to better determine if there were long-term trends in these structures, what might be causing them, and most critically, how to predict them. Overall, the global data would make it possible to test and refine various predictive models then under development. Summing up the special significance of the work, Carruthers claimed that "both the documentation of global auroral patterns . . . and the development of understanding of presently known features should be significant fruits of the far U.V. auroral photography effort."[27] In sum, this work would improve understanding of the interrelations between auroral activity and solar activity, and the disruptive interaction between the Earth's magnetic fields and the causative solar wind.

Carruthers went on to describe in detail how the Apollo observations would build on NRL's extensive efforts on OGO and OSO missions, which were ongoing and as yet unpublished.[28] His proposal was perfectly in line with NRL priorities, and he dutifully adhered to the suggestions of his supervisors, such as Talbot Chubb. But the instrumentation and the means were his alone.

As was typical of his proposals at this time, Carruthers restricted his responsibilities to the instrument. But this included training the astronauts in its operation, processing the flight films, and fully evaluating the performance of the instrument after flight. Grady Hicks and Talbot Chubb would concentrate on the "analysis of photographs and interpretation thereof."[29] Also typical of his many proposals, the instrumentation was hardly cookie-cutter. Carruthers sought improvements in spatial coverage and in spatial resolution, using his imaging systems instead of scanning photometers. He proposed adapting his Schwarzschild all-reflecting telescopic camera to collect and focus light with seemingly endless variations. For Apollo, he created a flexible design that could, with simple adjustments, be optimized for hybrid uses, either while traveling to the Moon, in lunar orbit, or on the lunar surface. Carruthers accordingly outlined in some detail these three possible modes of operation. He first laid out the "tentative plan for a lunar landing mission," where the instrument lands with the LM and remains attached to the LM on an extendable "porch" connected to the lunar descent stage.[30] The pilot would orient the lander so that the camera was in the shade of the LM with the Earth in view.

Earth would be the first target for Carruthers's camera on the lunar surface. It would be, of course, nearly motionless in the lunar sky given the synchronous rotation and revolution of the Moon. But since the Moon is in an elliptical orbit, it speeds up and slows down a bit in its motion around the Earth, which causes the Earth to shift slowly back and forth by about a degree over weeks. This meant that the astronaut would only have to adjust the camera once per day, minimizing the time required for its operation. The resulting observations would be a "continuous time-history of the auroral structure" in both the northern and southern auroral zones. In their proposal, Carruthers, Hicks, and Chubb repeatedly highlighted the importance of gaining a global view of the Earth's geocorona, along with associated auroral activity, linking the activity to solar activity.[31]

The second mode of operation, the "non-landed experiment," would also be directed to auroral activity, but at varying distances from the Earth and varying degrees of resolution. The only advantage of this mode, Carruthers admitted, was that the instrument did not have to be carried to the lunar surface. It was the landed option that he hoped for, "however, if technical considerations do not permit carrying the experiment down to the lunar surface, we feel that a non-landed experiment would still provide extremely valuable data."[32] Another reason for preferring the landed mode was the lack of stability for the telescope in flight. He did not think it would be possible to orient and fix the telescope during the flight with acceptable accuracy, "as even slight movements by the astronauts might disturb the pointing" unless the telescope was mounted independently on an inertially stabilized platform. This had long been a concern shared by numerous astronomers who preferred robotic to human-tended spacecraft.[33] Carruthers went on to consider other modes, such as observations from Earth orbit. Clearly, he was trying to cover all the bases; it was something of a fishing expedition.

By October, NASA turned him down. John Naugle, associate administrator for OSSA, explained that about one hundred proposals were received, about half were favorable, but less than two dozen were tentatively accepted, given the uncertainties of the program. Carruthers's proposal was evidently not received favorably. The Moon was the primary subject of scientific interest, there were other unnamed proposals of higher priority, and there were problems with fitting the instrument into the spacecraft. Naugle also claimed there were scheduling problems delivering the instrumentation.[34]

Naugle's negative reaction, especially his claim that non-Moon-related experiments were of lower priority, was possibly due to confusion caused by a major reorganization of responsibilities within NASA relating to the Apollo program. In May 1969, the MSC in Houston announced that its Advanced Missions Program Office would have management responsibility for lunar orbit and lunar surface experiments and, in September, transferred that responsibility to its Apollo Spacecraft Program Office. There were then a series of announcements issued for lunar orbit and lunar surface studies, and, by November, the office had issued an announcement detailing the science that was being proposed for geology, geophysics, geochemistry, bioscience, geodesy, cartography, lunar atmosphere, particles and fields, and astronomy. None of the few instruments listed in the astronomy category related to astronomical questions, but instead related to ultimate site selection and surface stability studies for an observatory, radio background noise levels, and improving knowledge of the lunar surface and atmosphere.[35]

Some of the instruments that were proposed and accepted were indeed astronomical in nature, including a small far-ultraviolet spectrometer by the esteemed Johns Hopkins physicist William Fastie to determine the presence, if any, and the extent of the lunar atmosphere. He also proposed a small UV camera to analyze the Earth's atmosphere above the ozone layer to "interpret unexplained features that appear in UV photographs of Mars and Venus."[36] At the time, there had also been at least one ambitious proposal from the Perkin-Elmer Corporation for an 80-inch segmented-mirror space telescope and eventually a lunar observatory sporting a 118-inch UV telescope on the Moon.[37] Just as plans for a Large Space Telescope were at the time in a state of "disarray," as Robert Smith has observed, so was thinking about what science would be taken to the Moon—and by what and by whom. Overall, this situation was symptomatic of the struggle over a vision for NASA's future in the post-Apollo era.[38]

## DUAL PROPOSALS IN OCTOBER

Carruthers, supported by his yearly funding for sounding rocket flights—first from NSF, then from NRL, and then from NASA's University Program through the late 1960s—enjoyed success after success. However, his larger proposals for placing instruments on Apollo-class missions, though they received strong encouragement from Friedman and Chubb, and endorsements from MSFC, were beyond the limits of his present funders and had failed now several times with Newell, Roman, and Naugle.

In April 1969, before she received Carruthers's proposal, but evidently aware of it, Roman made it clear that increased support for NRL would not happen; NRL would have to live within the allocation NASA provided. She stated emphatically, as we noted, that she would "leave it to NRL" to determine its own priorities. And, she added sternly, given the "difficulties" he was experiencing on his recent rocket sonde flights, as she understood the situation, his new proposals did "not sound reasonable." She ended her blunt letter on a hopeful note, stating, "I shall look forward to hearing more of your successes in the future."[39]

Even after Roman's cautionary advice and the Pioneer rejection, between April and October 1969, Carruthers submitted at least five highly detailed proposals to NASA for detector development, sounding rocket flights, and the Apollo mission noted above.[40] He also prepared four manuscripts for journal publication on his results thus far from rocket flights, indicating in some of them more proposals to come. Roman felt that Carruthers needed more guidance and direction from his superiors at NRL and, as we noted, warned him that NASA funding would be limited, and it might be best for him "to discuss relative priorities among various projects with your branch head."[41] Roman's warnings reflected her impression of dealing with astronomers. As she recalled, "Most astronomers were not only unused [to] such detailed management, but in fact tried to rebel against it."[42]

Unfazed and undaunted, enabled by Friedman's and Chubb's encouragement and support, Carruthers continued to develop prototypes and propose continually refined payloads for Apollo-related missions and sounding rocket flights, relishing the fact that each proposal justified his development of a new variation of his basic design in his laboratory. Excited by the successful landing of *Apollo 11* on the Moon in July 1969, Carruthers had also responded to another NASA Announcement of Opportunity for scientific projects on the lunar surface in two separate but related proposals, prepared in early October before he knew the fate of his Pioneer proposal and his first Apollo proposal for a lunar-module-based platform.

Friedman did his part, writing passionately to George Mueller, the associate director for manned spaceflight programs at NASA, that "what previously seemed to be very remote lunar-based astronomy has suddenly been brought much closer to realization." Friedman, of course, was thinking of a large array of X-ray detectors spread around the lunar surface, using horizon occultations to pinpoint the positions of X-ray sources. In his excitement, Friedman expressed only his personal passion: that a high-energy "x-ray observatory carries a higher priority than any other astronomical instrument proposed for the Moon."[43]

Friedman's uncharacteristically narrow appeal did not deter him from endorsing Carruthers's first Apollo proposals in October, which committed to sending small UV-sensitive instruments around the Moon and to the Moon, to be used on a deck mounted on the Lander, as we have described above, or removed from the LM and used directly on the lunar surface. With Friedman's approval, he would have the assistance of Talbot Chubb as principal administrator in charge of managing the project, dealing with the bureaucratic demands of a major NASA proposal. Echoing the scientific objectives expressed by NASA and by his peers at NRL, Carruthers once again proposed lunar surface instruments that would perform a variety of tasks, specifically wide-field observations of the UV spectra of diffuse background radiation and nebulae.[44] As W. C. Hall, associate director for research at NRL, explained, "These proposals are completely independent; although there is a great deal of overlap in the scientific objectives, the two studies are complementary."[45] One would be high sensitivity and low spectral resolution to make an initial survey of background radiation, and the second would be lower sensitivity and high resolution to resolve line structure beyond the Lyman-alpha region down into the extreme ultraviolet. Presumably, though not explicitly stated, these would be flown in consecutive Apollo missions.

One of his proposals also centered on longward UV spectra between 1,050 and 1,800 Angstroms to record galactic nebulae, the diffuse background radiation of interstellar, interplanetary, and possibly intergalactic mediums, along with some stellar spectra. Carruthers made the strong point that his effort was a means to "evaluate the potentials of the lunar surface as a location for a later astronomical observatory." It was a beginning. Though deeper into his proposal, he elaborated on a broad range of objects that could come under scrutiny, including the above and early-type blue hot stars, the Magellanic Clouds, planetary nebulae, galaxies, and even the newly discovered and highly enigmatic objects called quasi-stellar radio sources, or "Quasars."[46]

Carruthers now proposed yet another set of new and highly sophisticated designs for his image converters, camera, and spectrograph—highly imaginative and significant departures from earlier designs. We provide the details here to illustrate his complex thinking on the matter.

One spectrograph had a 2-inch concave grating working at f/10 that accepted light from an off-axis entrance slit. It then reflected a spectrum onto a huge (~5-inch) asymmetric concave photocathode image converter that again reflected electrons that were then accelerated by a solenoid array onto nuclear track film. There would also be two interchangeable gratings with coatings and rulings covering different wavelength regions. The astronaut would set up the instrument in the shadow of the LM, or in the shadow of a convenient lunar hill or rock, and then point the camera to several areas of the sky for unguided exposures of different lengths, programmed from 4 to 256 minutes. The spectrographic camera would see a square 5-degree field of the sky, so the pointing accuracy required was +/– 2 degrees. Pointing it to the desired object required circular protractors called "setting circles" on the two axes of a fork-mounted altazimuth mounting because the astronaut would not be able to see the stars through his protective visor.[47]

The second instrument was a more familiar electronographic Schmidt camera: a 3-inch aperture, a fast focal ratio of f/1.0, supported on a small portable pier with an equatorial mount. It was like those he was flying on

Aerobees but was modified to operate in two modes. Called an "objective-grating Schmidt Image Converter Spectrograph," it would operate in a wide-field mode covering some 20 degrees of the sky and also in a narrow-field mode of only ¼ degree for point sources. The latter mode would be activated by the astronaut flipping a lever to bring a "venetian-blind collimator" into the field of view in the direction of the grating dispersion, yielding low resolution but high speed. Carruthers felt that the low resolution would not be a hindrance "for this very first survey study" because the emission lines he was seeking were widely spaced, and there would be no detailed complex continuum spectra.

Carruthers argued that his designs were capable of addressing a wide range of tasks: his instruments could examine both diffuse fields and point sources, and, most important for an Apollo berth, his instruments were of the "smallest size and weight of the various types of instruments that might be considered for use in the 1050–1800A range." Further, these small devices, powered by his image converters, provide "a factor of ten to thirty times increase in sensitivity over direct recording."[48]

Carruthers maintained his usual step-by-step mode, offering up new, direct, and simple designs that hopefully, he argued, would provide more versatility. Especially clear in his proposal, he emphasized that his small, light, but powerful payloads were a logical first step on the Moon, and they would lead to something major. Surely, Carruthers was aware that there was strong competition for any berth on an Apollo lander. But his specific arguments and rhetoric hint at the possibility that he knew what he was competing against. Most clearly, his proposals for orbiters and landers covered the range of possibilities one might imagine that could be carried on the existing Apollo units, orbiters (the service module), or landers.

By now, both the Mariner V and Russian Venus probes had, en route, detected, measured, and mapped Lyman-alpha radiation in the galactic plane and in various discrete celestial objects and had results that supported Carruthers's own glimpses from his sounding rocket flights. As result, Carruthers reported, the high visibility of the emission line in the galactic plane indicated a deficit of absorbing atomic hydrogen in space "as much as a factor of 10 below" what was indicated from 21.1-cm radio mapping. This surprising result still needed confirmation and elaboration. And he emphasized that it needed observations by instruments far beyond the Earth's own night-sky glow. Extended observations from a stable platform on the Moon would be an ideal way to meet this need. The long exposures needed required only "slight compensation" in the position of the camera for the small emission regions, and none for the larger diffuse background radiation fields.[19]

To provide this "slight compensation" as simply as possible, when they were thinking of a landing far from the lunar equator, Carruthers switched from his altitude-azimuth design to an equatorial mounting on a small single column platform, complete with setting circles that could easily be calibrated by an astronaut pointing it to the Earth, Sun, or any pair of bright celestial objects with known positions.

The setting circles and equatorial mount were critical because the narrow field required locating and then tracking. Other than the Earth, the objects Carruthers hoped to examine would not be visible to the astronaut's protected eyes. The astronaut would have to set and level the instrument, point the polar axis of the instrument toward the lunar pole, calibrate the setting circles, and then begin the series of exposures with and then without the collimator. The equatorial design, moreover, would make it a bit simpler for the astronaut to point the instrument as long as the latitude of the landing point for the LM was known ahead of time, which most certainly would be the case. In fact, from the preliminary design offered in the proposal, the tilt angle of the polar axis appears to be fixed, and therefore notional. In distinction, Carruthers's original altazimuth design proposed in May for the hybrid mission was independent of latitude: a simple fork-mounting that was limited to short exposures.

Carruthers predicted that with the simplicity of the polar design for the lunar surface, astronauts would require no more than a week to practice the procedures in their suits in a suitable NASA vacuum chamber. He also hoped to be able to communicate directly with the astronauts during their observing periods to assist with any issues.

By the time he submitted his proposals through Friedman, he had built a prototype and tested it. Part of his proposal included at least one test flight on an Aerobee 150, which he predicted that, once approved by NASA, could take place within six months.

## ENTER THORNTON PAGE

A few days before Carruthers and Friedman signed and submitted their second Apollo proposal to NASA in October, and just before they also learned that the Pioneer proposal was turned down, Friedman received a letter at mid-month from Thornton Page, a Wesleyan University astronomer on detail to MSC in Houston. Page sent him a detailed plan to utilize Carruthers's camera in part of his own proposal for an ambitious lunar surface telescope. It was smaller than, but in the same mode as, those fancifully large telescopes imagined in 1965 by Kohlsman and in 1966 proposed by MSFC's Advanced Systems Office that envisioned a 40-inch altazimuth mounted reflector placed on top of a robotic LM.

Page, through MSC, had submitted his proposal to NASA in mid-August for a "Large Space Telescope (LST)," variously referred to as a "Large Telescope Facility (LTF)," a 20-inch reflecting telescope placed near the lunar equator for one year. It would be remotely operated from Houston and would serve a wide range of astronomical interests. Page was politically very savvy. He had recruited a "team of astronomers, astronauts, optical experts, and engineers" at MSC and the University of Texas to develop the instrument and plan for both far-infrared and far-ultraviolet studies of "planets, stars, and galaxies" to be conducted by Houston astronomers as well as guest observers.[50]

Page was a boisterous social mixer who was well connected in NASA (see figure 8.3). He certainly knew that deliberations at various scientific workshops in the mid-1960s placed a low priority on astronomical activities on presently planned Apollo missions and even that an astronomy panel headed by Northwestern University astronomy professor Karl Henize, later to become one of the first "scientist astronauts," believed, in the words of Donald Beattie, that "there was no intention to include an astronomy experiment on any of the Apollo missions."[51]

But Page also knew from his contacts in OSSA that by 1967, sympathies were growing for astronomy on Apollo. In his own proposal to NASA in October, he recalled that in these later meetings, there was "emphasis on site-testing with a small instrument." Subsequent to these meetings, observations from Earth-orbiting observatories, sounding rockets, *OAO II*, and "particularly the Apollo 11 lunar landing [made] the possibility of a large manned lunar observatory . . . more attractive."[52] If true, this warming attitude for astronomy on the Moon could have been partly due to Page's persistence and campaigning, as well as to Henize's efforts. Page had, in fact, recruited Henize for his cause, as well as about a half dozen astronomers, and both Henize and Page attended OSSA subcommittee meetings.

Page's new proposal still featured a 20-inch telescope but added another smaller instrument, "modified for the definite use of Carruthers's electronographic camera to be fabricated by NRL." Page then asserted that "it is my intention to work closely with Dr. Carruthers who will rank as 'co-principal investigator,' and I hope we can get the camera on Apollo 14." Page's claim that Carruthers was now involved is doubtful because a month after he submitted his proposal, Page wrote to Friedman: "We understand that you are proposing some observations of stellar spectra from the lunar surface. It will help our case for the LTF if we say that your project could be done with the 20-inch and UV spectrophotometer. Could it?"[53]

Page also claimed that he had tried to contact Friedman the previous week by phone, but failed. Still, Page knew what NRL was up to. He was the quintessential "insider" at MSC and in government circles, and he frequently participated in Roman's Astronomy Subcommittee deliberations. As we will note later, Carruthers later claimed that he was unaware of Page's proposal or his invitation to Friedman.

A brief diversion to introduce Page will illuminate the many complex ways NASA does business, and Page's centrality in that business. Born in 1913 in New Haven, Connecticut, and educated at Yale and Oxford in the 1930s, Page served in World War II in operations intelligence, part of a small group within the Naval Ordnance Laboratory that applied gaming theory to mine warfare.[54] After the war, Page continued to serve in the Operations Research Office, becoming deputy director in 1948 and remaining until 1958.[55] Page then became a professor of astronomy at Wesleyan University in Connecticut, taking multiyear leaves of absence at the Smithsonian Astrophysical Observatory (1964–1966) and at NASA's MSC (1968–1970s).

Page was the consummate networker and quintessential insider. As Homer Newell stated in his history of NASA space science *Beyond the Atmosphere*, "the best way of keeping close to the space science program was to serve on one of the NASA advisory committees."[56] Page was even closer; he had functional and advisory roles at NASA, United Aircraft, the American Institute of Physics, and the Smithsonian Institution, and he was an instructor for NASA's scientist-astronauts starting in 1967.[57] While at the Operations Research Office, in 1953, he became part of the "Robertson Panel," a CIA-directed committee that was pondering the reality of UFOs and their military implications. He even engaged J. Allen Hynek and Carl Sagan in public debates.[58]

Page's recruitment of Hynek and Henize, among many others, for his 20-inch telescope proposal was a strategy that responded to NASA's preference for "teams." But it was also part of his job at MSC; he was on leave from Wesleyan as a National Research Council research associate since July 1968. MSC hoped he would attract astronomers to MSC's programs to strengthen astronomy there, and he did so.

Page also reviewed space astronomy proposals for Nancy Roman's NASA Astronomy Subcommittee, evaluating them for how well their goals aligned with NASA's and MSC's missions. At one point, NASA even appointed him chairman of a committee on "Lunar Surface Experiments and Data" and charged him with making "plans for the use of optical instruments on the lunar surface."[59] In parallel, also with MSC's approval, he continued his own research on galaxy spectra, using data from Wisconsin's instrument payload on *OAO II*, which NASA was threatening to close down in the spring of 1971. Page, along with other well-connected names in astronomy, lobbied NASA in April to continue *OAO-II* because, he argued, with Carruthers's camera (now designated the "Apollo Science Experiment S-201"), "we are planning the targets for the UV Camera S-201 Experiment on *Apollo 16* to tie in with OAO measurements and must depend on OAO for follow-up on some of the S-201 findings."[60]

Page's role on the Roman committee bears particular notice, considering its low opinion of Carruthers's earlier proposals. In particular, Page had turned to Carruthers for assistance in building a suitable telescope and his open admission that he knew of Carruthers's proposal even before it was submitted. And, as we have also noted, Carruthers recalled that when he submitted his own proposal, he was unaware of Page's competing interests and, in fact, became aware later only through Friedman.[61]

Page was constantly in search of collaborators. His first efforts at attracting young astronomers to MSC did not bear fruit. Few were interested, and of those that were, none were acceptable to MSC management. Page's continued efforts, however, were successful. He attracted a talented specialist in 1969, the astronomer Guido Chincarini from the Astrophysical Observatory of Asiago, University of Padua, who was then at the University of Texas. Chincarini had experience with Lallemand electronographic image tubes from a collaboration with Merle Walker at the Lick Observatory, and Page felt he would be critical to adapt them to use on the Moon. He also

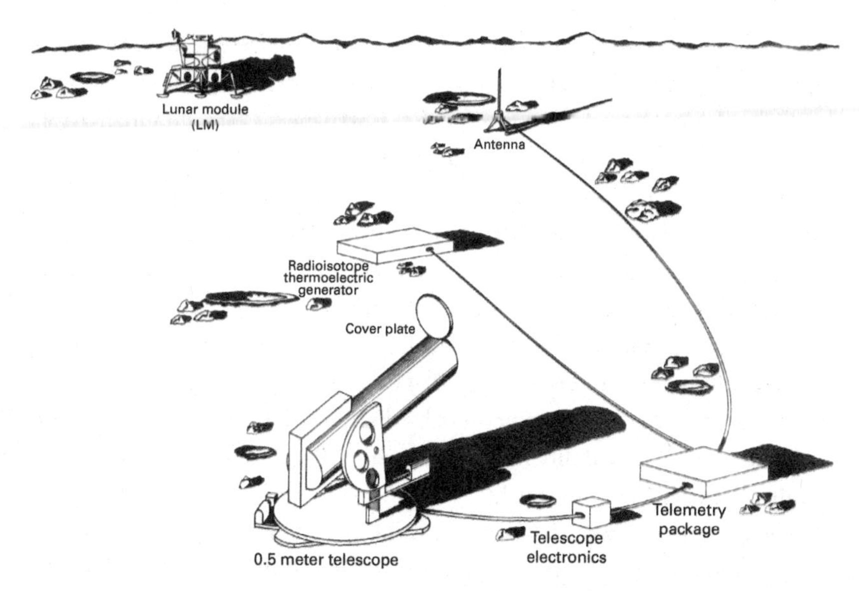


Figure 7.1
Thornton Page's concept circa August 1969 for a remotely operated 20-inch reflecting telescope placed near the lunar equator for one year. Folder "NRL, Geo. Carruthers Mat., S-201 Apollo Lunar Surf Camera," box 19, 1969–1974, HF/APS.

could help Page evaluate Carruthers's design, which Page realized was more efficient and easier to use than the Lallemand design and was sensitive to the far ultraviolet.

At the time Page wrote to Friedman, he had recruited Chincarini and was finally successful attracting, as we noted, a large team of MSC astronomers, including Yoji Kondo as the instrument PI for the large reflector; Stanley Freden as the administrator; and Karl Henize, Tom Giuli from MSC, J. Allen Hynek of the Dearborn Observatory, and several other connected astronomers. Page listed himself as a co-investigator, given his status as an NRC fellow and a Wesleyan faculty member.

NASA, however, soon nixed the 20-inch telescope.[62] Weighing in at 450 pounds with component dimensions as large as twenty-five square feet and requiring some sixty-three cubic feet for stowage on the LM, it was highly impractical now that the robotic instrumented LEM concept was long gone (see figure 7.1). But Page was never single threaded. Even before the LST was axed, he proposed a smaller version he called GALT for "Geophysical-Astrophysical Lunar Telescope," which could scan both the lunar landscape and the heavens. GALT was also nixed, given its expense and size. However, as we know, and as he stated in his "terminal report" as an NRC fellow at MSC, he also proposed what was, in fact, Carruthers's "small Lyman-Alpha camera," claiming it was his original proposal.[63]

Through the end of the year, various NASA committees evaluated all the proposals that had responded to its various Announcements of Opportunity. Over one hundred proposals passed preliminary screening for the orbiting package alone, and a similar amount for the surface experiments.[64] As with the Pioneer deliberations, in late November 1969, the Planetary Subcommittee, now referred to as "Planetology," deliberated over the proposals for three days, separating them by discipline and ranking them within each category. These deliberations show how Page's and Carruthers's proposals were eventually merged.

Thirty-two proposals fell under "astronomy," and priorities ran from "1" to "4" based upon the criteria we described earlier. A first ranking on November 24–25 rated Page's use of Carruthers's camera "1" by eight members. Carruthers's proposals for "Far-UV Spectroscopy of [Three proposals]" were rated "1" by all nine voting members, but a fourth proposal by Carruthers for "Diffuse UV Lunar Background" was rated "3" by seven members, and "1" by two members. After more discussion the next day, a new ranking rated Carruthers's first three proposals as "1" for his proposals to record the Earth's far-ultraviolet geocorona and the Small Magellanic Cloud, and his fourth a lower "3" rating. Page's proposal, called only the "Lyman-Alpha Ultraviolet Camera" using Carruthers's camera, was ranked "1," and the 20-inch reflector, given over to Kondo, ranked "4." In the notes taken by William Baum, a Lowell Observatory astronomer on the committee, Carruthers's and Page's priority "1" proposals were conspicuously combined and bracketed, with Carruthers indicated as the principal investigator.[65] Overall, Carruthers's proposal for the geocoronal work was ranked "1" by most members of the committee in the category. Yet even now, there would be compromise.

What did the bracketing mean? Somehow, Page and Carruthers had to join forces and work out a compromise proposal. Thus, "the far UV—camera/spectroscopy experiment is a new item which requires a development and qualifications program."[66] In late February 1970, Richard J. Allenby, assistant director for Lunar Science, asked Page and Carruthers to team up and issue a joint proposal, with one of them as principal investigator. Page had been in contact with Allenby since the previous April as a member of Roman's subcommittee and may well have suggested the merger since he was also a member of Allenby's long-term working group for possible future Apollo flights. Page quite likely anticipated Allenby's request and quickly drafted a joint proposal combining Carruthers's October 1969 proposal and his own from August.[67] By then, Page's 20-inch telescope was gone.

Allenby's request was not unusual for NASA, which typically formed teams for Apollo-scientific experiments.[68] Usually, there were far more proposals than slots.[69] How this merger came to be, according to Page, resulted from discussions between Friedman and Page, wherein, Page claimed to Captain Lee Sherer, "Dr. Friedman (NRL) agreed with us that I should be PI, partly because I am located at MSC Houston and have close contact with the astronauts here. It was (and is) my intention to stay here, and I bought a house with the understanding that I would be appointed to a position on the NASA staff."[70]

With his promised appointment as an MSC staff member, no longer on leave from Wesleyan, Page as principal investigator well knew that MSC was the most powerful political ally as the responsible agency. However, Page had also just been informed that due to a management turnover, his appointment was canceled. So Page now turned back to Wesleyan to administer the project, and "the University will become the world's first with an observatory on the Moon." Page also claimed, copying Friedman and others but not Carruthers, that by then, he had obtained MSC's full endorsement, as well as Roman's and other prominent astronomers, and that "we all hope that this minor matter of my non-MSC status will not jeopardize the experiment."[71]

In their combined proposal, "NASA Experiment Proposal for Manned Space Flight, Lunar Surface Ultraviolet Camera," dated April 25, 1970, Page was still listed as the principal investigator, and Anthony J. Calio, director of science and applications at MSC, as the principal administrator. By May 5, however, after considerable dithering, Page advised Friedman that the proposal could be "processed more rapidly" if Carruthers became principal investigator and Friedman became the administrator. Further, after a phone call, Talbot Chubb agreed that Page, as co-principal investigator, would take care of all the MSC-based actions, including trials, tests, and "interface" matters, "so that Carruthers will not be shouldered with excessive red tape as P.I."[72]

In mid-June, Friedman and Carruthers formally accepted Page's role to take care of all matters in Houston, and over the summer, they continued to negotiate over the details of who would be responsible for what.[73] By then, Carruthers and William Conway, a project engineer who was transferred out of NRL's Engineering Services Division to work with Carruthers, decided that both the camera and spectrograph would be supported on light collapsible tripods with altazimuth mountings, since it was expected that the landing site would be near the lunar equator. That was far from certain, however, given they were still jockeying for a slot on either *Apollo 16* or *17*.

Friedman worried about having Carruthers as principal investigator on such a complex and politically convoluted project. The political and managerial tasks would put too much unwanted pressure on him, given his personality and extreme focus on the laboratory. And Friedman was right. There is no record of Carruthers being actively engaged in any of the negotiations beyond providing technical reports on the many alterations to the payload.

Indeed, throughout all this confusion, Carruthers focused on developing the optical and recording systems, testing and calibrating them, and engaging in a series of *Aerobee* flights, the latest in March 1970 that refined his earlier observations and, finally, as we noted, gave a firm value for the hydrogen abundance in interstellar space. This latest success boosted Carruthers's status considerably after it was published in August, but it said nothing about his ability to manage or engage in what were complex and often touchy political matters. But just how he felt about the arrangement is unknown.

What is known is that even as the negotiations among Page, NRL, and NASA were progressing, Carruthers sought out alternative sources of support for other projects. He also moved beyond NRL and NASA to the Air Force, and even to NSF's Kitt Peak National Observatory in Tucson, Arizona. As we noted above, the creation of NASA had raised territorial issues with NSF, which already had a rocketry division at Kitt Peak.[74] Alan T. Waterman, NSF director, had been interested in consuming NRL's X-ray group, and his Kitt Peak director, Aden Meinel, a gifted and ardent instrumentalist, was enthusiastic about getting Kitt Peak involved in active space research. Meinel experimented with remote-controlled telescopes and was even thinking of a space telescope. After Meinel left Kitt Peak to establish his Optical Sciences Center at the University of Arizona, Kitt Peak enthusiasm waned, but his successor, Nicholas Mayall, still wanted to support Carruthers.

Hardly two weeks after his successful flight on March 13, Carruthers submitted a new proposal to Mayall to send a triple camera system up on an Aerobee to observe the Andromeda Galaxy before the end of the year. Carruthers would be a guest investigator at Kitt Peak, and NRL would provide the payload.[75] Mayall liked it and submitted it directly to NSF, but NSF Astronomy Section head, Robert Fleischer, was not enthusiastic and demanded changes to the proposal. It was resubmitted several times over the next few years while the project languished, until a flight was approved for October 1976, but under NASA cognizance.[76] Indeed, NASA had not taken kindly to NSF's intrusion. As Homer Newell confided in the astronomer Frank Edmondson, a high-level operator in the AURA (Association of Universities for Research in Astronomy) and Kitt Peak universes, "In view of our policy relative to space astronomy versus ground based astronomy we found it quite distressing when NSF chose to fund rocket astronomy at NRL and Kitt Peak, and to fund the AURA satellite telescope; but for the good of the field of astronomy we chose to not make a fuss about it."[77]

Despite Newell's claim that they made no fuss, responsibilities and control were shifted, and, given that NASA had to demonstrate its willingness to take these programs over, it worked to Carruthers's advantage.

Between the time of his initial proposal for Apollo and when *Apollo 16* finally flew in April 1972, Carruthers was constantly proposing to refine his detectors and maintain his UV program. He asked for additional Aerobees and for a larger rocket called Blue Scout, which could launch heavier payloads and even small satellites.[78] As always, Carruthers sought out new berths to try out new designs that might obtain data that would improve the observations of previous efforts. He proposed payloads for small satellites and for NASA's proposed Skylab flights. About 15 to 20 percent of his proposals were successful, which was a strong indicator of Carruthers's growing reputation. But there were always problems.

NRL Aeronomy branch head Phillip Mange laid out the darkening funding prospects, which was likely what prompted Roman's concerns. In April 1971, he advised Friedman that the lack of support from NSF and from NASA threatened support for Carruthers's staff member Chet Opal, stating that "we need to consider if we can or wish to support Opal in some other way."[79] Opal, a Hulburt fellow with a 1969 PhD from Johns Hopkins, became a key player on Carruthers's team and a candidate for a career appointment.[80] He had critical experience building UV payloads for sounding rockets and ground-based telescopes to measure molecular oxygen densities in the upper atmosphere. Fortunately, Mange was able to tap other laboratory resources to keep Opal supported, and his appointment was approved. This was one of many examples where NRL was able to provide management support and expertise that protected Carruthers and his programs, but it was never assured and was not sustainable. Even so, by 1971, Carruthers was promoted to the high pay grade of GS-15 (typically held by senior scientists, program managers, or section heads), which was a spectacular rise after just five years on the staff.[81]

Throughout this whole time, there is no evidence that Carruthers sought a role in the negotiations, letting Friedman and Chubb take the lead, as well as William Conway, who was by then designated project manager working directly with Carruthers. As NRL colleague Phillip Mange admitted, echoing the general opinion at NRL at the time, Carruthers "needed an individual, an associate who could represent him to the NASA oversight gang."[82] Conway also became the go-between for Carruthers and MSC technical staff, including Page.[83] Once on board, Conway worked closely with Carruthers on a wide range of tasks, both managerial and technical, paying close attention to details.

Throughout all this, Page remained an "effective front man" to the mission, which by June 1970 was designated S-201 and bound for *Apollo 16*.[84] Carruthers was aware of all this but was also constantly distracted by other projects, like Page's proposed follow-on GALT mission for *Apollo 17* and his many other continuing projects, including a major proposal for a Mariner mission. By December, Page was given a temporary appointment at NRL as project scientist, acting as NRL liaison, continuing to spend most of his time at MSC.

In this latest capacity, Page participated in creating the NRL "Statement of Work" for the S-201 package, completed in late July 1970. The "Statement of Work" defined the shared responsibilities of the scientific and technical support staffs, but also the costs and the delivery schedule. Carruthers would be half time on the project, lasting some two and a half years, and would have some liaison duties with MSC, attending Surface-Science Working Panel meetings with Page at MSC, identifying "scientific requirements, and providing scientific guidance, as required." Carruthers would organize the scientific effort overall, but "prefers to delegate authority to the Co-I for most of the support required at NASA/MSC." He would attend hardware tests at MSC and act as liaison for the staff constructing the hardware at NRL.[85] Carruthers would also be on-site at the Kennedy Space Flight Center, supervising the "assembly, calibration, and testing" to verify that the payload met all requirements. Then he would join Page at MSC for the launch, landing, and deployment sequences, remaining there throughout the rest of the surface mission. Both would work with the astronauts to ensure that all steps, from deployment of the hardware to the retrieval of the film cassette, were understood and properly executed.[86]

## DEFINING AND DEFENDING THE MISSION

Page ardently defended the mission. And throughout the rest of the year, others within NRL, at MSC, and at Headquarters, and even associated industrial contractors, also defended Carruthers's project even though constantly changing priorities and management changes within NASA put the mission in doubt, mainly from Headquarters critics who always felt that the main point of the landings was to conduct lunar science, not astronomy. Martin Malloy, who worked with Allenby in the Apollo Lunar Exploration Office at NASA Headquarters, filed a statement for the record in late November that made the case supporting the camera as a lunar surface experiment. Stating that Carruthers's camera was some twenty times more sensitive than UV film, he pointed out that "it was the most advanced light recording system in the world." Noting that the Russians were expected to attempt a similar set of observations from *Luna 17*, to be launched into lunar orbit in 1970, he added that "Carruthers' experiment may provide a spectacular discovery in UV astronomy to offset their lead."[87]

Soon, other Headquarters staff agreed. Norman Paul Patterson, who worked with Malloy, originally felt that "because the experiment does not have as its primary objective the study of the Moon, the necessity of conducting it on the lunar surface has been questioned."[88] But, he argued, the Space Science and Applications Committee gave it its "highest priority" and reaffirmed that it had to be conducted on the lunar surface. A third voice, orchestrated by Allenby and his staff, came from a manager at Bellcomm, Inc., a major contractor for Apollo science, who urged "the retention of the Lunar Surface Camera/Spectrograph experiment" on *Apollo 16*, giving many of the same reasons but adding that "Carruthers' electronographic camera is novel, and with film-return and semi-automatic operation, makes a productive auxiliary for a manned lunar mission." Beyond the astronomical targets, the Bellcomm manager, George Timothy Orrok, urged the importance of examining "the entirety of the Earth's geocorona" and expressed that the project itself was "clearly pertinent to the future use of the Moon as an observatory."[89] And Carruthers was the man to do it, as Orrok attested, pointing to his detection of molecular hydrogen.[90]

The many changes to the program made possible by the merged Carruthers-Page proposals, as well as Carruthers's creative refinements, expanded the problems that could be addressed. Among the myriad targets Page had proposed for his instrument were surveys of the Earth's auroral zone. But unlike Carruthers's earlier proposals, Page did not explicitly include imaging the Earth's outermost geocorona, which was NRL's highest priority. Sounding rocket flights could only briefly sample this uppermost region of the Earth's atmosphere, at heights greater than ten thousand miles, that interacts with the solar Lyman-alpha spectral line and faces the solar wind, all of which influence the ionosphere. Even low-Earth orbit satellites were still within its bounds and unable to explore it in its entirety. Now the goal was front and center justifying the camera: it needed a stable platform at lunar distance to view the geocorona in its entirety and to record how it changed over time in response to solar activity.

Some clarity emerged in June 1970 when Friedman and Carruthers formally authorized Page to provide all liaison services at MSC, acting both on Friedman's and Carruthers's behalf. All that was needed now was to process Wesleyan's proposal to NRL to support Page as project scientist for the next two years. Once NASA funding was assured, Friedman stated, the funding would be processed though the Office of Naval Research, and through Wesleyan to Page.[91]

Over the 1970 summer, Carruthers, Conway, and Page conferred by phone an average of three times a week to clarify shared responsibilities in the constantly changing project. By mid-July, the plan was for Conway to write up a complete "Statement of Work," along with a schedule and contract documents. Then work would begin at MSC designing the tripods and mounting, and recording rough dimensions for the cameras they would carry. Carruthers and Conway would then examine those concepts and refine them, and finally decide who would fabricate engineering models of the camera and spectrograph.

To meet scheduling demands, they pushed to have an engineering model available for testing by astronauts at MSC in its large vacuum chamber hopefully by September 1970. First, the optical systems of the two cameras had to be subjected to a series of radiation hardening and vibration tests.

Once all this was done, and deemed fully acceptable, a prototype mock-up would be delivered to the Grumman Engineering Corporation in Bethpage, Long Island, New York, to begin developing designs for interfacing with the LM, both for storage and access for deployment. One key question would be, could the cameras, mountings, and tripods, folded and stowed in a secure moisture proof cylindrical container some 32 inches long and 26.5 inches in diameter, fit in the LM?[92] The answer was no.

## TWO INTO ONE: THE CAMERA REDESIGN

Making the instrument package small enough to fit in the LM was only one of many pressing demands Carruthers faced. His initial proposals, combined with Page's, required two telescopic units: for imaging and for spectroscopy. This, of course, doubled the space requirements for the instruments and required a larger mounting or, as they were then proposing, two mountings, to say nothing of training the astronauts to use both. But why not combine them into a single unit?

Sometime in the spring of 1971, Carruthers and Conway decided to do just that: create a dual-mode system. This was not a new idea. They had both been involved with a project Chubb was leading for an "Optical Geophysics Satellite" proposed for NASA's fourth planned "Small Astronomy Satellite." It would use a UV-sensitive television system in both "narrow band imaging and spectroscopic modes" to monitor aurorae and emission in the upper atmosphere. The camera would be a combination of Carruthers's Schmidt system feeding a silicon-intensified vidicon. To operate in dual mode, the camera would receive light from an external back-to-back mirror/plane grating that could be rotated for spectroscopy or direct imaging. Carruthers soon modified this design: instead of the back-to-back mirror/grating design, he suggested a rotating array of mirrors and gratings on a triangular plate.[93]

For Apollo, Carruthers and Conway came up with a more rugged and efficient design that would be suitable for human operation. They put a single camera (containing the electronographic detector), film cassette, and Schmidt telescope on a pivot that was in line with the altitude axis of the system. Thus, the astronaut could either direct the camera to an external grating, for spectroscopy, or rotate it 90 degrees to view the sky directly for imaging, reducing the external reflection and obviating the need for two units (see figures 7.2 and 7.3). Carruthers chose the optically faster

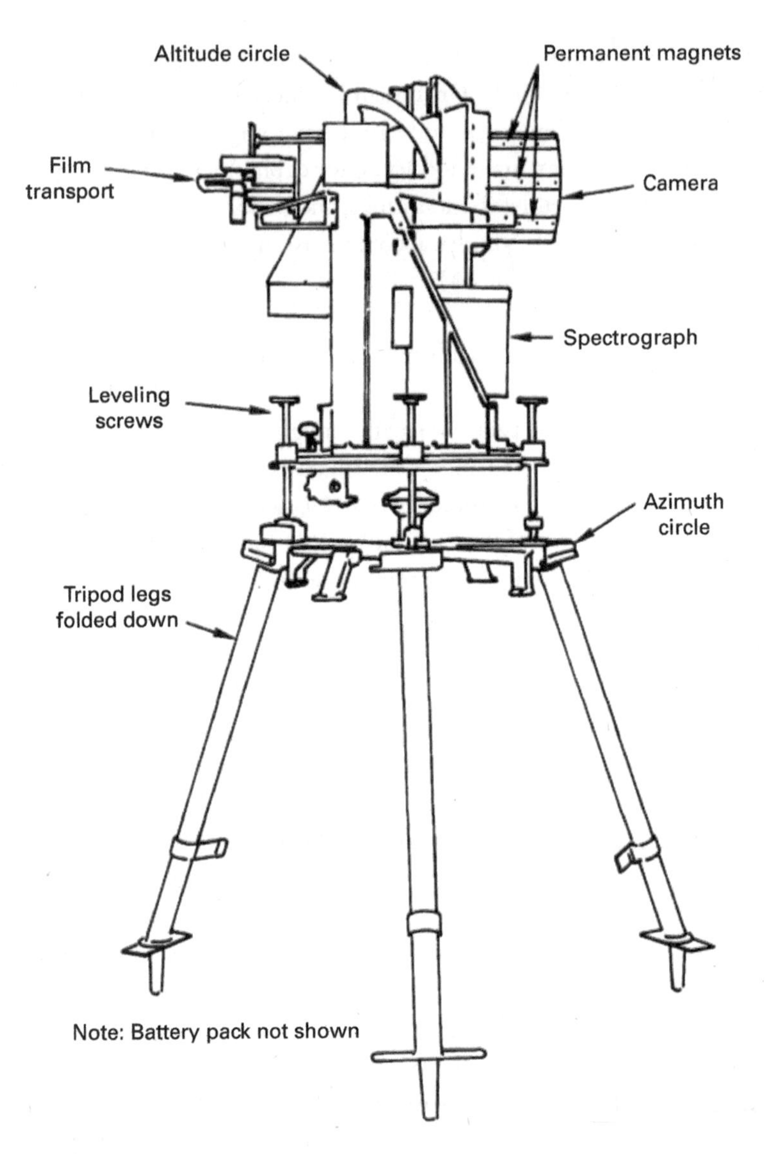


Figure 7.2
Schematic of the Lunar Surface Ultraviolet Camera/Spectrograph. *Source:* Carruthers, "Further Developments," from *Mission Science Planning Document Revision* (Houston: Manned Spacecraft Center, 1972), 2–112, figure 2–35b.

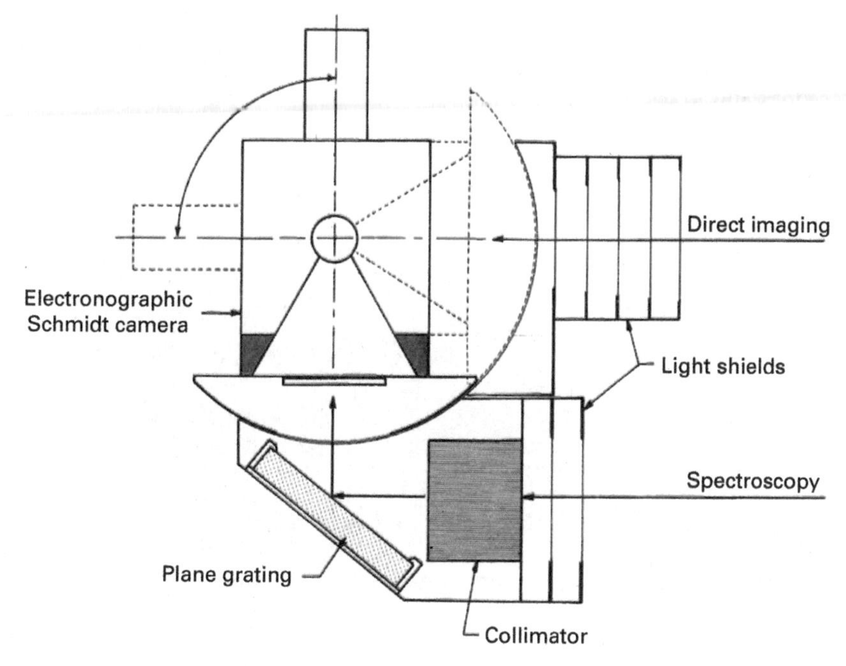


Figure 7.3
Schematic showing how the camera could be pivoted to produce images or spectra. The Schmidt camera (see figure 5.2) could be rotated to the vertical position (solid outline as shown) to look at the plane grating directly below for spectroscopy, where the field of view is restricted by an electroformed-grid optical collimator. The Schmidt camera could be rotated to the horizontal position (dotted lines) for direct imaging of the sky. *Source:* Carruthers, "Further Developments," figure 6 (figure 2 from Carruthers (manuscript), "Further Developments of Magnetically Focused, Internal-Optic Image Converters," paper presented at the Fifth Symposium on Photoelectronic Image Devices, Imperial College, London, September 16, 1971, folder "NRL, Geo. Carruthers Mat., 1972," box 19, HF/APS).

Schmidt—3-inch aperture, f/1.0, with a 20-degree field—as the sole camera and incorporated a motorized system to alternate UV filter correctors for different spectral bands, including and excluding Lyman-alpha.[94] And as planned from the beginning, they knew that the camera had to be designed to work in the cold vacuum of space, in the shadow of the LM (see figure 7.4), as depicted in their proposal.

NRL and MSC combined forces to build the first prototypes of the lunar camera system, which would utilize the altazimuth mounting and setting circles option since *Apollo 16* was now definitely bound for a spot near

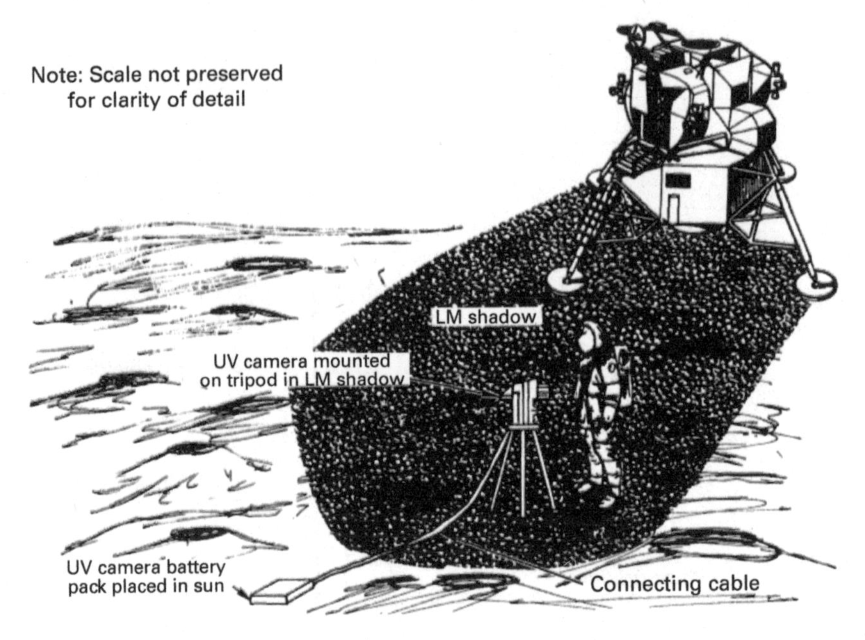


Figure 7.4
The planned positioning of the camera in the shadow of the lunar module. Note the wire connecting the power source. This is one of the wires Young tripped over.
*Source: Mission Science Planning Document Revision* (Houston, 1972: Manned Spacecraft Center), figure 2–35c.

the lunar equator. There was considerable confusion among the groups at times as to the details of the design—fitting everything together—not only between the Space Sciences and the Engineering Services Division within NRL, but with those engaged at MSC. Conway and Carruthers visited MSC numerous times to work out the problems.

NRL and the Naval Electronic Systems Command were now wholly behind the program, as was MSC, but there were severe scheduling issues raised at NASA Headquarters. Malloy, Patterson, Orrok, and Allenby campaigned to keep S-201 alive. In April 1971, Phil Mange, by then associate superintendent in Friedman's division, reported that S-201 was now definitely slated for *Apollo 16*, scheduled to fly in March 1972, but this put NRL seriously behind the delivery schedule set by NASA. There were "a number of engineering manpower and supply problems which now appear serious enough to threaten outright cancellation of the project." Conway had just met with his MSC counterpart, and their "intensive three-day review" resulted in a NASA representative threatening that the payload would be eliminated if the deadlines were not met. At the very least, NRL and MSC had to make assurances that delays would be minimal. Mange's point in his report was to alert Friedman that pressure had to be put on NRL's Engineering Services Division to step up the pace, finding ways to overcome the highly time-consuming "government supply procedures" and deal with industrial vendors who were "unwilling to shift production schedules to provide faster delivery."[95] Mange urged that this was a real threat and demanded that the project be given "highest priority" with the Engineering Services Division. Clearly, this was a management situation far beyond the finite boundaries Carruthers had lived in from his university days, the Illinois "do-it-yourself" style.[96]

There was no way to delay S-201. In July, Carruthers asked Rocco Petrone, Apollo program manager, if his instrument could fly on *Apollo 17*. Petrone bluntly said "no" and that the roster was filled—although he might appeal.[97] Problems persisted through November to the end of the year, including humidity control and power supply issues. Mange sent Conway and Carruthers to MSC again to break the logjam. The humidity bag, which would contain the experiment in flight, leaked. And the power supply was erratic. These issues, plus problems with the integration pallet, called for quick analysis and solutions to meet NASA standards for Apollo. The bag, pallet, and power supply were, however, NASA responsibilities, but they all had to work together to satisfy qualification requirements, and so having NRL staff there facilitated communications. On November 9, Mange, responding to a telex from MSC, assured higher-ups at NRL that MSC was "proceeding in a sympathetic evaluation of the exploration." The telex, however, bluntly warned that if the problems were not solved quickly enough, S-201 would be removed from the roster.[98]

By this time, it was no surprise that the media was taking notice, informed and encouraged, no doubt, by dutiful NRL and NASA public affairs offices. Carruthers was featured in January 1970 in the *Chicago Daily Defender*, which pictured him with an Aerobee payload and a hip caption that read "This Is His 'Thang.'"[99] And in late November 1971, as we noted above, *Washington Post* reporter Thomas O'Toole chimed in announcing "Black Scientist Develops Apollo Instrument."[100]

Carruthers worried that O'Toole's colorful reporting would add to the tensions NRL was having with Houston MSC staff. The article was reprinted and distributed in Houston, and MSC staff, including Page, surely would have noticed the sole focus on Carruthers with no mention of MSC or Page.

In bold red block letters, Carruthers told Mange, "Phil—In view of the recent resurfacing of old skeletons in Houston, and NASA feedback regarding same, I felt compelled to put this matter to bed once and for all. G.C."[101]

Carruthers wanted to distribute a disclaimer, and NRL approved. He first wanted everyone to know that the O'Toole article was not an official NRL news release, although O'Toole was given permission by NRL to interview him. He was annoyed that the article was mainly about him, not about the project, and that "the many personal details regarding the P.I. were very embarrassing to him: many details were not true, only half true, and some things which were true were gathered by the grapevine, and not by direct quotation as inferred." Further, Carruthers claimed to have repeatedly cited "the contributions of people other than the P.I.," but O'Toole omitted them.[102]

Conversely, newspaper accounts of the mission in the *Hartford Courant*, local to Wesleyan, featured only Thornton Page. The first feature, in July 1969 during the flight of *Apollo 11* and before Page approached NRL, dealt, of course, only with Page's 20-inch telescope and his vision of future lunar exploration. The second, in November, again featured Page and the 20-inch telescope, noting it was the same diameter as Wesleyan's classic refracting telescope on campus. The extensive article also described the creation of the MSC out of a huge Texas ranch and humorously described Page's "now nicely renovated" office there, fashioned, it claimed, out of one of thirteen bathrooms in a classic mansion that was converted into the headquarters. And in a third feature in April 1972, after the merger, titled "Astronomer's 2½ -Year Project to Be First Moon Telescope," Carruthers was barely mentioned in the "joint project." As the piece ended, "In about two weeks, Page will know if his 2½ years of planning will pay off."[103]

Underlying these contrasting public notices was continuing friction between MSC and NRL, at all levels. Final assembly, testing, and delivery of the instrument payload were underway, but fingers pointed in all directions when deadlines were missed. The overall *Apollo 16* schedule was on track: Grumman had delivered the LM ascent and descent stages in May 1971, and the command and service modules came soon after from North American Aviation. By September, Boeing produced the Lunar Roving Vehicle and the Instrument Unit arrived soon after.

Finally, after solving lingering problems like humidity control, S-201 was ready, so everything was in place for full integration and testing to start. As all parties raced to meet deadlines, there was some slippage of the

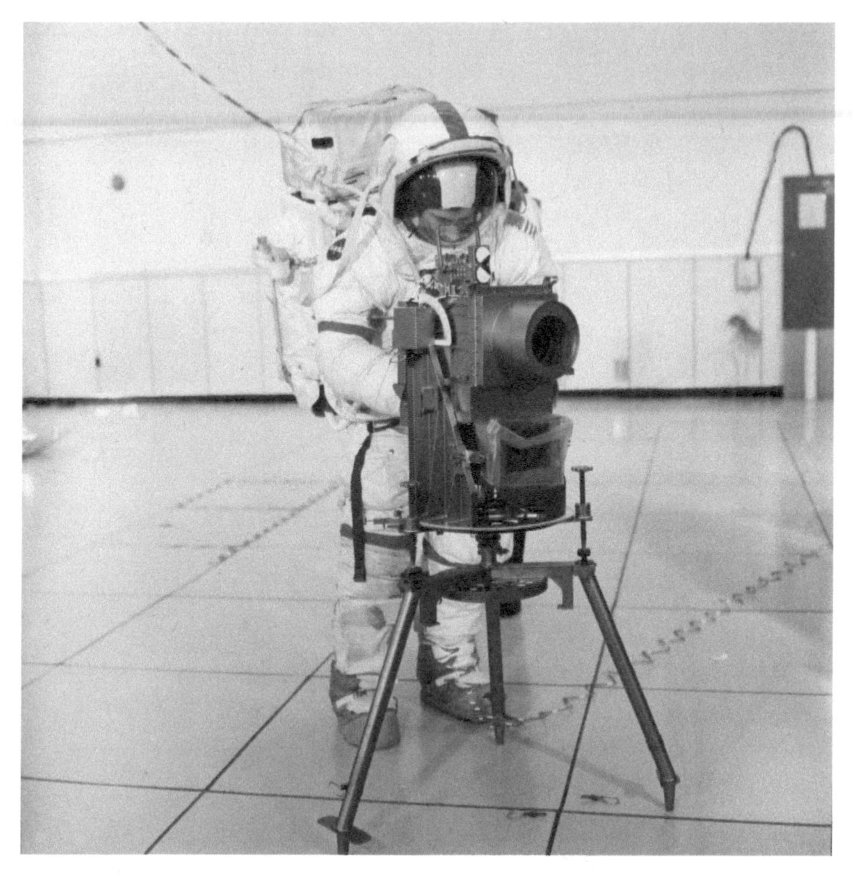

Figure 7.5
John Young practicing the steps he would take on the Moon using a backup model of the camera. *Source:* https://www.hq.nasa.gov/alsj/a16/ap16-KSC-71P-628HR.jpg.

launch, from March 17 to April 16, due to problems with the space suits and the launch system. But by then, the instrument units, including the camera, were stored, sealed, and ready.[104]

The thirty-hour countdown began on April 15, and the astronaut crew boarded on the morning of April 16. John W. Young was mission commander, Thomas K. Mattingly II was the command module pilot, and Charles M. Duke Jr. was the LM pilot. Young was among the most experienced of the corps, and the other two were rookies. Young would have chief responsibility for the UV camera/spectrograph, after training in its use with Carruthers (see figure 7.5).

# 8

# ON THE MOON, AND AFTER THE MOON

After attending the launch at the Cape on April 16, 1972, Carruthers and Page flew to Houston when *Apollo 16* landed on the Moon four days later. As recounted in chapter 1, the crisis of *Orion*'s balky propulsion system delayed the lunar landing for some six hours. This delay would shift the positions of all the celestial objects they wanted to observe, except the Earth, of course, because the Moon was in synchronous rotation (its rotation rate closely equals its revolutionary rate around the Earth), keeping the Earth more or less stationary in the lunar sky. Hence, as we noted, Carruthers and Page were wholly occupied recalculating the altitudes and azimuths of the regions they wanted to observe.

After a scheduled rest period and preparation for the first extravehicular activity (EVA) by the astronauts, some sixteen hours after landing, Young removed the camera from the LM pallet and placed it into the LM's shadow, as deep as possible yet far enough away from the LM (some ten feet) to view the Earth and most of the sky (see figure 8.1). The lunar surface within the shadow, however, was not level. There was a small shallow crater, and Young's first challenge was to level the tripod. "The only way I could get the camera level was to really step down on two of the legs, push them clean out of sight in the dirt."[1] He then placed the battery in the sunlight to keep it warm and connected its cable to the camera. The cable presented his second challenge, having to remember to step over it as he moved back and forth. On several occasions, in his suit and with restricted vision, he tripped, fortunately only dislodging the battery and not the camera. Finally, he was confident enough to deploy the camera, removing the pins to activate the battery and free the altazimuth mount, following procedures he had practiced time and again in Houston.

In the next few hours of his EVA, Young managed to align the camera. At 120:33 hours into the mission, after catching his foot on the cable again, he pointed the camera straight up to view the Earth, delighted

Figure 8.1
The lunar camera in the shadow of the lander *Orion* with Charles Duke in the background. *Source:* https://images.nasa.gov/details-as16-114-18439~orig.jpg (3880 × 3924).

that, after moving the mount to the coordinates sent from Houston, the 2-degree-diameter Earth was right in the center of the 20-degree Earth sight finder field. Still, his first attempts to expose the field first in the camera mode and then in the spectroscopic mode failed. "I fouled up that procedure and had to do it all over again."[2]

Young then reset the camera, correctly initiating the programmed sequence of exposures so that he could then attend to other matters. The electrical activity in the camera itself emitted a weak VHF signal, which told the astronauts that it was working. During the first EVA, some five hours long, Young periodically returned to the camera to set it on another target, following Houston's instructions.

As Page recalled in an autobiographical statement, his "main contribution was selecting 15 targets in the lunar sky that would be above the horizon at the scheduled mission time and calculating pointing angles for the astronaut [John Young] to set on the camera's altazimuth circles."[3] As we have noted, of course, the six-hour delay required that Page and Carruthers hastily recalculate those pointing angles. This became the moment of truth, but they would not know for many weeks if they had done it correctly.

Up to this point, the camera behaved perfectly. But Young soon had trouble turning the camera: "That son-of-a-gun really was hard to turn in azimuth; and I didn't think I was going to be able to turn it."[4] The azimuth motion got so stiff that turning it dislodged the camera tripod, and the whole thing had to be reset and re-leveled. This was frustrating and slowed the procedure considerably, partly because Young had initially set the azimuth to point west and not north—not what Houston had specified. Young only realized this when Houston's coordinates for objects pointed it straight at the LM! Throughout all of this, Houston kept tabs on the astronauts' heart rates. Duke was in the 120s and 130s, whereas Young remained in the 90s.

Young made three visits to the camera, setting the camera on specific objects like the Earth as well as areas of the sky selected by Page. His first field on each visit, NRL's highest priority, was always the stationary Earth, imaging and taking spectra of its outermost atmosphere. Then he turned the camera to the center of the Milky Way in Sagittarius and then to bright regions in Cygnus. Next were areas far from the plane of the Milky Way (including fields in the constellations Fornax and Cetus) and then beyond the Milky Way (the Small and Large Magellanic Clouds), where Carruthers and Page hoped to detect both interstellar and intergalactic gas and dust. Overall, he covered some ten fields in exposures ranging from one to thirty minutes for images, and three to two hundred minutes for spectra.[5] Eleven targets had been planned out of the fifteen Page had selected, but "no data were recorded for the eleventh target because the camera had run out of film, as expected."[6] In all, some "209 exposures on 9.727 m (383 in.) of film" were exposed in the three EVAs. Even though the film roll length was designed for shorter EVA times, more exposures of the Earth were taken than expected.[7]

In addition to the race to adjust predicted target settings due to the delayed landing, there were also frequent colorful remarks from the astronauts meant to be humorous, but they added to the pressure. As Duke blurted out at one point, "Man, that [UV] camera is some contraption, John." And, as he later testified, the only things he could see through his visor were the Sun and Earth. "The UV camera was just looking up into the heavens all the time, to me; and I don't know what they were looking at. We didn't take the time to dark adapt."[8] Carruthers recalled that "we could actually hear them talking about our instrument," but he could not talk with the astronauts directly, only indirectly through mission control.[9] We do not know how he reacted to Young's jocular snippets. But as we noted in the first chapter, when Heckathorn tried to speak with Carruthers in the Houston mission control area during the flight, he was met with silence.

In subsequent EVAs, the UV camera had to be moved as the LM shadow shifted, requiring leveling and recalibrations each time. Due to the irregular shape of the LM, it was difficult to predict when parts of the camera would be exposed to the Sun. This raised problems because when parts of it were exposed, they quickly heated up. During the second EVA, Page and Carruthers asked the Houston controllers to have the astronauts change targets. The flight director refused, saying, "We're wasting time, here." But he soon relented when they found that the film cassette was about to be exposed to the Sun if they stayed with the original plan.[10]

The azimuth motion was getting worse and worse. At one point in EVA-2 (143:19), Young suggested that he simply pick up the entire camera and rotate it. The Sun became even more of a problem during EVA-3, although most of that EVA was to travel on the rover to selected sites on the lunar surface to gather rocks and soil. With the closeout of the third EVA, spending a total of twenty hours on the lunar surface and seventy-one hours overall on the Moon, the astronauts returned to the command and service module via the lunar ascent module, delivering their precious cargo of data, including 96 kg (211 lbs.) of lunar rock and soil and S-201's cassette with its exposed roll of film.

One can only imagine the tension Carruthers and Page experienced during the days of the flight. They had spent countless hours over the past several years writing and revising proposals and ensnared in bureaucratic debates over who would be responsible for what. But even after the astronauts returned and the film cassette was retrieved, there would be many more weeks of worry and tension, on many levels.

### CONTINUING TENSIONS AFTER THE FLIGHT: TECHNICAL AND MANAGERIAL CONFLICTS

The technical problems experienced during the *Apollo 16* mission, in lunar orbit and on the lunar surface, raised considerable debate within NRL and between NRL and MSC. The balky azimuth motion was eventually determined to be due to an assembly error at MSC (noted below), but there were also clashes between the technical and scientific groups, and between NASA's and NRL's management styles.

Outwardly, in the following months, the media hailed the flight, and Carruthers was hailed, mainly by the Black press.[11] But within NASA, and especially at NRL, there were many concerns about how the mission was organized and who would get the data and when. In his initial 1969 proposal, Carruthers stated explicitly that he expected to "process the film and make post-flight evaluation of instrument performance, as well as analyze and interpret the data."[12] In May 1970, however, Page told Friedman that the film would be processed first at MSC.[13] After the flight, though, the data were split. After removal and processing at MSC, the film record was immediately duplicated (both positives and negatives), and Page eagerly took on the task of data reduction. The film cassette also went through testing as well, remaining at MSC for quite some time.

In May, Friedman asked MSC to return the film cassette for radiation tests at NRL, but Christopher Kraft, MSC's director, replied that would not be possible for some time. Preliminary analysis of the cassette cannister's frame advance mechanism and frames in the film showed evidence of variations in the amount and nature of the film advance between frames, and this was puzzling. But fortunately, the data were hardly compromised. Kraft promised to send it to NRL for their own tests, but, oddly and unnecessarily, he reminded Friedman that there was an agreement between the Smithsonian's National Air and Space Museum and NASA whereby all excess Apollo hardware went to the Smithsonian, which was then responsible for loaning the items for study or "display."[14] Friedman, of course, wanted the mechanism for evaluation as well as display, but the agreement did not apply to the scientific data—the film roll—which they retained.

Young's struggles with the azimuth motion of the mount had to be evaluated, which took some time. And meanwhile, fingers pointed in all directions. Carruthers's design was a major target, but MSC engineers had constructed the flight version of the mounting, making some modifications, creating confusion over the orientation and calibration of the circles. During the development and testing process, there was also friction between those who constructed the mounting and the dual camera/spectrograph at MSC and NRL. There was also tension within NRL between its Engineering Services Division(ESD) and with Mange and Conway, though much of it was directed to Carruthers. It reflected continuing conflict between the different divisions and institutions, which was typical for these huge projects, as we noted before, citing Homer Newell.

In July, ESD issued a strong critique in a lengthy unsigned and undated memorandum, later attributed to Sam Cohen in NRL's ESD management office. Cohen singled out Carruthers as the main problem. The "Principal Investigator wanted the least possible change" in the camera design. He alleged as well that Carruthers did not allow for proper management review, subject to a "formal Management Plan" set in place by the Marshall Space Flight Center. At NRL, ESD was responsible for project management services, but for the S-201 project, the "project engineering function and the reliability and quality assurance function reported to the same supervisor." Indeed, William Conway's transfer out of ESD to report directly to Carruthers raised objections within ESD as well as at MSC. Cohen claimed that it inhibited the management function "in the event of a controversy by a conflict of interest on the part of the supervisor." Due to this fact, Cohen asserted, MSC found unacceptable the overall "reliability and quality assurance" at NRL. Management was "too small," deficient by at least 30 percent. The critique also alleged that the science part of the mission was improperly organized and "under its present scheme of organization, NRL cannot possibly comply with NASA rules for program management."[15]

The critique compared NASA and NRL management. NASA was "geared to operation in an industrial type of environment," which "is directly antithetical to the NRL philosophy of management which aims at maximizing informal integration." So Carruthers's "initial unwarranted assurance that the pre-existing design of the camera should not be restudied" was based, the critique continued, "on the apparently pragmatic but professionally naïve judgment that one does not quarrel with success."[16]

Cohen admitted that it would not be right to "divorce a scientist from control of his own project," but the "more scientifically oriented he is, the less satisfactory it will be to leave the scientist in control of decisions which are clearly of an engineering production nature." In sum, he called for more independence for ESD and, when problems arose, that they be considered by "an unbiased but competent independent observer."[17]

Conway, who took the lead for this aspect of the work, reacted strongly, saying his transfer was beneficial for better communication and "in fact, strengthened the management aspect of the program." Further, as the program manager, Conway felt that he did "exercise exactly the functions spelled out in the Management Plan" that Cohen had copied from NASA's management statement. And third, most critically, he countered Cohen's assertion that Carruthers was too rigid. Carruthers had conferred constantly with Conway, who was well versed in the technology, and he also worked with "numerous NASA personnel of various backgrounds" remaining open to modifications. "To the knowledge of this writer, the Principal Investigator never prohibited any review of camera design."[18]

Phil Mange was far more incensed and protective. He argued that the "project succeeded as much as for any other reason because the principal investigator and his program manager had the background and fortitude to resist NASA's 'management by committee' tactics when major problems arose." He added strongly that they persevered and "made their judgements stick . . . in the face of harassment by committees of as many as a dozen persons." He also reminded everyone that Carruthers was an engineer as well as a scientist.[19]

The anomalies during the mission were all eventually explained. There were design problems, but also assembly and performance blunders. Astronauts packed deeply into their helmets could not see their feet, of course, and Young got caught more than once in ground cables, which led to the loss of a heat flow experiment.[20] Another cable was broken when he deployed the shade for the cosmic ray detector, which compromised its performance.

The balky azimuth motion for the UV camera was a technical blunder in Houston. A later evaluation concluded that the exposed 12.5-inch ball bearing ring that provided the horizontal merry-go-round-like azimuth motion had not been contaminated by lunar dust. It was packed with a "waxy, low-outgassing grease which stiffens appreciably at temperatures below 50° F." Someone had applied a sealant rather than a lubricant, and since the camera was always in the shade by necessity, the grease stiffened. Somehow this system was not tested in a thermal vacuum chamber.[21] The azimuth problem was, therefore, not due to any design issues caused somehow by Carruthers's alleged intransigence, as might be implied from Sam Cohen's critique, but by the unfortunate choice of lubricants by an MSC technician.[22]

In addition to flight operations, *Apollo 16* addressed over sixty distinct research problems in thirty-one general categories—most related to exploring the landing site, collecting lunar samples, performing seismic, magnetic, X-ray, gamma ray, and cosmic ray surveys, biomedical experiments, ad infinitum.[23] It would take weeks, months, and even years to fully analyze the S-201 observations, supported by repeated funding from NASA and elsewhere.

## THE MOST NOTABLE EVENT OF THE YEAR, FOR NRL

The key result, of course, was that the camera worked, and the film was returned to Earth. In his annual report to NASA, Herbert Friedman singled it out as the achievement of the year for NRL: "Perhaps the most notable event to take place in the program of Hulburt Center activity during fiscal year 1972 was the successful deployment on the lunar surface of the NRL far-ultraviolet camera/spectrograph on 21 April 1972 during the *Apollo 16* mission."[24]

He cited the images of Earth's far-ultraviolet geocoronal glow from hydrogen and oxygen and its auroral ring, seen "for the first time in full distant perspective." He also cited observations of UV fields including the Magellanic Clouds that offered the chance for comparison with Earth-based photographic maps.

Deeper into his report, however, in a priority listing of eighteen NRL achievements, Friedman marked the top five X-ray observations from Aerobee flights; sixth through eighth in order were from *Apollo 16*, and the rest were solar and radio discoveries. Carruthers's achievement was highlighted at the outset, given its obvious public visibility, but it was still too soon to draw conclusions from the observations. The fact was the camera worked: "The Far-Ultraviolet Camera/Spectrograph was successfully operated from the lunar surface during the *Apollo 16* mission, April 21–23, 1972. Nearly 180 frames were exposed during the mission, and essentially all of the scientific objectives were achieved. Data analysis is now in progress and is expected to require about one year to complete."[25] (The reported number of frames exposed varied.)

In fact, it would take years. After the film cassette from S-201 was returned to Houston and then eventually to NRL, and the dickering continued over who got what part of the data, the matter was partly settled by making Page an NRL research associate to analyze S-201 film. These visits continued through the decade as Page collaborated with Carruthers and his staff, analyzing data from Apollo and later from a flight of the instrument on Skylab to observe Comet Kohoutek (see chapter 10) and finally, new versions on space shuttle missions. Initially, Carruthers, soon with Robert Meier and later adding Harry Heckathorn, concentrated on NRL's central interest, the Earth's geocorona, whereas Page was primarily interested in the deep sky fields.

## THE EARTH'S GEOCORONA

The photographic frames brought back illustrated the behavior of the Earth's airglow and polar auroral zones over the three EVA days. The images and spectra were duplicated and converted into positives and prints, and they were visually examined for structural changes and chemical identifications. Just who did the work is not known, as the earliest publications of preliminary S-201 results were authored by Carruthers and Page with no acknowledgments. Carruthers, however, was first author on the instrumentation and the Earth observations, whereas Page was first author for an early report on the Large Magellanic Cloud.

They reported their preliminary findings at the August 1972 meeting of the American Astronomical Society in East Lansing, Michigan, and soon published preliminary results in various places, including the September issue of *Science*. Their *Science* article was limited to an analysis of the images and spectra of the geocorona, which they acknowledged were "necessarily preliminary and qualitative." They displayed six direct images and four spectra, describing the "fine structural details" of the coronal arcs and how they played out at the Earth's shadow boundary (see figure 8.2). These observations offered tantalizing hints of the physical processes at play, such as "the spatial distributions and relative intensities of emissions" of the atmosphere's major components, which confirmed sounding rocket observations that

Figure 8.2
Earth's geocorona. The coronal arcs are parallel to Earth's equator, extending from the day/night terminator to the limb of the Earth. *Source:* Courtesy of Naval Research Laboratory. NRL Photograph 76604(4)A.

the polar auroral zones were "largely due to atomic oxygen and molecular nitrogen, rather than atomic hydrogen." Overall, they highlighted that their results "gave the spatial distributions and relative intensities of emissions due to atomic hydrogen, atomic oxygen, molecular nitrogen, and other species—some observed spectrographically for the first time."[26]

In an interview years later with NASA historian Glen Swanson, Carruthers expressed his feelings quite clearly: "Well, the most immediately obvious and spectacular results were really for the Earth observations, because this was the first time that the Earth had been photographed from a distance in UV light, so that you could see the full extent of the hydrogen atmosphere, the polar auroras and what we call the tropical airglow belt. All of these were revealed in pictorial form for the first time, so that's something you don't have to wait for data analysis to show people."[27] Indeed, the image has become iconic. NASA's "Preliminary Science Report" for

*Apollo 16* highlighted a colorized geocoronal image on its cover, along with Carruthers's camera in the shadow of the LM.

In January 1973, Carruthers and Page appealed to NASA for more funding for data processing and analysis, funding sufficient at least to support staff physicist Robert R. Meier, "who is expert on the Earth's upper atmosphere and geocorona [that] will allow completion of a three-dimensional theoretical model of the atmosphere and geocorona, valuable to other NASA projects, and reduction of photometric data on the 500 stars, nebulae, and galaxies on the S-201 frames."[28] Meier's job was to derive new theoretical models of the airglow and geocorona based upon the S-201 observations.

Indeed, Meier was up to the job.[29] Meier trained under Thomas M. Donahue at the University of Pittsburgh, an expert on airglow studies using data from sounding rockets and other means.[30] Meier had analyzed data from previous rocket flights that measured sodium in the upper atmosphere, and his thesis put the parts together, describing the steps to ready a payload for flight from Wallops Island, Virginia, and then to analyze the data. Meier supplemented his data with NRL's far-ultraviolet observations from rockets and satellites to model the Earth's upper atmosphere.[31]

In 1966, following Carruthers and others, Meier obtained a Hulburt postdoctoral appointment at NRL exploring the physics of the geocorona and was then hired as a research physicist two years later. He collaborated with Phil Mange, as well as NRL physicist Dianne Prinz, analyzing observations from sounding rocket flights and from OGO-4. He was well prepared to interpret the spectra from *Apollo 16*.

Working with Page, however, had its moments. As Meier and others recalled, he was definitely a character, of elite bearing and with habits that annoyed those around him at NRL, especially his insistence that his dog accompany him around the laboratories at NRL (see figure 8.3). Heckathorn even sensed that Page could be somewhat dismissive of Carruthers at times.[32] When the author asked Carruthers to reflect about Page during an oral history interview, he demurred. Given the politics and pressure dealing with NASA, he was content to let Page handle the logistics and the data reduction. Page was also essential to keeping lines of communication open between MSC and NRL.

Through the mid-1970s, Meier worked with Carruthers, Mange, and others in refining their analysis of the S-201 geocoronal observations.

Figure 8.3
Thornton Page at his desk at the Van Vleck Observatory, Wesleyan University, Middletown, Connecticut. *Source:* Courtesy of Wesleyan Library, Special Collections and Archives.

They compared their observations with existing OGO-4 data and supplemented them with data collected later from Skylab and *Dynamics Explorer 1*. These efforts constantly improved connections between ionospheric structure and its response to solar activity, allowing them to construct more sophisticated and functionally more useful models of its structure and dynamics. Meier and others continued refining their analyses of the geocorona's structure, composition, and extent over the next several decades, and it continues to this day, with NRL scientists like Christoph Englert and Charles Brown collaborating with space scientists around the world.[33]

This first reconnaissance of the whole of the Earth's geocorona might well be considered a fine example of a poetic 2018 reminisce by *Apollo 8* astronaut Bill Anders. He was one of the first humans to leave the Earth's orbit, which was a highly emotional experience, exemplified by his famous

"Earthrise" photo as they orbited the Moon in 1968. Thinking back, he mused, "We set out to explore the Moon and instead discovered the Earth."[34]

## STARS, STAR-FORMING REGIONS, AND THE LARGE MAGELLANIC CLOUD

The S-201 far-ultraviolet observations of star fields, nebulae, and galaxies were, of course, not the first glimpse of the ultraviolet universe. As we know, Carruthers's sounding rocket missions, as well as those by Princeton and others, and especially the second Orbiting Astrophysical Observatory, which by then had been in operation for about two years, had been mapping the UV universe. But few reached the Lyman-alpha line, and S-201 went far deeper, recording the spectral signatures of objects and regions well beyond Lyman-alpha and at least 2.5 times fainter than *OAO II*'s maps.

Page was far more invested in these observations than he was in the geocoronal work. Known as a "galaxy man," he focused on stellar and galactic fields in his original proposal. So Page wanted to take the lead in the stellar work. Meier recalls an early meeting in Chubb's office where Carruthers, Page, Opal, and Prinz deliberated over "who would be doing what on the data analysis."[35] Carruthers wanted Meier, Opal, and Prinz to take the lead. Prinz, in particular, had experience operating devices called densitometers, which converted photographic images into quantitative measures of photographic density as a function of position on the frame.

Carruthers wanted the data analysis done as quickly as possible to get it published and to be able to provide convincing arguments for flying improved instruments on future human-tended flights, like Skylab or the planned *Apollo-Soyuz Test Project* (*ASTP*). Meier recalls that Prinz hastily performed preliminary densitometry at NRL, which informed their first reports. But apparently Page soon took the lead.

Page visually inspected the fields, comparing them to photographs of those areas by the Palomar 48-inch Schmidt and star catalogues such as the compilation created by the Smithsonian Astrophysical Observatory. But this first pass left too many potentially interesting objects unidentified on ground-based images. So, as he explained in an autobiographical recollection, with computer support from Richard Hill of Lockheed Electronics in Houston, Page, like Prinz, also used a "microdensitometer" (densitometry with high positional and photometric accuracy) to make precise measurements of the optical densities of the exposed film, correlating them with position on the film. The images were original negatives, so measuring the darkening profiles across the film quantified the position and brightness distribution of the source. This could be the brightness of a star, or the structure of an extended object like a nebula, or the complex structure of its spectrum.

To achieve greater spatial and photometric resolution, Page took his copy of the film roll to Pasadena to use a newer, more sophisticated, and higher-resolution computer-controlled measuring machine, a Boller and Chivens PDS [Pacific Data Systems] microdensitometer. Once again, Hill used the converted digital data to produce contour maps of extended objects like the Large Magellanic Cloud, the North American Nebula, the Andromeda Galaxy, and the blue hot stars in Orion. This somewhat improved the digital data, and it was an all-important step to correlate more of the sources with known stars and extended objects with ground-based surveys. In the late 1980s, as Carruthers pointed out in retrospect, for the S-201 reductions, "perhaps the greatest impediment to proper analysis and interpretation of the far-UV results is the lack of, or inadequacy of, ground-based identifications, photometry, and spectral types for the objects detected in the far-UV."[36]

Carruthers's point here reflected common thinking among astronomers at the time, and over time. Radio sources identified in the 1950s had to have optical identifications to be "accepted." Observations by infrared balloon and satellite-borne instruments had to be confirmed by ground-based telescopes.[37] Therefore, a full analysis of the far-ultraviolet S-201 images and spectra most of all required that they be correlated with objects in those fields seen by large ground-based telescopes for deeper analysis. And to be sure, making correlations between the observations at different wavelengths and in different media vastly expanded knowledge about the objects. The act of correlation was at the center of observational astrophysical practice.[38] The challenge was to survey those fields in the near ultraviolet (longward of 3,000 Angstroms, the blue and visual portions of the spectrum known as the UBV system) with a ground-based telescope and the most sensitive detectors available. That is what Page initially tried to do. But more had to be done.

In the early 1980s, Heckathorn led an effort to improve Page's results. He analyzed the S-201 nuclear emulsion films using a new and far more powerful computer-controlled Perkin-Elmer PDS 1010A microdensitometer at NRL, thereby making them fully accessible for more detailed computerized analysis. Heckathorn concentrated on operating the intricate microdensitometer, while Opal attended to its computer control system.[39] At the same time, with Opal, Heckathorn traveled to the McDonald Observatory in Texas to use their 36-inch reflector with a visual electronographic camera built by Paul Griboval to close the identification gap. They did find some confirmations, but not many. They also needed access to the southern skies, and searched at the Cerro Tololo Observatory in Chile (which had unexpected benefits—see chapter 12).

## LOOKING AHEAD

In 1984, Carruthers and Page summarized their star field findings and looked to the future. They concluded that their imaging surveys thus far had detected previously unknown or unrecognized "hot objects" showing that their UV brightnesses were far larger than heretofore known. They also argued that since the spectra of known objects now extended into the far ultraviolet, the present system of classifying stars by their spectra needed to be revised. And third, they felt that their data, including spectra and brightnesses, could be useful for the "mapping of interstellar extinction over extended areas," another way to assess the unseen material between the stars. But predictably, they concluded, in a section on implications for future surveys, that "the S201 survey, of course, leaves much to be desired in terms of the utility and objects of an 'ultimate' far-UV survey."[40]

Throughout this time, there was no question about where Carruthers's priorities lay. Whereas Page, Heckathorn, and Meier focused on analyzing the data, Carruthers planned larger and more refined variations of his cameras, which involved experimentation and testing in his laboratory. He was in a hurry also because he wanted to refly a modified version of S-201 on *Apollo 17*, due to launch in a few months. In May 1972, well before they even received the processed film from *Apollo 16*, according to Capt. Earl Sapp, NRL director, the success of the first flight "excited Dr. R. A. Petrone and many scientists in the Washington Area, including the PI [Carruthers] and the Co-PI [Page] of NRL." There were two flight-qualified units in storage at MSC, as well as the prototype model in the custody of NASA Public Affairs. Sapp then added, "Our engineers, who designed and fabricated all these models, consider that minor modifications might reduce weight and improve performances of the UV camera." But, he warned, "There is barely time enough to accomplish this before Apollo 17 launch." Accordingly, he called for an immediate decision to fly again. But as Apollo program director Petrone and Kraft had originally stated, the roster was filled.[41] Others within NRL, however, like Herbert Rabin, associate director for research in NRL's Space Science and Technology Division, still pushed the matter. In the end, however, Carruthers finally decided that it was not feasible, given that *Apollo 17*'s landing site was much farther north, requiring more fuel consumption, a reduced payload, and an equatorial mounting. And, as associate superintendent Phil Mange also explained, expanding on Carruthers's advice, the "politics of payload assignment rule out any chance of reflight for the lunar camera. Moreover, flight arrangements for *Apollo 17* were made before the flight of *Apollo 16*."[42]

Carruthers knew there were practical and political limitations, but he was more interested in new designs than reflying old equipment. Nevertheless, *Apollo 16* had changed his life in profound ways: it made him a public figure, and a target for romance.

# 9

# ATTENTION: PUBLIC AND PRIVATE

Carruthers's energy during the preparation and flight of *Apollo 16* and its aftermath were not confined to *Apollo 16*. During this time, he also published some seven academic and refereed papers on topics including laboratory studies with K. L. Bromberg and papers with Chet Opal and others on the uses of electronic imaging devices in astronomy. He and Opal, with others, also submitted more than a half dozen new proposals to NASA and to NRL and took part in institutional proposals to NASA, NSF, and the Advanced Research Projects Agency for a wide range of projects. These included refinements of his electronographic systems with the significant addition that, for robotic missions, film recording would be replaced by some form of television sensor, an electronically amplified scanner such as the secondary electron conduction vidicon from RCA or Westinghouse or the newer silicon intensified target being developed by the Naval Ordnance Test Station at Inyokern, California. These electronic detectors could transmit their data to ground stations and so did not require recovery.

But after *Apollo 16*, there were new demands on Carruthers's time and energies. He was no doubt aware of the public attention he was receiving. But he worried how this attention might again sour relations between NRL and NASA. Throughout this tense period, on at least one occasion we have noted, Carruthers expressed concern over the ramifications of the O'Toole article in the *Washington Post*, fearing that it would only raise the hackles of MSC people. He must have been aware of Cohen's critique, but let others react to it. When everyone was pointing fingers, however, or celebrating after the fact, Carruthers stuck to his laboratory and desk, cranking out reports and publications, proposals for future missions with his camera, and constant improvements to the design utilizing various recording media. Characteristically, he never sought, but never shied away from, public exposure during and after the flight. But in the midst of the pressures of the flight, when CBS's Special Events Unit secured a mock-up of the camera via NRL's public affairs office a day after the LM landed on the Moon, William Conway represented NRL.

Before, during, and after the flight, NRL public affairs got numerous requests from the press, especially after dramatic visuals were released by NASA in June showing the Earth's geocorona and the Large Magellanic Cloud.[1] Soon, Carruthers and his story started appearing in newspapers beyond Washington; in Chicago, Hartford, New York, and Boston. As Richard Paul and Steven Moss have observed, newspapers and magazines appealing to Black audiences had initially been critical of NASA calling "into question the amount of money spent on space exploration."[2] But they were now eager to take special notice of Black achievers at NASA, Baltimore's *Afro-American*, Chicago's *Daily Defender*, and Brooklyn's *New York Amsterdam News* all hailed the moment.[3] *Ebony*, in particular, made a special effort to highlight Carruthers."[4]

Much of this public attention came in requests for Carruthers's appearances in schools and on tours. Starting in May 1972, NRL's public affairs office began to receive inquiries asking specifically for Carruthers's presence as an inspirational speaker for Black achievement. One of the first was from Lydia R. Thaxton, writing on behalf of "the alliance of blacks for Children's motivational learning opportunities" in Harlem. Confirming an initial request earlier that month, she added:

> We wish to honor him in a public ceremony so that all our thousands of students can see, if not meet him for his historic and pioneering contribution. . . . As the first black to design any instrument placed on the Moon, he is a perfect "role model" for our youngsters.[5]

As the *New York Amsterdam News* dramatically reported on its front page in December, after Carruthers visited a local school:

> There were hushed whispers among the small fry and some of the older children as a modestly dressed soft-spoken Black man stepped up to the front of the auditorium of Public School 243, Dean Street and Troy Avenue, Brooklyn on Friday, December 8. The children were seeing in the flesh the man who made history on the Moon. No, he did not take the "one giant step for mankind" by walking on that satellite, but his contribution was far greater in scope, for it will be used to give men insights into a heretofore unknown phase of space travel.[6]

Carruthers had crafted an engaging slideshow featuring scenes from *Apollo 16* and screened a dramatic NASA film highlighting the mission. He also spoke of his own work, describing how he imaged the Earth's UV geocorona and why that was important. After his talk, he and the group moved to the school's gym to mingle and answer questions. The gym, the auditorium, and the entrance to the school were festooned with welcome signs and drawings of rockets and student-built spaceships. Kids could also touch a scale model of the LM, provided by Grumman Aircraft, and there were displays from General Electric and NASA. Carruthers's entourage included Grumman staff, City College of New York faculty, and personnel from Central Brooklyn Model Cities.

This initial attention blossomed. Francis Redhead, executive director of the Caribbean House in Brooklyn, invited him to appear at yet another New York City event but on a larger scale "aimed at inspiring the youth of our black and other minority groups." Redhead was then, among many titles, Deputy Permanent Representative, Permanent Mission of Grenada to the United Nations. The program he represented in Brooklyn had as its motto "One Sun, One People."[7] He was soon to become Carruthers's father-in-law.

Caribbean House was established by a "group of prominent Harlem leaders born in the West Indies" and in the late 1950s became an activist community organization providing social space and community service broadly in many venues. In 1969, Redhead, one of its founders, explained its origins: "We feel the time has come when we have to get away from island grouping and speak in terms of regional association. Caribbean House will bring about cohesiveness and solidarity that we are dreaming about."[8]

The mid-1970s were an especially sensitive time for Grenada, which was on the verge of independence from Great Britain. Redhead had become envoy to the United Nations before independence and in 1974 became its consul general after independence. His charge was to keep relations as positive as possible with other nations, and to seek unity in the Caribbean.[9]

Carruthers may have been aware of Redhead's background and motivations but seemed unconcerned by the political ramifications of such invitations. He was willing to accept such requests, urged, no doubt, by the NRL Public Affairs office, Chubb, and certainly Friedman. As quiet and shy as he may have been informally, he had already gained a reputation, attested to by Friedman and others, of being an eloquent speaker among his peers, as O'Toole of the *Washington Post* reported in 1971: "Something happens to George when he's addressing his peers on astrophysics. . . . He gives beautiful lectures."[10] But how would he fare with children and the public? The

December 1992 *Amsterdam News* account of his performance at a school in Brooklyn answered that question.

As NRL public affairs officer James Sullivan recorded, "George was well-liked and well received by everyone." Redhead had also upped his request. In addition to meeting and mixing with his Caribbean House children, he and the noted Caribbean broadcaster and journalist, Frank Rojas, requested that Carruthers also join a planned celebrity tour of the Caribbean: "It would be especially beneficial, he [Rojas] says . . . since most of the citizens of these countries are black." Sullivan also relayed Redhead's formal invitation to Sullivan's NRL boss. Copying no less than President Nixon in his request, Redhead stated that it would be a goodwill tour headed by two astronauts (Thomas Stafford, *Apollo 10*; and Jack Swigert, *Apollo 13*), along with Ruth Bates Harris (at the moment, NASA's equal opportunity officer).[11] Others named were Isaac Gillam IV, a rising star as a NASA program manager, and other Black figures as "success symbols" working in NASA. He also mentioned that they were still looking for "One Chinese" and "One Indian Space Engineer." Redhead above all emphasized that:

> Dr. George Carruthers, the black astro-scientist, could be the fastest, surest, and most impressive manner that can be employed, not only of overcoming the apathy and defeated attitudes of black and other non-white children in the undeveloped islands of the Caribbean, but such a tour will straighten the bonds of friendship. . . . Grumman aerospace also very interested.[12]

Throughout this period, Carruthers received numerous awards and honors for his achievements. After the United Nations held a small reception in his honor, likely orchestrated by Redhead, NASA held a "special ceremony" at the space agency's headquarters where administrator Dr. James C. Fletcher presented Carruthers with NASA's Exceptional Scientific Achievement Medal.[13] By the end of the year, among countless invitations, Carruthers agreed to address students in the Cleveland Public Schools Supplemental Educational Center and was invited back to Brooklyn numerous times, not only to the Caribbean House but also to organizations like the Science Motivational Program at District 16, Brooklyn.[14] The fact that he had won a patent for his invention was noted frequently by the popular press.[15]

Certainly, race played a role in Carruthers's sudden prominence in this era of increased effort to promote racial and civil equality. After all, there were hundreds of people producing dozens of creative payloads for NASA missions to the Moon. But there were very few Black faces among them. The organizations he was identified with—NRL and NASA—wanted to demonstrate their support for racial equality. The latter was then being criticized for exclusionary employment practices. And it is understandable that dedicated and focused promoters like Redhead and the Black press would hail his accomplishment as they did others who had made a mark and became "role models with stories of African Americans working directly or indirectly with the American space program."[16] But the more important question is, as we have seen here, and will see also in later chapters, what did Carruthers do with his new found fame? As we shall see, far from seeking or exploiting his success for personal gain, he used it to "give back" to the community he evidently cherished. He accepted his responsibility as an active and engaged success symbol and wanted to use it to make students appreciate that the success he had attained was not unique and would be even more accessible to them in the brightening future. As he recalled in a 1992 interview:

> Certainly students are interested in knowing how I got to where I am at the present time. Of course, there are some changes with time. For example, many of the obstacles that African Americans faced in the fifties and sixties are no longer obstacles now, although the obstacles are not totally absent. Certainly if we give examples of more difficult times in the past and point out that they actually have it easier now, then maybe that would give them some incentive to do better.[17]

In various interviews, Carruthers claimed casually that color was not a distinction when he interacted with others, and race was not an issue. This was not a unique view among Black inventors and scientists, but from what we see here and will see later in this biography, through his actions, he was sensitive to the fact that young Black people needed to be encouraged to consider careers in science and technology.[18]

One thing is sure: it was also his fame that led to his marriage to Sandra Renee Redhead.

## A QUIET COURTSHIP

Possibly the best insight into Carruthers's character and view of life came from his correspondence with his future wife, Sandra Redhead. Some of Carruthers's visits to Brooklyn and New York City, enabled by Francis Redhead and others, were coordinated by Redhead's daughter Sandra, who was active in Caribbean House programming. Evidently, they struck up a relationship in these visits, as well as on the Caribbean tour, and by late May 1973, the gossip column in the *New York Amsterdam News* reported that "Sandra Redhead, pretty daughter of Francis Redhead of the United Nations, exchanged marriage vows with Dr George Carruthers a NASA scientist."[19]

Born about 1950 and raised in Grenada, Sandra graduated from St. Joseph's Convent Secondary School and then moved with her family to Brooklyn to assist and be active in her father's projects, to work at the Eastern Caribbean Tourism Association in Manhattan, and to continue studies at both Brooklyn and Hunter Colleges. In 1969, she placed third in a local beauty contest—the "Miss CARIFTA"—held in Brooklyn. James Earl Jones presented her with a red rose at the event.[20]

George and Sandra's courtship consisted of passionate letters and postcards back and forth, with weekend visits either to Brooklyn or Washington. Their contact began at a June 21, 1972, event honoring Carruthers for what was declared to be, by the president of the Borough of Manhattan, Percy E. Sutton, "'George Carruthers Day' as America's distinguished Black Space Scientist."[21] This two-day event was carefully orchestrated by Francis Redhead with NRL and Roscoe Monroe, then the assistant director for equal employment opportunity and an outreach activist at NASA. The event included celebrations in Brooklyn, Manhattan, and Harlem. The culmination was a Harlem-Brooklyn Community Space Program event that honored Carruthers, as well as *Apollo 16* astronaut John Young and *Apollo 10*'s Thomas Stafford, former chief of the Astronaut Office. The coordination came from the Caribbean House, and Sandra was deeply involved.

In the wake of all this excitement, Sandra's letters were at first formal but warm and friendly, asking Carruthers if they could meet again when she visited Washington, DC, in a few weeks. She would be escorting a group of young visitors from England to see American highlights and wanted them to see Carruthers's laboratory where, she hoped, "you could show us some of your work?"[22]

Other visits followed and became much more personal. In late August, Carruthers wrote, "To say that I miss you already would be an understatement." Continuing in an uncharacteristically emotive voice,

> It was not until this morning that I started to realize the full significance of this weekend. First, I am incredibly lucky to have had such a wonderful and beautiful person as you as my companion. Secondly, however, at least at first, I found it

> difficult to express my feelings in words, as I have always been rather shy at such things in the past and this was a totally new experience for me. I think that it was not only because of your love for me that I was finally able to begin to overcome my apprehensions and to realize that I love you also. Please believe I always will.[23]

Carruthers also started sending Sandra catalogs from Washington area colleges and advising her about possible careers in science. These were evidently responses to her ardent hopes and wishes for being a part of his life. The first books he sent were Carl Sagan and Iosif S. Shklovskii's stimulating *Intelligent Life in the Universe* and a lavishly illustrated NASA text curiously called *This Island Earth*, edited by Oran Nicks, designed to get her acquainted with space science and with what were then and now the most provocative questions a scientist could ask.[24] Carruthers was frank: "I would, of course, like to see you take up some field of science (even if only as a minor field or as elective subjects) since we would then have them in common, and would be better able to understand, and perhaps to be able to help out in, each other's work."[25]

He emphasized that it would be her decision and that he did not want to push. Given her present work in organizing travel and tourism events, he added, possibly courses in meteorology, geology, Earth resources, and geography might help her be able to "write some of the travel literature that you now distribute." But overall, "the main thing is to finish [college] and to maintain a broad background of basic subjects."

Most important, he warned that his own work "keeps me quite heavily occupied," suggesting, among other things, that he hoped she would find her own interests after they married. He was most ardent that she broaden her horizons because, he admitted, "I do feel a strong urge to learn about non-scientific subjects, and regardless of what you decide to study in school I will be very happy."[26] Up to now, Carruthers had signed off with "best wishes"; now it was with "love."

Sandra passionately responded that his suggestions gave her "the guidance I needed concerning my further education." She was enjoying the readings he had sent and looked forward to more area college catalogs. In September, her father was once again making plans to feature Carruthers in upcoming community activities in Brooklyn, and as always, Sandra ended her letters with repeated expressions of love and longing. But she was also frank about her possible future: "What I would like you to understand is that I am only interested in science courses in school, Moon pictures, etc. because I care for you and love you. I feel that having some idea of your work would help me to know you better. In fact, the only reason I said I'd pursue that field of study in school is to be interested in what you were doing."[27]

She was already active in music, piano, ethnic dancing, drama, and English literature, and she seemed torn between them and taking on science to be a part of Carruthers's life. So she needed his help to "pursue any field in science to a degree level," which, obviously, would cement their relationship: "Sometimes all I can think of is loving you for the rest of my life. You are the most important in my life, next comes my school interests, hobbies, etc." She ardently wanted to see him again very soon, but alone, "by ourselves and not with two hundred other people."[28]

Something in her last letter, and a series of long phone calls, made Carruthers step back a bit and take a breath. He wrote a long letter about what it all might mean to be married to him. Lacking a scientific background does not mean she should, or should not, take a course of study to let her become familiar with his world. "This is, in fact, the same thing I tell the high school students in attempting to encourage more minority young people to go into science and engineering," he admitted. Then he added, "However, I would not in any way expect you to become a professional scientist." She should explore a wide range of possible interests and possibly look at those sciences—like geography, geology, cartography, and social geography—that could be useful to appreciate space science but would also be useful in her present interests in travel and tourism.

He then, most significantly, stepped back to help her appreciate what she faced, and in so doing provides the most intimate portrait of the man thus uncovered. Expressing compassion for both racial and gender barriers, Carruthers speculated,

> Perhaps you are wondering why I am placing so much emphasis on your education and career. I feel that women, as well as men, be they married or single, have a right to do things they feel are interesting or worthwhile. However, it almost always requires a good education and hard work to attain this goal. I have seen too many people drop out of school at an early age only to be stuck in jobs whose only benefit to them is the take home paycheck, or (in the case of married women) with full time housework.
>
> If we should establish a permanent relationship, I think that you would be much happier with something interesting and worthwhile to keep yourself occupied, since among other things I will probably always have to put in considerably

> more than the normal 40 hour week on my job, and there will be many occasions, when we are preparing for major space launches, when it is necessary for me to work 16 or 18 hours a day, 7 days a week in local preparations, and then be gone on field trips for one to several weeks. Since, in such space flight experiments, many thousands (or even millions) of dollars of the taxpayers' money (and many man-months or years of effort) can go down the drain as a result of a single mistake, you see why we must put in such long hours in the final preparations and check-outs of the experiments. Hence, you can also see from this why if you had a basic background in the applicable science subjects you would be better able to understand what I am trying to do and why, and also you would be able to help out with some of the less technical aspects of my work, allowing us more time together.[29]

After this clear and frank statement, Carruthers assured her that "last but not least," he appreciated her interests in music and nonscientific subjects and did "not in any way suggest you drop them in favor of science-related subjects," adding, "I, also, am interested in music and a number of other subjects not related to science. Our common interests in such things would allow us things to do together when we have some free time from our work."

Indeed, Carruthers felt that there were many places to see and things to do in the United States and elsewhere "that we would both like to visit when we have more than just a weekend off." He was more than open to looking for ways to be together and away from work, and he much wanted to discuss these matters with her next time they were able to meet "in private."[30]

Carruthers felt strongly that he had to clarify just who he was and what she should expect from him, as well as to break through her sense of urgency to make plans. In mid-September, he alerted her that he planned to visit again on the way to Amherst to give yet another talk. But he cautioned, "I think it best that we keep it secret from everyone else, so that we will have a chance to talk things over alone without any obligations for me to see anyone else." If she preferred, they could have that talk in DC and maybe even take in a performance at the Kennedy Center. He looked forward to this meeting, urging her, "Please believe that I love you very much and will love you always."[31]

This effort did little to subdue Sandra's passion and urgency. She frequently expressed her "pain of missing you." In late September, after another weekend in Brooklyn, she cried out,

> I realize how much I want to be with you and I am considering June 9th, 1973 as the date of our marriage. In that way maybe I can attend summer school in Washington and you can help me with Math. Among other things we will have to let the priest know our plans six months before. I am quite honored to be in love with a wonderful person like you and I will want you to be very happy with me. My love, I want to give my whole self to you; just you.[32]

And as an afterthought, she thanked him for helping her with her transcripts, signing off "I love you, Tiger."

Sandra pushed to tell their parents their plans, so in early October, they broke the news to their parents as well as to Carruthers's uncle Ben, and by the end of the month, the deed was done. As Sandra proclaimed, "Things should be easier for us now that we have told our parents."[33] They formally announced their engagement at the end of November and were quietly married on May 19, 1973, at a Manhattan courthouse. His brother Gerald and his family as well as Ben Carruthers attended the ceremony, and it was followed sometime later by a small but festive family reception at a Manhattan hotel (see figure 9.1).[34]

Sandra obviously was enamored by George and his achievements. After their marriage, she collected newspaper stories and notices of awards and honors, and she prepared an elaborate scrapbook covering George's accomplishments from 1970 to 1973, presenting it to him with "my love."[35]

True to form, Carruthers neglected to tell his NRL colleagues that he got married. Only when the October issue of *Ebony* magazine noted it in passing in a feature article on Carruthers's work, suggesting that "few of his coworkers even knew about it," did the word get out.[36] Indeed, NRL's Robert Meier first heard of it some months later when Randy Taylor, a young Black chemist in the division, ran breathlessly down the hall at NRL waving the magazine. Meier called out, "What do you have there, Randy?" And Randy screamed, "George Carruthers got married! And nobody knew!"[37]

Carruthers felt that Sandra's continuing education was paramount. During their courtship, scouring course catalogs from local universities, Carruthers felt that the University of Maryland in Greenbelt, in the northeast suburbs, "offers the widest choice of courses."[38] In November, with George's help drafting her text, she wrote to the University of Maryland Office of Admissions for entrance in fall 1973, declaring a major in geography. She would be a transfer student from Hunter, where she completed a two-year program in 1971 with a major in English and minors in Spanish and music. She worried that Maryland required transfer students to

Figure 9.1
Sandra and George Carruthers on their wedding day. *Source:* Courtesy of Gerald Carruthers and Carruthers Collection, box 1, folder 2, Archives Department, Smithsonian National Air and Space Museum (NASM 9A19879).

have a least a B average and admitted that she did not meet that level of performance, assuring the admissions officer, however, that her subsequent work and experience had helped her better "define my interests and to greatly increase my scholastic motivation."[39] She also admitted that she was deficient in science and would like advice on summer courses to get ready. Influenced by George, she declared that "Earth resources surveys from aerospace vehicles" was "an area in which the University of Maryland offers courses of far greater relevance than do the other universities."[40]

Sandra may have taken classes at Maryland, but by fall 1973, her course notes were from George Washington University. They included "Resources and the Environment," "Central Place Theory" (a way of interpreting the distribution and nature of human populations), "Urban Geography," and courses in physical geology.[41] She received her bachelor of arts degree from George Washington University in September 1976.[42]

By then, they had moved across the Anacostia River to a modern townhome at 337 O Street SW, near the waterfront. Built in 1962, it was the first integrated co-op in DC, known as Riverpark Mutual Homes. It was farther from NRL and required that he ride his bike across the Frederick Douglass Memorial Bridge. But this was still a relatively short commute and was closer to areas of the city, like George Washington and Howard Universities, which would benefit Sandra.[43]

Her schooling and Carruthers's unending work had prompted her father as well as Carruthers's uncle Ben to worry in mid-1975 that George and Sandra were not getting enough "recreation and rest." Nevertheless, referring to himself as "Dad," Redhead continued to organize programs in Brooklyn and elsewhere featuring Carruthers and his work—mainly appearances, popular talks, and displays.[44]

Because, or despite the fact that, they were busy, Sandra and George did not seek out his NRL colleagues for socializing. Robert Meier met Sandra Carruthers only once, in the late 1970s at a weekly social hosted by Chet Opal. "George would never come there on a Friday night . . . or anything like that," but Sandra wanted to meet his colleagues, hoping that one of them might sponsor her to join the American Geophysical Union, which Robert Meier agreed to provide. On another occasion, Meier heard that Sandra had called NRL looking for George only to learn that "he was at White Sands, flying a rocket. She didn't even know that."[45]

Figure 9.2
The brothers and their wives during a 1973 New Year reunion, photographed by their mother. George was the last of his brothers to be married; their youngest sister Barbara remained single. L to R: Anthony and Alica, Gerald and Corine, George and Sandra. *Source:* Courtesy of Gerald Carruthers.

Meier's recollection that Sandra was interested in joining the American Geophysical Union supports another cousin's memory that Sandra wanted to find "common interest" with George.[46] But there is no evidence that she completed any postgraduate coursework or ever worked professionally during their marriage. Another cousin felt that Sandra would have liked to get out of the house and find a job. But the situation was more complex than anyone realized.[47]

Sandra's cousins' contact with George left them with the same impressions that his NRL colleagues related. On occasional family gatherings at their home on O Street, as well as in Brooklyn and Chicago (see figure 9.2), George "had little interaction in our conversations. It was an ongoing observation that he only spoke up if it was space related conversations."[48]

As cousin Suzanne Phillip recalled, "He sat quietly while Sandra and I did the girly gossip, and she would tease and say something about the planets and watch him come alive."[49] Indeed, others like Hattie Carwell recalled that Carruthers "was not much on small talk unless it was about space." She added poignantly, "On space matters, he could talk for hours. You could set a clock by his punctuality. His sense of humor was dry (much like English humor). He could tell 5 jokes before you realized that he was telling jokes. After that, the jokes did seem funny."[50]

# 10

# HUMAN SPACEFLIGHT POST-APOLLO

Carruthers, as we noted, hardly skipped a beat after *Apollo 16*. Even during the weeks and months of proposing an instrument for an Apollo berth, building it and using it and then working with the data, he was writing reports on past missions and proposing flights for other new missions. Between 1969 and 1972 he authored and coauthored, with Richard C. Henry, papers about their 1969 observations of the far-ultraviolet brightnesses of the stars that provided evidence for discrete interstellar absorption lines in bright blue stars, and the presence of general absorption by diffuse gases and dust (extinction) in the Orion Nebula region. Other papers in professional journals and conference proceedings with Hulburt fellow Stephen V. Weber as well as his senior colleagues Chubb, Byram, and Friedman garnered many dozens of citations.

After *Apollo 16*, Carruthers continued to collaborate with Page and others proposing for upcoming piloted missions, as well as robotic probes to the planets, such as Pioneer orbiter to Venus, which flew in 1978. For these, however, he now proposed to modify his electronographic cameras to amplify signals for a variety of electronic sensors to telemeter the data back to Earth. But his heart remained in human-tended and sounding rocket flights, which made physical return of the data possible.

## THE SKYLAB MISSION

Carruthers and his NRL colleagues, pushing for berths on more hoped-for lunar missions, originally discounted the proposed Skylab that would be confined to low-Earth orbit. But by 1973, with the cancellation of an extended Apollo program, Skylab was the next opportunity to fly a film package. So Page and Carruthers once again joined forces to propose for both Skylab 4 as well as the only remaining Apollo mission, *Apollo-Soyuz*. By then, Page had resigned from Wesleyan and lived in Houston, in a complex working arrangement with both NASA and NRL.

As we have seen, the space station eventually called Skylab went through many different designs and scenarios before it flew in 1973.[1] Colorfully put by Harry Heckathorn, NASA's plan by the late 1960s was to "hollow out the second stage of a Saturn rocket and [turn] it into a space laboratory."[2] There were two primary designs in contention, called the "wet workshop" and the "dry workshop." The wet option would be boosted by a Saturn V first stage and a fueled Saturn IVB second stage. The spent IVB would achieve orbit for outfitting as the "orbital workshop."[3] The dry version would be outfitted on the ground. Between July and October 1969, the "dry" option won out. It was favored by von Braun and the astronauts to improve "the probability of mission success and crew safety."[4]

Officially designated "Skylab" in February 1970, it soon included the OWS's cluster of large solar telescopes (now called the Apollo Telescope Mount, or ATM) that would be attached on one side of the Skylab main unit by a "Multiple-Docking Adaptor." Most of these observatory-class instruments, from NRL, the High-Altitude Observatory, Harvard, and the Marshall Space Flight Center, as we have seen, had been part of the canceled robotic mission called the Advanced Orbiting Solar Observatory (AOSO) in the mid-1960s and now were being modified for human-tended operation.

In addition to the ATM suite, which would operate on all missions, there would be dozens of smaller experiments and instruments for stellar astronomy, space physics, Earth resources, life sciences, the material sciences, and for monitoring the spacecraft environment. In addition, by early 1972, there were slots for about twenty experiments by advanced secondary school students. Many of these were automatic, but clearly, the ambitious program ensured that the astronauts had plenty to do.[5] As with Apollo, there was great competition for berths for the smaller instruments in the orbital workshop.

When the space station was approved, and an "Announcement of Opportunity" was issued in the spring of 1971, Carruthers immediately proposed an electronographic Schmidt camera/spectrograph for a far-ultraviolet sky survery on the first Skylab mission. He had already built an upgraded version of his S-201 package for such a platform, arguing that a full sky survey would serve as a guide for later Skylab missions carrying larger and more elaborate telescopic payloads.

Carruthers had been proposing for berths on some form of human-tended mission since 1966, and as the design of the space station changed, due to the often-heated interplay of the primary NASA centers and Headquarters, numerous contractors, and the astronaut corps itself, his plans had to change too. Carruthers's first proposals were not approved, as we have noted, but in March 1973, a celestial surprise created considerable excitement that allowed Carruthers's S-201 payload to reach space once again. A recently discovered comet was nearing the inner Solar System, and his camera was a perfect instrument to observe it, given its wide field and deep UV sensitivity.

On March 7, 1973, astronomer Lubos Kohoutek of the Hamburg Observatory in Bergdorf, West Germany, found a faint fuzzy object on a series of photographs; it was moving slowly through the constellation Hydra. It was a comet, and by April, preliminary orbital calculations by the Smithsonian Astrophysical Observatory at Harvard predicted that it would approach and pass around the Sun in late December and would be bright. Even though Smithsonian astronomers rightly warned, "These comets are tricky things. . . . They don't always do what you expect them to do," the media had a field day.[6] Because it was bright enough to be caught while still so far away from the Sun, some predicted it to become the "Comet of the Century" as it neared and then whipped around the Sun.

Since the comet was discovered still far from the Sun, giving ample time to prepare for it, NASA had to respond. Astronauts had already flown to Skylab in late May. Their second visit would be in July when the comet was still far from the Sun, and the third and last visit was planned for October (*Skylab 4*), still two months before the comet would be close to the Sun and become most interesting to study, given the vast amounts of water ice, gas, and dust expected to be expelled from the comet's head by then. Accordingly, in late July, NASA delayed the last Skylab to November for its planned seventy-day flight to better cover the unfolding event. As NASA associate administrator Dale D. Myers admitted, quoted in the *Washington Post*, "It really looks like the kind of thing we can't pass up." Still, there was some dissent. A delay would be costly, but "it will be a terrific opportunity to observe a phenomenon we might never have an opportunity to study again."[7]

Proposals poured into NASA to observe Kohoutek, from sounding rockets, from the ground, and from Skylab. There was enormous competition and quite a bit of duplication, especially many proposals for UV observations, including with the present telescopes on Skylab's ATM, but also with smaller wide-field instruments brought to the station for the event.

In late June 1973, advised by a wide range of NASA staffers including public affairs specialists, NASA administrator James Fletcher personally invited Carruthers's team to modify their originally rejected proposal to fly his camera on Skylab, to now record the comet. He also invited Carruthers to take part in the planned 1975 Russian-American "*Apollo Soyuz Test Project*" (*ASTP*). According to Herbert Rabin, NRL associate director for research, alerting NASA in mid-July, "We are preparing a crash program to ready this camera for *Skylab 4*."[8]

Indeed, within a week, Page, Conway, and a newly married Carruthers initiated the modifications and proposed for *ASTP* with an addendum for Skylab. But even with Fletcher's invitation, there was doubt about its inclusion. In early July, Conway reported darkly that NASA Headquarters and MSFC were still deliberating how to reassign existing Skylab payload instruments, including ATM, to the comet, and "we weren't mentioned in the message. . . . It looks like they assign rather low priority to any experiments not already on board or at least existing as backup for Skylab."[9] By mid-July, however, NASA/Goddard astronomer Steven Maran, the manager for Operation Kohoutek, informed the Houston center, now renamed the Johnson Space Center (JSC), that they would add Carruthers's UV camera. This still needed Headquarters confirmation from John Naugle.[10] Clearly, a lot of wrangling was going on; there were also suggestions to utilize the ATM's solar telescopes, as well as other airlock payloads.

Even with Fletcher's directive, Carruthers learned from Page, NASA would not make a final decision until August. Regardless, he urged that "shop work must start immediately."[11] They had already established a timeline two weeks earlier to create, on a "crash basis during the next 90 days," a mock-up within ten days for stowage tests, a training mock-up by mid-September for the astronauts, and a flight model by mid-October. The problem was that JSC did not agree to fund the program until July 23, and even then, it was informal.[12] By the end of August, the NRL package was formally funded in an addendum to a NASA "Mission Requirements" document.[13]

While he and Page campaigned for the Skylab flight, Carruthers and Opal proposed to fly an Aerobee from White Sands to view Kohoutek using an existing payload they had built for an Air Force program, a four-camera array. Again, NASA rejected it at first, but in late June reinstated it, given the public attention. The Aerobee 150 flight would take place when the still-hoped-for bright comet nucleus was partly eclipsed by the Earth and not directly visible, allowing its expected huge but faint coma to be isolated for detailed study.

Page's and Carruthers's primary goal was Lyman-alpha imagery of the development of the comet's head and coma over time, as it approached and receded from the Sun. They also planned for a wide array of observations ranging from the Earth's airglow to finding gas and dust within clusters of galaxies that they could not reach from the Moon.

The structure and extent of Kohoutek's huge gaseous halo ("coma") were especially important observations to make, and there were many proposals to do so. It would be highly valuable as confirmation and hopefully as an extension of observations by the first successful Orbiting Astronomical Observatory (OAO-A2) of three comets in the late 1960s and early 1970s. NRL agreed to make Carruthers's camera compatible with two Skylab mounting systems so the astronauts could use it either in the antisolar scientific airlock using an articulated mirror system or on an EVA (extra vehicular activity) with NASA's "Standard Universal Friction Mount." NRL worried that the airlock option, the one chosen by NASA, was less preferable because a small external mirror that collected the light would also vignette the field of view, reducing it from 20 degrees to only 7 degrees. As with *Apollo 16*, Page would work with the astronauts while Carruthers and Conway would liaise with scientists at the centers, including Steven Maran and Bill Snoddy.[14]

Brian Marsden at the Smithsonian Astrophysical Observatory had already provided them with approximate positions of the comet from October through January, which they then mapped onto starfield charts. Of course, the sensitive camera could not be pointed at any spot within 15 degrees of the Sun, so "with great caution" for fear of damaging the optics and detectors, they plotted out ten dates and positions when they hoped the faint comet could be safely recorded.

## THE FLIGHT OF SKYLAB 4

Over the next few months of feverish preparation, not everything went smoothly. The camera optics and mechanics were not the problem, but there were issues with humidity degrading the quantum efficiencies of the detectors in the human-tended environment. In August, after a critical design review, Page, through Conway, appealed to Carruthers to look again and report on the problem, and provide suggestions to fix it. Carruthers had done the same for the Apollo system. His experience there told him that there was not sufficient time to perform more tests. He replied immediately that the types of tests requested for Skylab were "impractical . . . especially on the time scale of the SL-4 mission." His intimate knowledge of the tests convinced him that it was fruitless. Furthermore, his original tests in 1971 more than satisfied him that as long as the relative humidity was 25 percent or lower in the Skylab airlock, the detectors would be fine. This was just one of many examples of how Carruthers's experience and intimate knowledge of the instrument helped to avoid time-consuming wrong turns.[15]

But this was far from the only problem. New designs for the electronics, especially a set of integrated circuit boards from MSC, had not been thoroughly tested, again due to the time limits. Further, the airlock option was mechanically complex and was misbehaving, and as feared, tests showed that when the camera assembly was drawn back into the airlock and pressurized, it would cloud over which degraded UV reflectivity.[16]

Additional worries arose in November. Based upon analyses of OGO-4 data, there was growing concern that perturbations in the Earth's ionosphere caused by natural and, more critically, artificial events, like the passage of missiles and their exhaust gas plumes through the region, could influence the observations. Though not articulated here, the ability to detect these perturbations was of value to military reconnaissance but might well reduce the observing time devoted to the wide-field astronomical observations of Kohoutek's brightness and structure. Carruthers knew of this problem because he had already proposed, in line with a continuing interest of the Navy, a mission to study the effect of exhaust plumes.[17]

Meetings throughout this time at JSC, reported by Page, identified the many changes in procedure that would give the astronauts sufficient time to add Kohoutek observations when working in the airlock and on EVAs. There were many difficult decisions to make regarding when to observe the comet as it moved rapidly across the sky, and these affected all the other observations that were already planned. Steven Maran called an "Operation Kohoutek Planning Workshop" in October—and there would be many others to coordinate observations. Carruthers attended some of the final meetings just before launch at the Kennedy Space Center but devoted most of his time to buttoning up the payload.[18]

Even after *Skylab 4* launched on November 16, 1973, NASA still felt the need to justify the addition of the camera to reinforce the importance of observing the comet in an already seriously overloaded mission. "Mission Requirements Change 8," issued in early December 1973, highlighted S-201, enumerating all the tasks it was then doing onboard: to confirm data from *Apollo 16* for the Earth's tropical airglow bands, to search for lunar atmospheric hydrogen emission and for interstellar material in the Large Magellanic Cloud, and to search for evidence of intergalactic hydrogen in clusters of galaxies. But most of all, in a specially highlighted section, it would observe Comet Kohoutek.

Still being touted as becoming "one of the brightest comets of the century at an early time," a NASA release noted that it "has prompted the scientific community to prepare for extensive observations of the comet." And observations from Skylab instruments would be unique: "It will be possible to verify the prediction of huge clouds of hydrogen-emitting Lyman-Alpha radiation around the comet." Not only might the observations provide clues to the origin of the Solar System the notice claimed, but there would also be great popular interest: "Next to eclipses of the Sun, comets are the celestial phenomena that have most aroused human curiosity since earliest times."[19]

Throughout this period, Page constantly sent Carruthers detailed notes and transcripts of proceedings from JSC, asking for input. Page, designated principal investigator now because he was actively associated with JSC, also managed complex interface issues, performing mainly procedural and political tasks. Snoddy and Richard J. Barry of JSC assisted Page and Carruthers in this project, concentrating on the technical issues and assisting in the analysis.

During the flight, Page constantly reported the voice recordings of astronauts Gerald P. Carr (Commander, or "CDR"), Edward G. Gibson (Science Pilot, or "SPT"), and William R. Pogue (Pilot, or "PLT") that were transcribed and issued as "Dump Tapes." They were a running account of the astronauts' work with the various payload instruments and at one point reported on problems with Carruthers's instrument, including the motor drive for film advance, and with access to various fields of the sky at specific times that were hindered by the orientation of the airlock or blocking by portions of the station. They also provided a continuous record of the objects being observed. Page added commentary in his constant telexes to

Carruthers, emphasizing "the complications in S-201 operations on EVA." Most critically, the astronauts could not tell if the camera was in the shade or being exposed to sunlight. As Gerald Carr reported, "I'll go out front and look at the front of the camera and tell you whether its shaded or not. You know . . . and if the Sun line starts coming onto it, I'll holler at you, and you can snatch it in."[20]

The astronauts' banter as Young operated the camera was colorful, but worrisome to those in Houston: "Watch that power cable!" and "You got to keep an eye on the attitude error, Bill," and even "Oh Jesus. Hey don't do that." These were sprinkled through the chatter between the astronauts on board with astronaut Story Musgrave in mission control. But the most troubling commentary early in the mission was about the camera's film advance motor. Bringing the camera film cassette inside and pressurizing the airlock, Pogue took it apart, checked the motor with a power source, and then put it back together, evacuated it, and tested it. He was advised, "During S-201 ops, press ear against the canister and listen for the ¼ second film advance motor noise." But Pogue heard nothing, so he asked the Commander for advice, who advised him to listen not for a "click" but for a "clunk," which Pogue tried to do: "I will strain my little ear bone and see if I can't hear that clunk this time." The film advance was indeed erratic, partly due to constant problems with the power supply, which drew even more colorful language from the astronauts, and concern from Carruthers and Page.[21]

After the mission ended and the astronauts returned the film cassette, negotiations between NASA and NRL, reminiscent of the *Apollo 16* experience, intensified over who was responsible for what, especially how the film cassette would be handled and who would handle the data. The overall atmosphere in the aftermath was, as usual, tense. The astronauts themselves were assigned so many tasks that one of them, scientist-pilot Edward G. Gibson, described the first four weeks as a "fire drill." Adding the Kohoutek assignment at the last moment did not help.

Carruthers's camera was not the only instrument pointed at Kohoutek. ATM's coronagraph, a photometric camera, and other instruments were recruited. One of the main issues was how to point the major instruments at the comet since the ATM was designed to lock onto the Sun.[22] Despite the heavy workload, with frequent instrument repairs and competing priorities, the astronauts managed to record over five hundred photographs with the S-201 camera in both EVA and airlock mode. On thirteen occasions, between November 26 and February 3, they pointed the S-201 camera directly to Kohoutek and obtained some 126 images, recording its constantly evolving structure—namely, the development of the cometary halo of hydrogen as it approached the Sun, and its decline thereafter.[23] In the midst of these observations, Carruthers and Opal had also collected both image and spectroscopic data from their Aerobee 150 sounding rocket flight on January 8, 1974, observing the comet from an altitude of 120 miles over White Sands after it passed around the Sun. In addition, between January 11 and 17, 1974, NRL sent a team to photograph the comet from Big Bend National Park in Texas, for "ground truth." It was an impressive three-way campaign, and everyone anxiously awaited the results. Of course, beneath all this effort, the question was, were these efforts complementary, leading to conclusive results?

After the cassettes were returned and inspected, they were examined at JSC. Page was asked about anomalous features in some of the photographs that they had also seen on *Apollo 16* images and now on Skylab results. Were there problems with the film transport, or with the optical system? He responded that there were problems with a barrier membrane in the camera separating the film transport from the focal plane of the telescope, on both the lunar and Skylab flights: density variations causing the brighter images to appear wavy. Brighter objects also seemed to have small tails; as Page explained at a March meeting in Austin titled "Electrography and Astronomical Applications," the camera for Skylab had no shutter, so exposures were started and ended by advancing the film, which took one quarter of a second. This was acceptable for fainter objects, but "a very bright object such as the comet may actually show as a streak on the film moving in."[24] Left unsaid was the frustration of the overall experience.

Page, as principal investigator, dealt with the dickering between JSC and NRL over priority for holding and analyzing the films. There were issues on both sides. He worked feverishly to prepare and present preliminary results at several conferences in the spring. He was also not shy to state that there was considerable frustration getting control of the films from NRL, citing "administrative delays," and, in fact, his first reports were made before all of them were secured and reduced. The frustration peaked in May when Page angrily warned, "Since things seem to disappear in NRL without a trace, I'll stop sending the scans until specific ones are requested."[25]

Despite these continued rough edges, by then, Carruthers had also received enough data, mainly negative prints, to jointly prepare a preliminary discussion with Page of their Kohoutek cometary halo observations. Page formally wrote up the results and presented them at conferences, while Carruthers shared the images with local schools and universities. Within the next three years, Page produced some five publications on Kohoutek—two of them coauthored with Carruthers and other NRL staff. Snoddy and Barry independently reported on the S-201 observations from Skylab.

Initial impressions of the data from both the rocket flight and Skylab in March were that the "the Skylab data is of somewhat lower quality than the rocket data."[26] But the data from both flights were consistent enough to report their findings with confidence. In all, Carruthers and his colleagues, including Page, Conway, Opal, Prinz, and Meier, published about a dozen articles and abstracts and gave many talks based upon their observations. There was much to report: delineating hydrogen, carbon, and oxygen production rates through sublimation in the comet's nucleus, and the spectra of stars and interstellar gas in the Magellanic Clouds.

All this analysis work required support, and NRL could not support all of it. Herbert Rabin did his best, making NRL's pitch for funding from the Johnson Space Center's Skylab Office in March 1974. He claimed, "These data far exceed the preflight expectations and will require 18 months' or more analysis."[27]

By December 1974, they were able to "derive a preliminary, qualitative history of the development of the cometary hydrogen coma and the associated hydrogen production rate."[28] By late 1975, they were sufficiently satisfied with their overall results to suggest continued UV and infrared observations of comets from future flights of the Space Shuttle.[29]

A 1976 report, led by Meier, was a "final analysis" of the Lyman-alpha data. It made a detailed comparison of both rocket and Skylab data with theoretical models that verified their former conclusion that the hydrogen production rate as the comet neared the Sun was relatively constant until the comet was at closest approach. Their results were consistent with Fred Whipple's "icy-conglomerate model of the comet nucleus," but they also warned that it was still possible that something other than water was the source, and more work was needed.[30]

Meier and his colleagues were not totally confident about their conclusions because both the Skylab and rocket observations had some problems.

After the observing run outside the airlock, and then after the camera was repressurized, the astronauts detected an electrical discharge from the critical photocathode, which compromised its sensitivity and confused subsequent recalibration. Further, the reflectivity of the articulated mirror assembly decreased over time, as tests had indicated it might. But the full EVA observations were not affected. The rocket observations were also compromised because the focusing magnet degraded during initial vibration testing. They had to reduce the voltage, resulting in less sensitivity. In the end, however, both missions returned valuable and consistent data.

Page's final report from his comparative analysis of *Apollo 16* and *Skylab 4* observations, which he conducted mainly under NRL auspices, was not finished until March 1977. It incorporated and summarized some nineteen papers dating from 1972, as well as a popular book Page had written with his wife, Lou Williams Page, *Space Science and Astronomy: Escape from Earth*, which appeared in 1976. Still highly descriptive, it made the point that primary observations had yielded information on the Earth's tropical airglow belts, the general shape of the geocorona, and the relative strengths of far-ultraviolet spectral lines of major constituents like helium, oxygen, and hydrogen that were consistent with models prepared by Meier and others. Similarly, their Kohoutek observations fit models of the comet's huge hydrogen halo and the evaporation rates of hydrogen from the comet's nucleus.[31]

Although Meier and his team reached their final conclusions with caution, the results stood the test of time. Years later, this work was cited by William M. Jackson, an astrochemist at the University of California, Davis, who in a support letter for Carruthers's 2011 nomination for the President's Medal of Technology and Innovation, noted, "The spatial resolution of the images that he had obtained [of Comet Kohoutek] were excellent so that they could be fitted with a detailed model involving the photodissociation of water." Once again, this clinched Whipple's dirty snowball theory for comets.[32]

Much good science came from Kohoutek's passage, but the popular press and the public were disappointed by its failure to brighten as predicted. One theory suggested that as a "primordial comet" in its first visit to the Sun, it was far denser than expected and therefore performed less ice-to-gas sublimation as it tore into the inner Solar System, rendering it far fainter to the eye than expected. Another suggested that sublimation took place much earlier. But as Johns Hopkins physicist William Fastie summed it all up (he led a team performing sounding rocket studies of the comet), "Comet Kohoutek, also known as the Comet that Couldn't, was a disaster for public viewing, but the coordinated research that was performed will be most significant."[33]

Given Page's leadership and Conway's support, Carruthers experienced less bureaucratic pressure during *Skylab 4* because he knew his instrument so well and was through this drill before. And, frankly, he was comfortable letting others take the lead because he now also had experienced management assistance from Conway.

This protection allowed Carruthers to do what he did best, proposing yet other projects. Earlier, in December 1973, even with *Skylab 4* still in flight, he proposed a high-resolution UV spectrometer to inspect the Venusian atmosphere for a planned 1978 *Pioneer* probe and orbiter. Along the way, the instrument would also sense interplanetary gases and monitor the solar UV, far-ultraviolet, and extreme-ultraviolet spectra, as well as examine interstellar gases in those spectral ranges. Then, in January 1974, he also proposed to participate in the scientific definition of planned space shuttle missions for AMPS (Atmospheric, Magnetospheric, and Plasmas-in-Space) payloads, with Opal as co–principal investigator. For both, he also suggested a series of sounding rocket flights to test out the "remote sensing" payloads that were designed to better understand the dynamics of the thermosphere, exosphere, and magnetosphere.[34]

The Space Shuttle was the newest platform on the horizon for human-tended spacecraft experiments and profoundly changed the opportunities for flying payloads, and for humans flying in space, like Carruthers.

# 11

# PREPARING FOR THE SPACE SHUTTLE

## A PLETHORA OF PAYLOADS

After years of debate over the development of a true space transportation system, based upon decades of speculation in fact and fiction, in January 1972, President Richard Nixon announced that the next step in the nation's space program would be a fleet of reusable space vehicles. They would provide access to low-Earth orbit, serve as platforms for launching satellites into higher orbits and space probes to other planets, provide the means to construct a permanent space station, and carry a wide range of scientific instruments into space for short-period observations from a large open bay or from free-flying craft that could be released and recaptured for return and reuse.

Given his visibility, Carruthers was invited to participate on an early space shuttle scientific definition team and by March 1974 had teamed up with scientist-astronaut Karl Henize, who was leading a proposal for a new deeper ultraviolet sky survey. Henize had recruited several scientists as coinvestigators to create an advocacy group and told Nancy Roman that he would help to meet the costs with his own "personal (astronaut) research budget."[1] It could well be through his contact and collaboration with Henize that Carruthers decided to apply to be a mission specialist in 1977, which we explore later. But first, we look at Henize.

Karl Gordon Henize, born in Cincinnati, Ohio, in 1926, was initially a farm boy like Carruthers and had a similar childhood interest curve. He received his PhD in astronomy from the University of Michigan in 1954 and then held a series of postdoctoral positions at Mount Wilson and the Smithsonian Astrophysical Observatory, where he engaged in satellite tracking, and ended up at Northwestern University's Department of Astronomy. After NASA abolished its age limit for astronauts in 1967, Henize was selected as one of the first scientist-astronauts, a category historian

Matthew Hersch has shown was never fully assimilated into NASA culture.[2] Henize was assigned to provide astronaut support for *Apollo 15* and then for the three *Skylab* flights. He finally flew almost two decades later, on *Space Shuttle Challenger* in July/August 1985, providing scientific support for an array of instruments on *Spacelab-3* in the shuttle bay.

Henize's first detailed articulation of his ideas came in an internal document to the University of Texas in June 1973, but it would be another year before he recruited and organized his workforce, the group that included Carruthers. Henize envisioned his camera as a "modest budget" project for low-cost and repeated access to space. Since the shuttle was expected to start flying by 1978, there was little time to waste planning for mission payloads.

Carruthers became a coinvestigator in Henize's formal proposal to NASA in April 1974, along with Texas astronomers Harlan Smith and James D. Wray and astronomers from three other institutions. The proposal described in detail a "two-color survey" (comparing brightnesses in three wavelength bands) of the far-ultraviolet sky for what was then called a "shuttle sortie mission," a new class of standardized payload that could be used by many observers for a variety of goals. Henize wanted to create an array of instruments that could make a deep survey of the sky to "supply the world's astronomers with similar quick look data (similar but deeper and more UV sensitive than the classic Palomar Sky Survey)." [3] To prepare for the spaceborne instrumentation and its execution, they asked for a two-year photographic prototype study starting with a ground-based sky survey, which they described as a "finder telescope" for what was then being called the Large Space Telescope.

Henize claimed that Carruthers's electronographic designs were the best option to reach deep space efficiently and that they might be the detectors of choice for the shuttle sortie missions and ultimately the Large Space Telescope. The ground-based study was a first step, testing curved photocathodes for the proposed fast all-reflecting folded Schmidt camera, dubbed a "large electrograph," on an altazimuth mount in the shuttle bay. With a primary mirror in the range of 24 to 30 inches, it would be the largest space telescope equipped with one of Carruthers's cameras yet flown.

Shuttle sortie payload carrier missions had three planned goals: (1) to boost robotic planetary and interplanetary missions; (2) to launch, recapture, and return robotic missions in low-Earth orbit; and (3) to provide a platform for instruments exposed to space.[4] Sortie missions were also proposed to provide either pressurized laboratories, open platforms, or short-term free flyers during shuttle missions. In 1974, these options merged into what was soon called "Spacelab," a reusable space laboratory for a wide range of interests (astronomy, biology and life science, and the material sciences) and for developing manufacturing techniques in zero-g. Spacelab was a cooperative venture between NASA and the European Space Agency and first flew in the early 1980s.

Another mode of carrying instruments that Henize and Carruthers proposed was a short-term free-flyer mode for astronomical studies, named SPARTAN (Shuttle Pointed Autonomous Research Tool for Astronomy). Promoted by the Navy and Air Force, NASA justified it as a "low-cost" replacement for Aerobees and other sounding rockets. It did provide for longer look times and higher-altitude perches and was also a way to protect the instruments from the local shuttle environment, where outgassing and a persistent UV glow compromised observations. It could also obtain greater pointing stability as an autonomous free-flying subsatellite. The first one was carried aboard the shuttle *Discovery* in June 1985, the year the last Aerobee was flown from White Sands. Although it provided many hours of exposure time, astronomers did not fully accept the change because they preferred the fast turnaround they had enjoyed from sounding rocket flights—especially important when training graduate students.[5]

This tension had a long history. Riccardo Giacconi, a leading X-ray astronomer who was successful obtaining larger and larger berths, spoke for Friedman, Carruthers, and many other astronomers when he complained to the General Accounting Office in 1977 that NASA was focusing nearly all the space sciences on the Space Shuttle. He was repeating, in part, but ardently amplifying, concerns raised by the 1973 National Academy of Sciences' summer study, "Scientific Uses of the Space Shuttle," which concluded that NASA's planned mode of performing science in a "shirt-sleeves laboratory environment," such as with Spacelab, "was an extremely inefficient use of the money." Free flyers were marginally better, but still, "in astronomy the need is for permanent observatories." Spacelab and the free flyers could be useful as test beds, but there were far too few in planning at the time to meet the needs of astronomy without sounding rockets.[6]

Carruthers's reaction to the death of sounding rockets is not known, but he lost no time modifying his Aerobee designs and experimenting with many options that these new observing platforms promised. This mode also perfectly fit his design philosophy: using film requiring retrieval. But he had never built anything that could serve a telescope larger than 6 inches, and this new opportunity, being part of Henize's project, gave him all sorts of ideas. The system would use film, but it could be in a larger format. He stuck with his all-reflecting Schmidt camera, building laboratory prototypes that at first used permanent magnets to focus and amplify the electron beam. But as he did before, he turned to a "super-conducting solenoid to further improve the electron imagery."[7]

Carruthers proposed two different camera systems for a two-color sky survey on a sortie mission. One was for the 1,800–2,800 Angstrom range and another, with interchangeable filters, for 1,050–1,600 and 1,250–2,000 Angstroms. Each, he predicted, would have "about 10 times the overall detection efficiency of the fastest vacuum-ultraviolet photographic emulsions." They would also have a linear response to the intensity of the signal, better resolution, a wider dynamic range, and insensitivity to the visual region of the spectrum.[8]

One new challenge he faced was the curvature of the all-reflecting Schmidt optical system field, a characteristic of classic Schmidt cameras. It was negligible on the small aperture f/1 systems for *Apollo 16* and *Skylab 4* but would not be a problem for a larger aperture system with a large film field. To achieve uniform definition over the field, the photocathode had to be curved. And it would be significant because his photocathode was huge: 8-inch with a 98-inch radius of curvature.

Carruthers proposed a stepwise process where he would enlarge his existing prototypes incrementally. During the first year of this effort, he would refine his present camera with a 4-inch-diameter cathode using 2.4-inch film, as well as test various photocathode formulations with colleagues in Texas, ramping up the field strength first with focusing magnets, and eventually with the superconducting solenoids.

## SIMULTANEOUS NRL PROPOSALS

Characteristically, Carruthers's collaboration with Henize paralleled his work with Opal on their proposal for the AMPS (Atmospheric, Magnetospheric, and Plasmas-in-Space) payload for Spacelab, which NASA was now describing as a "constantly evolving . . . orbital national laboratory."

By November 1975, "some 60 instrument candidates and 80 possible investigations" had been proposed.[9]

To meet the competition, Carruthers and Opal modified their electronographic camera once again to make it even more powerful and to sharpen its view. They added a combination of a photocathode and an amplifying "microchannel plate" feeding a magnetic focusing system recording spectra from an external curved objective grating. As Carruthers explained in one of his more accessible articles on electronic imaging for *The Physics Teacher* in 1974, although his present camera had 100 percent photon efficiency, some efficiency was lost when other output recording schemes were added, like lens-coupling amplification or a television camera, and other means of intensification became necessary. So from the early 1970s, he started experimenting with what are called "microchannel plates" to amplify the signal. A microchannel plate, as Carruthers described it, is a tightly bound bundle of 10-to-20-micron–diameter glass tubes with interior metallic coatings that could reflect and amplify photoelectrons via a voltage gradient. They were first used in X-ray astronomy, which required small incident angles of reflection. But then they were applied at longer wavelengths not only to collimate the beam but also to amplify it through 100 percent reflection (low-angle reflection called "grazing incidence" where no absorption or transmission takes place) off the interiors of the tiny microchannel tubes. These combinations typically provided electron gains of about ten thousand and photon gains of one hundred for a resulting amplification of one million.[10] Carruthers had tested similar systems on sounding rocket flights but wanted to develop larger versions. And with colleagues, he adapted closed-circuit television units that would allow the astronauts to point, control, and monitor the system.[11]

And as if AMPS was not enough, there were three more proposals. In March 1974, he proposed for continued analysis of data from his S-201 instrument on *Skylab-4*. Page would take the lead with Robert Meier, Chet Opal, Dianne Prinz, and Carruthers assisting. And in April, NRL asked other funding sources in the military to support Carruthers's continuing use of sounding rockets. Then by the end of the year, Carruthers teamed up with Opal and Meier as well as Warren Moos and Richard C. Henry at Johns Hopkins to propose a Lyman-alpha explorer mission. They planned for a mission "in the same class as the *International Ultraviolet Explorer (IUE)* launched by a Delta rocket that would carry it into a geosynchronous or higher orbit."[12]

The *IUE*'s design and function were well defined by the time Carruthers made his proposal. He knew there would be a backup payload, a full-sized operating telescope, and possibly duplicates of portions of the spacecraft. Backups for these missions were common practice at the time, and it was also not uncommon for people to propose reuse of the backup hardware. Carruthers explicitly made this suggestion, proposing that his instrument would complement the *IUE*'s mission by providing "a complete sky survey to the faintest possible limiting magnitudes."[13] He proposed two telescopic cameras: one was a wide-field Schmidt with interchangeable corrector plates for direct imagery and wide-angle surveys of the UV sky. The second telescope would be a narrower-field and higher-magnification Ritchey-Chrétien reflector.

Both cameras would utilize his electronographic image converters, but this time—and a significant departure for Carruthers—their output would be recorded by SEC (Secondary Electron Conduction) photoelectric imaging systems so the data could be telemetered to the ground. The two telescopes would be independently mounted so two programs could be carried on simultaneously, with the narrow-field telescope commanding the orientation while the wide-field telescope serendipitously covered the sky. Both would have direct imagery and spectroscopy modes. The Ritchey-Chrétien, the responsibility of the Johns Hopkins contingent, would have an effective 14-inch aperture, whereas the Schmidt would be only 4-inch clear aperture, something with which NRL was already familiar. As with the planned *IUE*, this Explorer would be operated by ground-based observers in real time.

Carruthers's proposal typically went into considerable detail over how these observations would complement or extend the work of present and future proposed missions, from the *IUE* itself to the *OAO Copernicus* that had launched in 1972. But most of all, Carruthers wanted to repeat and extend his own Earth-directed studies of the geocorona, followed by related studies of interplanetary gas and dust.

Once again, NASA rejected this proposal, which led to some of the same back-and-forth quarrels. But Carruthers was not phased; in fact, it gave him more time through the 1970s to collaborate on continued data analysis and projections with Page, Opal, Heckathorn, Meier, Prinz, and many others, leading to some seventy-seven publications and presentations from 1972 to 1980 based upon data from *Apollo 16, Skylab*, and sounding rockets.

This latest refusal did not dampen his enthusiasm for producing a continued series of proposals from the late 1970s through the 1990s. On one hand, as we have been emphasizing, Carruthers's never-ending infatuation for instrument development paralleled his practical need to keep his dwindling staff employed with the external funding he was seeking (see below). That was a requirement of the job and was reflected in laboratory space allocation. However, there had been (and would continue to be) far more rejections than acceptances, which was the typical experience of space scientists proposing to NASA in the 1980s and 1990s. Reacting to the ills of stagflation, a double whammy of high inflation and little economic growth, combined with the rocketing price of oil and other drains on the economy, the Carter administration did not give the space program high priority, and the situation grew even worse under Reagan.[14]

Carruthers's colleagues sensed that he took these rejection notices stoically and would not burden people with his anger, if it existed, since it was a fate shared by many. But he was overheard at least once saying that "NASA is a four letter word."[15] He evidently was smarted by Roman's stern cautionary advice to limit his countless proposals, but he continued on undeterred.

Younger colleagues like Tim Seeley sensed and shared his frustration: "You could tell he was upset, I think. I remember that." Seeley then went on to relate his experience with NASA: "You can work on a project for years, and it can get cancelled. . . . That's just the name of the game in the space flight projects."[16] And despite the many frustrations, former Hulburt fellow Richard C. Henry recalled, "Looking back, how could I have been so lucky? Working with George Carruthers was a dream."[17] And Carruthers definitely had his own dreams.

## APPLYING TO BE AN ASTRONAUT

A HistoryMakers interviewer asked Carruthers if he ever wanted to be an astronaut "or a space traveler like Buck Rogers." He brightly responded, "Yes, I did, and in fact, I applied, after I got out of school, of course, a couple of times. I didn't quite make it."[18] The earliest application that has been found dates from 1965 when the "scientist-astronaut" category was announced. The only record found of this was a telegram on January 16, 1968, rejecting him as a mission specialist, but he made it to the preliminary interview stage.[19]

Nixon's 1972 announcement of the shuttle program created new pressures for replenishing and expanding the astronaut corps. Finding itself in a "goldfish bowl" from pressure by the media and Congress and having not issued a call for new astronauts since the late 1960s, NASA now actively sought out candidates irrespective of gender or race and highlighted the importance of training in math or the sciences at least to the bachelor's level.[20] The call, NASA's eighth (called "Group VIII"), was issued on July 8, 1976, and would be open for a year. To reach a broader pool, NASA publicized the openings widely in both mainstream and minority outlets like *Ebony* and *Jet*. By June 1977, JSC had received over 20,000 applications, and some 5,680 passed the preliminary screening. The listing was then weaned to some 150 semifinalists who were brought to JSC in groups of about 20 for a week of intense interviews and testing before there was further weaning and a far longer training and testing period.

Carruthers eagerly responded to the 1976 NASA announcement, which now focused on a new category of "mission specialist" to complement pilot and commander and, since 1965, payload specialist and scientist-astronaut, and to provide a wider range of expertise and responsibilities that did not require flight experience. And, most welcomed, it was also the first to call explicitly for women and minorities.[21]

Carruthers had good reason to be heartened by the new category, and it was shared by many people, including those who had worked with him. In an extensive Bellcomm contract analysis, "Astronomy on the Shuttle Sortie," widely distributed in November 1971, N. Paul Patterson, a former Hulburt fellow who had worked with Carruthers and then at NASA Headquarters, used his experience with preparing for what became the S-201 payload to argue strongly before the President's Science Advisory Committee that the person who designed and built a payload was the best person to use it onboard a shuttle. Using Carruthers's camera design as an "illustrative example," citing its successes to date as an indication of its versatility during sounding rocket flights—measuring the extent of atomic hydrogen in the interstellar medium, discovering interstellar molecular hydrogen, and "recording unusual characteristics in the spectra of stars" in the far ultraviolet—he speculated on discoveries possible from a sortie mission extending for days. He then described how the experimenter who conceived of, designed, and built the instrument would also be the ideal "observer" on the sortie mission: "The particular advantage of this experimenter's presence onboard would be his experience as a pioneer in the development and use of this new type of instrument in space, his motivation to acquire significant results and his sensitivity to the subtle balance of factors which determine the ultimate quality of his data."[22]

Carruthers's desire to fly and operate his own instruments in space also fit Friedman's agenda. Friedman strongly urged his staff to propose sending instruments into space on the shuttle, implicitly encouraging them to consider flying with their instruments. He petitioned NASA to make "realistic plans" to send small payloads on shuttle sorties. And, in fact, on the same date as Carruthers's draft proposal, Friedman told Philip E. Culbertson (director, mission and payload integration) that "within our Space Science Division at NRL, I have urged my colleagues to prepare informal proposals that might be worthy of acceptance for some of the earliest shuttle test flights."[23]

The 1976 call explicitly for mission specialists further expanded the connection between astronaut and instrument. Payload specialists were focused solely on the instrument, but the mission specialist would have broadly defined duties before, during, and after flight. NASA also made the call to meet the anticipated need to staff the forthcoming Space Shuttle, which was being popularly hailed as a transportation system that would "allow ordinary citizens to fly into space."[24]

Carruthers responded to the call in early August, insisting that if selected, he wanted to remain an employee, on detail, from NRL "as [was the case] for the military candidates."[25] His "Personal Qualifications Statement" was for "Mission Specialist," but he made it clear that he was also applying for the existing category of "Payload Specialist," which was not part of the call. He claimed that he would accept a pay and grade reduction to take the position and would be ready to move to the Johnson Space Center by December 1978. Sandra was about to graduate from George Washington University, and he hoped she could become absorbed in her own career in his extended absences.

Carruthers's proposal, vita, and attached personal statement offer many additional insights into how he viewed his interests and his qualifications. He listed his training, his major projects from his one dozen sounding rocket flights, *Apollo 16*, *Skylab 4*, his contributions to NASA and Space Science Board advisory committees, his awards (eight by then), and his "selected" bibliography consisting of some fifty-one publications with three in press.[26]

He described what he was hoping to do in an appended personal statement blatantly titled "Mission Specialist/Payload Specialist." This left little question where his priorities lay: while at Johnson in training, he hoped "as time permits" to continue his research and his development of electronic imaging devices, both having "potential applications in future Shuttle missions."[27] He emphasized his training and expertise in science and engineering, and his many ongoing projects, including hoped-for sounding rocket flights and berths on future shuttle launches of Spacelab, either in its pressurized capsule mode or on its exposed platform in the shuttle's bay. He envisioned attaching his camera to the recently dubbed STARLAB, a 40-inch general purpose f/15 Cassegrain telescope proposed by a large team of prominent astronomers at NASA and in academe, operated by scientists in orbit and on the ground.[28] In closing, he hoped that his work "would be coordinated with that of his associates at NRL" as well as at JSC who might want to use the imaging camera that he would continue to develop "on a time-available basis."[29]

Carruthers listed Karl Henize (one of the proposers for STARLAB); astronomer C. Robert O'Dell, then at MSFC as Space Telescope project scientist; and, of course, Herbert Friedman and Talbot Chubb as supporting names. Friedman gave Carruthers the highest ratings, "exceptional" in all aspects of technical competence and in professional integrity. Henize and Chubb rated him slightly lower, but only in the "theoretical research" category. All three whose responses have been found rated him "one of the very best" for "Over-all Fitness for Professional Work as an Astronaut Candidate." Chubb identified him as "the world's leader in the development of far UV electronographic cameras," warmly adding that "he has an excellent sense of humor and is fair and honest." Henize personally noted that "his modesty, unusual for a man of his race and stature, makes him an easy person to work with and get along with." He added that this modesty did not keep him from "standing up for what he thinks is right in a technical or management situation." Friedman summed it all up, stating he has a "very stable personality."[30]

Carruthers was one of the few score candidates who passed all the preliminary screenings and was in the sixth group of semifinalists that arrived at JSC on October 3, 1977.[31] There was major competition in this group, such as Sally Ride, a physicist who would become the first American woman astronaut, along with astronomer George "Pinky" Nelson, who had just completed his PhD at the University of Washington and was then at the Sacramento Peak Observatory. There was also astrophysicist Jeffrey Hoffman, a Harvard PhD then working at the Center for Space Research at MIT. And Carruthers was not the only Black candidate: those who were chosen from Group VIII included Guion Bluford (USAF fighter pilot and aerospace engineer), Frederick Gregory (military engineering and engineering test, and a fighter and helicopter pilot in the Air Force), and Ron McNair (a karate champion and recent PhD graduate in physics from MIT).[32] The competition was intense.

The candidates were subjected to an exhaustive battery of physical and psychological tests and mental evaluations. Nelson had the impression that the real testing and evaluation, beyond the extensive physical exams and mental examinations, took place throughout his time there during social gatherings and on the baseball field.[33] In an interview with the author, Nelson also recalled that NASA was as we've noted looking more for generalists among the scientists, to handle a wide range of responsibilities as mission specialists, and not for scientists focused on a specific payload.[34] Still, among the astronomy candidates who knew Carruthers by reputation, Nelson personally felt that Carruthers was major competition, recalling in an interview, "How can we compete [laughs] against somebody like this?"[35]

Nelson's impression about NASA looking for generalists aligned with R. Thomas Giuli, an astronomer at JSC in the Physics Branch who was then program scientist on the *Apollo-Soyuz* Test Project. Giuli was a semifinalist in Group VI in the 1960s and wanted to fly as a payload specialist on *Skylab 1*. He put it this way in 1977: "if he got the job, his mission would be to . . . operate and be responsible for conducting many experiments in a large number of disciplines, and . . . [he] must be trained in general scientific procedures. What he will not be [on *Spacelab 1* at least] is an expert in a particular experiment field, at least not necessarily."[36]

These descriptions contrast with how Carruthers described himself. In his proposal, he focused more upon what a payload specialist had to be, rather than a mission specialist. Everyone knew, as he also stated numerous times in his application, that he was proposing payloads for future missions. His reputation for being an expert instrument-centered scientist was well known among the other candidates, and to those evaluating the candidates.[37]

So it is easy to speculate that this preference was an issue during his interviews at JSC and may have reduced his competitiveness for mission specialist. Carruthers knew well that the call for Group VIII was not for payload specialists, as they were selected separately as part of the selection process for payloads themselves. Still, his presentation made it pretty clear that this was his preference.

Carruthers's mixing of his interest in being either a mission or payload specialist also reflected the confusion at the time over what these designations meant. Up to the mid-1960s, astronauts were military pilots, or test pilots. A few had science training and, as academic candidates, still had to take USAF jet pilot training. The first call for scientist-astronaut was in 1965, but in the colorful words of historians David Shayler and Colin Burgess, socially "they were judged to rank below test pilots, jet pilots, astronaut wives and even astro-chimpanzees in the flight hierarchy."[38]

During his visit to JSC, Carruthers underwent many demanding tests—physical, medical, and social. Shayler and Burgess described candidates as "Guinea Pigs." Among the psychological tests, the candidate was subjected to penetrating interviews while sitting on an uncomfortably wide and high chair that left their feet and arms dangling. In another test, they had to squeeze into a 36-inch-diameter "rescue ball" described as "a claustrophobic's nightmare."[39] Not surprisingly, Carruthers endured these challenges.

But was he an appropriate candidate? It is quite clear that Carruthers had strongly stated his interests in flying payloads that fit Giuli's description. He had already, at the end March 1976, with Opal and Conway, proposed to enlarge and modify their sounding rocket Schmidt camera/spectrograph for use on Spacelab in the 1980–1982 timeframe. It included a closed-circuit TV system that the payload specialist would use to home in on the desired fields. The payload specialist would also monitor critical electrical components and replace filters, gratings, and other components to make the system capable of performing a wide range of observations. As Carruthers had pointed out, "The sounding rocket instrument uses only a single fixed optical configuration, but in Spacelab missions the flexibility of interchangeable configurations is highly desirable."[40]

Carruthers never stated explicitly that he would prefer to be a payload specialist, but he made it pretty obvious. His constant reference to specific problem areas and training requirements the payload specialist needed to address, and his frequent reference to proving the feasibility of his designs and making them compatible with Spacelab, should have made that fact obvious. He knew well that the first Spacelab flights would be engineering test missions, and so he emphasized this as his primary goal: converting and testing his camera to be useful on Spacelab.

One might well suspect that Carruthers's complex interests and ongoing desire to continue his research mitigated his prospects to be a mission specialist. Further, specialists were informally assumed to be career NASA employees. Carruthers was not in this category.

## THE OUTCOME

Carruthers passed the preliminary exams, which got him as far as semifinalist, but in the second or third rounds of testing, as one source indicated, a "medical issue disqualified him."[41] Indeed, although his medical review was redacted in a copy of his restricted file released by NASA under the author's FOIA request, it did not delete Carruthers's lengthy response to the decision.

In his original proposal, Carruthers claimed that he had no "known health problems and currently use[d] no medications." However, ophthalmic testing revealed a condition that made him unacceptable as an astronaut. He knew of this condition, commonly known as "lazy eye," diagnosed by his own doctor some ten years prior as "amblyopia." His vision in his left eye deviated from that of his right eye. In his rebuttal, he stated that when he used just his left eye, he did not see any difference in "image quality and sharpness of detail" but then added, frankly, "Yet I cannot read print of usual size with the left eye alone."[42] But he countered that his peripheral vision "was equally good in both eyes" and that his left eye improves under low-light-level conditions. He could still detect fine detail, "particularly isolated details," and, moreover, he had never sensed any difficulty in perception of depth, except in "artificial test situations" like ophthalmic procedures. Carruthers defiantly demanded, "If I am automatically disqualified because of the left eye deficiency, I would be interested in knowing the reasons for the requirement of 20-20 vision in both eyes."[43] But the decision was made. And, in fact, it was not a unique call. At least two other scientists who were semifinalists in Group VIII were rejected for the same reason.[44]

Carruthers spent over a week at JSC, and his presence was advertised (see figure 11.1). But it must have been a stressful time for Sandra because it might change their lives for years, with long separations. And Carruthers also made clear in his application that he wanted to remain active at NRL

Figure 11.1
NASA publicity photo showing Carruthers and a model of the Space Shuttle during his week of testing and evaluation at the Johnson Space Center. NASA JSC s77-28828. *Source:* Courtesy Carruthers Collection, NASM Archives. GC/NASM.

somehow, even remotely—so further limiting his attention to Sandra. Carruthers took the eventual rejection (first by phone and then telegram) stoically, just as he would NASA's increasing rejections for payloads later in the 1980s and 1990s. But his inner feelings are not known, though as we have seen, prompted dark humor at times.[45]

## BUT WHY DID HE APPLY?

Why Carruthers applied bears deeper examination. As we have noted, when he applied in 1976 and visited Houston, Carruthers already had his hands full with his many continuing projects and responsibilities, including his wife, who in the late 1970s was becoming a concern (see chapter 14). If he were selected, it would mean many weeks and months away from home, and up to a year in active training. But as Karl Henize stated bluntly in his commentary supporting Carruthers, he "has been strongly motivated to be an astronaut ever since I have known him."[46]

One can well imagine that given his expressed interests in space since childhood, Carruthers's motivation was rekindled when the requirements were liberalized. He was sensitized in his contacts with Henize, and it is also possible that others he was in contact with, like Roscoe Monroe, the NASA EEO director who had worked with Francis Redhead to promote events featuring Carruthers, would welcome his selection to be an astronaut. Even Friedman would have been sympathetic; it would promote visibility for NRL's scientists and their programs. Carruthers's motivations were far more focused, but they aligned with a young Black candidate who saw in space travel, in the words of Lynn Spigel, "a utopian hope that space will release her from earthly burdens . . ." like housing discrimination.[47] Other than his Aerojet episode, Carruthers did not face those kinds of problems, but, as with his trips to White Sands, it could have been the ultimate escape for him.

Carruthers, however, did not want to sever connections to NRL, and Sandra must have had her own worries since the separation could last on and off for several years if he was selected. It is still not surprising that he applied, again given his close contact with Karl Henize and their joint proposals for shuttle payloads. But, though evidence is lacking, it must have left him, and those close to him whom he consulted (if anyone), with mixed feelings.

Beyond his personal passions, he likely felt encouraged to apply when he was invited to join various subcommittees and panels of the National Academy of Sciences' Space Science Board that exposed him to growing issues surrounding NASA's treatment of science.[48] In late 1974, he became part of an assessment effort, starting in November and lasting through March 1975, of how science could be better treated and made more useful on shuttle missions. And after all, as humorously noted above in retrospect, the NASA "scientist-astronaut" was regarded as a test object below "chimpanzees." Also, their duties and relation to the other astronauts were far from clear. And few who were selected actually flew.[49]

One of the panels Carruthers joined conducted a space shuttle passenger study that would "review the effectiveness and value of NASA's scientist-astronaut program in the past, and to make recommendations for the future."[50] Their mission was to advise, with the Space Shuttle on the horizon, on what types of astronauts were required to manage it as a "Space Transportation System." And especially, how could science be better served and serve the mission? The panel collected testimony from a wide range of people and recommended that "NASA continue and expand the scientist-astronaut corps in consonance with space shuttle science and application needs." The scientist-astronaut, however, would be recast as a mission specialist: "that crew member serving as coordinator of overall orbiter payload operations."[51]

After attending early meetings of the panel and reviewing minutes and draft conclusions in February 1975, Carruthers seemed more concerned with describing the requirements for, and duties of, a "Payload Specialist." At one point he reflected Giuli's view, that "it will be very rare that a Payload Specialist will fly on the shuttle to operate only 'his' experiment." And further, he also felt that the payload specialist should be connected to the NASA center most involved in the specific payloads on board.[52] But, understandably, he also argued that payload specialists had to be selected who would be "the most qualified" not only to operate the instruments but also to adjust them and even repair them in orbit, and "would not necessarily be a NASA employee, but would be one of the investigators on the particular payload."

The panel report included Carruthers's commentary as an appendix. Though impassioned, it was also rather confusing, making conflicting suggestions. One can only guess that this odd behavior, for someone who normally prepared and delivered highly focused and coherent arguments, is an indication of the complexity of his thinking on the subject. His laboratory work at NRL was at the very center of his being, and he could not imagine being separated from it. And we again speculate that he must have been concerned that this would separate him from his wife for extended periods of time.

Carruthers's insistence that he remain connected to NRL is not surprising. At the least, as we noted before, when his number came up for the draft, he knew how critical he was to the health of his programs at NRL. He knew that he was considered an irreplaceable employee at NRL at the time by his colleagues and bosses. However, NRL was supportive of having one of its own in the astronaut corps, as it was with physicist Dianne Prinz and astrophysicist David Bartoe, who applied in the same round as Carruthers and eventually became, indeed, Navy payload specialists planning to conduct solar observations.[53] And resistance to losing Carruthers to the draft did not mean that other more stimulating appointments would not have been welcomed.

This episode in Carruthers's life and career has no simple summation, but it has revealed in dramatic fashion the depths of Carruthers's passions for spaceflight, and for his science. In this instance, there was considerable conflict of interest by all parties involved: in NASA's policies and priorities for who and what to fly, in Friedman's arguments on the one hand to keep him close at NRL, and on the other, to see him fly, and finally, in a deeper sense, within the man himself, to the extent that we can speculate, who passionately pursued spaceflight and space science in all possible ways.

## PROPOSALS FOR SPARTAN—SUCCESS AT LAST

One long-term proposal that did eventually bear fruit, but only after two stages of rejection, was likely an outgrowth of Henize's initial efforts to mount the large pallet-borne STARLAB telescope in the shuttle bay for a wide-field survey. Henize's continued lobbying with astronomers pushed NASA to establish yet another working group in 1982 "to assess the scientific need and objectives for such a telescope, and to study its feasibility and implementation."[54] Goddard Space Flight Center contracted with the Perkin-Elmer Corporation to do a feasibility study, confining the program to a single shuttle orbiter pallet. The now 30-inch aperture Schmidt would have at least a 5-degree field captured by a large electronographic camera. The goal was to produce a survey of the UV sky to at least twenty-seventh magnitude, some twenty-one magnitudes (~250 million times) fainter than the eye can see on the darkest of nights.

Carruthers joined Henize proposing wide-field, large-format electrographic shuttle pallet-based cameras that were a variation of his usual design, this time most significantly employing a semitransparent photocathode. In a draft statement for a symposium in London in September 1978, he predicted that it would be useful for a wide array of missions in addition to STARLAB, such as the proposed Spacelab wide-angle telescope and other variations. His design could be modified to be sensitive to either the far ultraviolet, the middle ultraviolet between 2,000 and 3,000 Angstroms, or the near visible, from 3,000 to 10,000 Angstroms. As always, he hoped his inventive hammer would reach all possible nails.[55]

Henize's persistence and dream of an all-sky survey was never realized. His smaller instruments were flown on Gemini and *Skylab*. He finally flew as a mission specialist on *Spacelab-2* ([Space Transportion System] *STS-51-F*), managing an all-pallet array of some thirteen experiments. There was an all-sky survey instrument, but it was for the infrared. None were Henize's, whose major contribution was to operate a new instrument pointing system and manage the experiments.[56]

Carruthers, however, remained focused on the idea and took the lead. When the Henize proposal failed after several years of dickering, Carruthers reproposed a scaled-down version, again with Henize, along with Opal, Heckathorn, and Adolf Witt of the University of Toledo. Instead of a large pallet system, they would propose a smaller system as one of several payloads on a free-flying SPARTAN. As described in "Ultraviolet Wide Field Imaging and Photometry: SPARTAN-202 Mark II Far Ultraviolet Camera," the system would be "similar to but larger (and having higher resolution and point-source sensitivity) than similar cameras we have used in numerous other space flight experiments." Called the Mark II to distinguish it from the S-201, this Schmidt would have a 6-inch aperture operating at f/2 and could reach stars five magnitudes (100X) fainter than S-201 (see figure 11.2). Versions of the Mark II had already flown some five times on sounding rockets, and the design was firmly set. As he would on several occasions and in keeping with astronomical practice, Carruthers promoted this survey instrument to prepare for the recently renamed *Hubble Space*

Figure 11.2
By the mid-1980s, Carruthers was mentoring students, giving them hands-on experience developing and testing his payloads. Here, (r to l) Donna Stockton and William Glascoe from the University of Maryland are with the SPARTAN payload. Photograph P-2513(7). *Source:* Courtesy of Naval Research Laboratory.

*Telescope*, using it to identify potentially interesting targets for the telescope and to calibrate its sensors.[57]

He proposed an initial flight for testing, to be followed by berths on "Spartan Space Station or Space Station programs" for full sky surveys. By then the camera had also been tested by Harry Heckathorn at the University of Texas McDonald Observatory.[58] Ground-based observing was, in fact, one of Heckathorn' s contributions to the team, given his training and background.[59]

Like Carruthers and his contemporaries at NRL, Heckathorn built telescopes in high school and was attracted to physics and then astronomy after he joined an astronomy club in his hometown of Edina, a suburb of Minneapolis.[60] Heckathorn came to NRL with practical knowledge of spectroscopy and electronography based upon his PhD research experience at Northwestern using a fiber optic–fed electrostatically focused Westinghouse image tube to study galaxies. He moved to MSC's Astrophysics Section on a joint postdoctoral appointment with the University of Houston, continuing this general line of research on galaxies with Yoji Kondo at NASA and Guido Chincarini, now at the University of Texas, using a Kron-type camera at McDonald Observatory. Collaborating with Chincarini exposed Heckathorn further to electronographic techniques.

In 1975, Heckathorn was invited to give a colloquium at NRL. He was soon offered a senior postdoctoral appointment and then a staff appointment, taking an office next to Carruthers in September 1976.[61] At NRL, engaging in rocket flights to observe nebulae (star-forming regions and galaxies) with Carruthers and Opal, Heckathorn also continued to experiment with what he called "electrographic detectors," even though Carruthers preferred "electronographic." They were the same technology, but it was a choice in how one interpreted the elements of the technique, involving for some, "electrons," and for others, "photography." He teamed up with Carruthers, Opal, and others to make comparative tests of different designs. He participated in rocket flights and in SPARTAN missions and assisted in the reduction of observations, including the *Apollo S-201* data with Page. He also continued to explore microchannel plate intensified electrography and by the mid-1980s, developed computer programs to simulate, calibrate, and delineate the structure of the images, making them far more accessible to analysis.

With the modifications made possible by Heckathorn's expertise, Carruthers submitted numerous proposals and redesigns for SPARTAN flights through the 1980s and was poised to fly an instrument on *SPARTAN-C* in December 1986. But then the *Challenger* disaster quashed all hopes for years. What Carruthers and his team finally flew from his years-long campaign was a small set of pallet-mounted Schmidt cameras on STS-39 aboard the Space Shuttle *Discovery* in April 1991. It was the Department of Defense's (DOD) first unclassified mission. Carruthers's shuttle bay payload, one of about a half dozen on the Air Force pallet, had goals similar to his *Apollo 16* instrument, such as upper atmospheric phenomena and a series of wide-angle far-ultraviolet fields of various nebulae. A scaled-down version of Henize's deep all-sky survey, Carruthers' cameras also recorded twelve 20-degree star fields, including the Magellanic Clouds. Of particular note, they recorded "vast interstellar clouds" around the head of Scorpius, the Scorpion.[62] During the mission, the cameras also made observations for

DOD to help better understand the shuttle's environment in orbit and how it influenced the outer atmosphere, from the infrared to the far ultraviolet.[63] The instruments were funded jointly by NRL, the Air Force Space Test Program, and various corporations. As Carruthers ironically explained in his 1992 NRL Review,

> A need exists to determine the natural background against which targets of DoD interest would have to be detected. Also, a need exists to directly measure artificial sources of potential DoD interest in the near-Earth space environment.[64]

By then, Carruthers had expanded the scope and nature of his support staff. Even though his formal NRL staff dwindled, through his outreach efforts Carruthers now had helping hands he had once nurtured and was now carefully mentoring in his laboratory to further his many interests.

### STUDENT ASSISTANCE -1-

As we will cover in greater detail in a following chapter and in appendix A, by the late 1980s, students from the DOD-sponsored Science and Engineering Apprenticeship Program (SEAP), which Carruthers helped to establish, were aiding and assisting Carruthers and his coworkers at NRL and gaining real work experience in his laboratories. This program supported high-school students and university-sponsored co-op students in a wide variety of activities in the 1980s, from recruiting students for summer work at NRL to participating in and producing a range of educational videos. Among early students who worked in Carruthers's laboratory were Donna Stockton (a co-op graduate student from the University of Maryland) in the summers of 1984 and 1985 and William Glascoe (then a cadet at the US Air Force Academy) in the summers of 1985 and 1987 (see appendix A). In his "Final Report for 1985," Glascoe, who had also worked on *SPARTAN-202* with Stockton, described the "High Resolution Shuttle Glow Spectrograph" (HRSGS) as "intended to measure the Shuttle's mysterious glow in the visible spectrum." Glascoe's role was to "give Dr. Carruthers and Donna" a detailed listing of the types of atoms and molecules that one expected to be present at the shuttle's orbital height. He and Stockton did far more than search the literature. They also set up gas discharge lamps and electric torch flames and other hot sources in the laboratory and examined them with prototypes of their cameras.[65]

Carruthers was anxious to get Glascoe back in the summer of 1989, enthusiastically reporting that the "High Resolution Shuttle Glow Spectrograph project is now moving full steam ahead. . . . This summer we hope to put it all together and try it out!" But that was not all. If he joined them, Glascoe's summer would be filled with other projects: another UV camera payload and the Far-Ultraviolet Imaging Spectrograph for *SPARTAN-281*.[66] During their summers working on projects like these, Stockton and Glascoe knew that there was no guarantee of when the camera would fly, which was all dependent upon funding priorities and politics. It was a practical real-world lesson into the realities of space science.

Carruthers's payload finally flew in *SPARTAN-204* on the Shuttle *Discovery* (*STS-63*), launched on February 7, 1995. It was also the second "Shuttle-Mir" mission and the first to be piloted by a woman, Eileen Collins. Carruthers's instrument was the sole occupant of SPARTAN, and it was deployed by Bernard A. Harris, a Black mission specialist. Supported by the US Navy and Air Force, what Carruthers called "FUVIS" (for "Far-Ultraviolet Imaging Spectrograph") had originally been proposed for the *SPARTAN-202* flight in 1984, which slipped constantly, and as with all other shuttle payloads, it was delayed by the *Challenger* disaster in January 1986. As Nancy Roman reminisced, "That catastrophe cancelled all shuttle launches for three years" and delayed everything, including the *Hubble Space Telescope*.[67] Carruthers provided a detailed report on the mission for the *NRL Review* as well as a constant stream of talks and papers from 1988 to 1996. In the NRL report, he stressed that more wide-field surveys were needed for DoD applications "to directly measure artificial sources of potential DoD interest in the near-Earth space environment."[68]

## WITHER ASTRONOMY?

*Challenger* deeply affected Carruthers' work, and also his personal feelings for the astronauts and for his own fate. In December 1985, just before *Challenger* disintegrated, Carruthers was in Houston working on one of his planned SPARTAN payloads, helping to determine how best to integrate the camera into its mounting and train the astronauts in its use for a later flight. When in Houston, he occupied an office that was temporarily vacated by the teacher-astronaut Christa McAuliffe, since she was then at Cape Canaveral. When *Challenger* was destroyed and its crew killed, Carruthers had just returned to

Washington, but it hit him hard: "I will always remember being in Christa McAuliffe's office two days before she flew."[69]

Even before *Challenger*, NASA's "overly optimistic" promise of a cost-effective and efficient route to space was a deep concern for many; the program was definitely not meeting the needs of astronomers, who now were limited to piggybacks on shuttle missions. Just a few of the frustrations: Increasing competition for access, requiring longer proposal and preparation times, and orbits lower than some experiments needed. Hardly any of the recommendations of task groups and surveys sponsored by the National Academy of Sciences, like the Field Committee in the late 1970s, were met yet. Thus, the frustrations and delays Carruthers faced were hardly unique.[70]

By the mid-1990s, however, though funding still squeezed new projects, the climate for space science and astronomy was brightening. The *Hubble Space Telescope*, anticipated for decades, finally launched by the shuttle in 1990 and repaired in the first shuttle-servicing mission in 1993, was now observing the universe with great fanfare. The dual giant 394-inch Keck telescopes were operational in Hawaii, and the *IUE* was still working beautifully. And most critically, the ASTRO (Autonomous Space Transport Robotic Operations) missions on the shuttle, yet another variation of the shuttle payload universe, used the full pallet to hold a set of large fully pointed UV telescopes and spectrographs that were gaining much positive attention.[71]

Carruthers was not part of ASTRO because, directed by Navy priorities, he was preoccupied now assessing shuttle glow. After many modified proposals with evermore lengthy detail, Carruthers teamed up (possibly prompted by NASA) with Reginald J. Dufour of Rice University, John C. Raymond of the Harvard/Smithsonian Center for Astrophysics, and Adolf Witt of University of Toledo to repeat and extend the study of how the shuttle influences the space around it as it travels. They prepared and flew a modified HRSGS (HRSGS-A) on SPARTAN along with FUVIS on *STS-63* in 1995.

These instruments, as before, were designed to record both natural and artificially induced sources of visual and UV radiation. The results of these were paired to assess the artificial contribution, definitely a Navy interest. The natural sources were nebulae, supernovae remnants, and the diffuse galactic background radiation. And the artificial sources were, beyond Carruthers's shuttle glow studies, for gaining experience in the ability to monitor ballistic missile signatures. It was a wide-field but low-resolution device. But not the wide-field high-definition deep survey that Henize and Carruthers had fought to build and fly for so long.[72] It was reported to have had a nominal flight lasting some forty hours and was successfully retrieved. However, no publishable data were reported, either in a NASA database or in Carruthers's subsequent open publications. Carruthers did speak on FUVIS results at the 186th Annual Meeting of the American Astronomical Society in Pittsburgh in 1995.[73]

Throughout these years, Carruthers and Henize had teamed up a good number of times instigating cooperative multi-institutional ventures for shuttle missions. After retiring from the Astronaut Corps in 1986, Henize continued to work at JSC, following his interests in wide-field surveys, this time turning his attention from the heavens to near-space, estimating just how much orbital debris was accumulating in the region from low-Earth orbit up to geosynchronous altitudes. This collaboration ended, sadly, when Henize died from a pulmonary edema in 1993 while trying to meet another challenge—conquering Mount Everest. For Henize, Everest was both a challenge and a mission: he carried passive sensors on his body to monitor how his skin reacted to the high-altitude conditions. His death was a loss for his family, for astronomy, and definitely for Carruthers.[74]

# 12

# CHANGES IN MANAGEMENT AND TECHNOLOGY

## FRIEDMAN'S RETIREMENT

In his oral history with the author in 1992, Carruthers neatly summed up his professional history after he achieved permanent status at NRL in 1966:

> From that point on, there were no major changes in direction in the sense that we continued to further develop the technology that was used in the sounding rockets which led to the *Apollo 16* camera and which continues today with our recent shuttle flights. The only major change is that in the last few years, we have been going away from the film recording devices to charge coupled devices, CCDs. But the basic technology is still pretty much the same as we developed in the mid-sixties.[1]

Indeed, by the 1990s, solid-state detectors like the charge-coupled device (CCD) had improved to the point that they were replacing film recording. As already noted and as explored in the next section, they were small and lightweight, they required less power, and their data could be telemetered. Improvements including larger collecting areas, higher resolution, and greater sensitivity over a wide range of wavelengths made them especially attractive for many applications. Although Carruthers was willing to replace film return with electronic sensors of various types that were electron sensitive, he remained reluctant to dispense with his vacuum tube–based electronographic amplification systems, using them as the initial imagers and amplifiers.

Although there was consistency in his research style and direction, the world around him, and at NRL, was certainly changing. Herbert Friedman retired in December 1981, after what had "been a difficult year," he reported to a colleague. Arthritis and double hip replacement surgery that year just knocked him out. Friedman remained as consultant "emeritus" a few days a week. Most of his focus, however, turned to the National Academy of Sciences, where he had long been an active member.[2]

Friedman's retirement stimulated and also reflected changes in the working atmosphere at NRL due in part to evermore complex relationships with NASA, especially in the Space Division. In an interview with historian Richard Hirsh, Friedman described working with NASA at first as cordial, but:

> As time went on, of course, space science became big science, and you just cannot spend tens of millions of dollar[s] on a mission [and] not have it organized in a rigorous engineering [way] with full accountability for everything you do. So I guess I was becoming sort of old fashioned and out of tune with the times, by continuing to do what was a kind of individualistic . . . program here. Also, in the first decade, if you liked to compete, it was the right kind of environment. We enjoyed it, and we succeeded. We had a very high success ratio in getting funded for what we proposed. Now, as time went on, it all became much more complex.[3]

And in a manner that poignantly reflects Carruthers's experiences, and in some ways may well reflect the growing rate of rejections in the 1980s and 1990s:

> The program became much more expensive, and NASA established rules for proposals, rules for review, conflict of interest positions, and one of the things that has bothered me for years is that you don't have the straightforward opportunity to come before the selection committee and answer their criticisms face to face or criticize your competitor's approach and debate with him before a committee. Everything is done sort of second hand.[4]

His greatest frustration was that NRL "was treated, I think, somewhat shabbily in the non-solar areas." Most of the NRL staff who migrated to NASA were from the space science section at NRL and were certainly anxious to continue in that field. UV missions, Friedman felt, were therefore "not offered to an open competition." In retrospect, overstating the problem a bit but reflecting his early frustration with lack of access to the three missions in the OAO program, Friedman felt that: "We never had a chance to bid on any of those, even though we made the only ultraviolet photometric measurements preceding them in the rocket program. So as a result, we never pushed to get into the NASA ultraviolet rocket astronomy program."[5]

NRL of course was not the only team flying sounding rockets prior to the NASA years, but it was the largest and most visible. In sum, his remarks clearly express how NRL perceived the effect of Nancy Roman's negative opinion of NRL science.

In Friedman's wake, the transition was not easy. Who could succeed him? The competition seemed to be between Phil Mange and Talbott Chubb to become interim superintendent. Chubb had technical expertise but was less motivated by management responsibilities. Even though Mange had been at NRL since 1960, section head for aeronomy for a decade and then associate superintendent for an equally long time under Friedman, he feared that he lacked the technical knowledge to prioritize decisions across the many disciplines the superintendent had to deal with.[6] He was trained as a theorist at Penn State and was a student of the famous Belgian physicist Marcel Nicolet.

The third candidate was Herbert Gursky, then at the Center for Astrophysics at Harvard. Gursky fit both the technical and managerial requirements and felt ready to face all the challenges Friedman had described, both internal and external, heading the NRL Space Division. He was a man of considerable energy and productivity, and he was respected for his views on gender equality.[7] Born in the Bronx in 1930, he received his PhD in physics from Princeton in 1958 and by 1961 was working in the Space Research Division of American Science & Engineering, Inc. (AS&E), a recently formed consulting group in Cambridge, Massachusetts, that concentrated on X-ray detection technologies. X-ray astronomer Riccardo Giacconi and cosmic-ray physicist Bruno Rossi led the company, and the three of them collaborated on projects in X-ray astronomy, establishing partnerships with MIT and at major observatories to fly X-ray sounding rocket payloads. Giacconi and his crew attained great notoriety, and eventually a Nobel Prize for him, from an Aerobee flight in June 1962, supported by the US Air Force, that detected the first nonsolar X-ray sources.[8] The culmination of their work at AS&E was the dedicated X-ray satellite *Uhuru*, launched in 1970.

In that year, Gursky followed Giacconi to the Smithsonian Astrophysical Observatory in Cambridge, where he held increasingly more senior positions, ending up as Harvard professor in the practice of astronomy and associate director of the Optical and Infrared Astronomy Division. He was an aggressive and capable builder. Like Friedman, Gursky was an X-ray specialist and a proponent of new technologies, especially solid-state detectors for a wide range of applications. By the time he arrived at NRL in

1981, he had written seven papers either exploring or utilizing the CCD. He came to NRL as an agent of change.

One of his first actions was to reorganize. Mange reported to Gursky as associate superintendent. Gursky was a builder and wanted more assistance and expertise in engineering. He asked Mange to come up with a solution, but Mange recalled he failed and Gursky felt he "hadn't done what he wanted me to do." So Mange took a position at the Office of Naval Research (ONR), but after a year, Gursky called him back to work as liaison to other Navy laboratories and sources of funding, especially ONR.[9]

Through the 1980s and especially after the *Challenger* disaster in 1986, NASA's support continued to dwindle, and ONR funds were stagnant and diminished by inflation.[10] It was a time of retrenchment for space astronomy at NRL. When Carruthers's proposals failed to gain sufficient support, his staff, including Heckathorn and others, were either reassigned or left altogether. NRL's UV group was hit the worst.

For example, as Carruthers's revenue from grant overhead dwindled, Heckathorn's abilities at computer simulation made him more attractive to other projects. So Heckathorn was reassigned to develop a proposal for a strategic scene generator model, a computer program that could identify images that missile defense sensors would categorize as "battlespace targets and backgrounds."[11] These included, most poignantly, the ability to detect and recognize rocket plumes of missiles headed toward the United States, an application of Carruthers's earlier efforts detecting the UV signatures of missile launches.

Accordingly, as the new head of the division in early 1980s, Gursky approached Heckathorn, saying, "Harry, I want you to get involved in . . . the SDIO, the Strategic Defense Initiative Organization." Heckathorn agreed to the transfer, but only after he had finished his observing in Texas for Carruthers's project. He was not unhappy with the transfer because it seemed exciting, and he was allowed to continue assisting in data reduction from Carruthers's payloads. But the reason, he knew, was that by then, despite his continued efforts, Carruthers was just not bringing in enough funds to support his staff: "He was in tough straits, because his payloads were sitting on the shelf."[12] And sadly, Chet Opal, who had continued to work with Carruthers through the 1970s and the 1980s, still actively collaborating after he moved to the University of Texas McDonald Observatory in 1984, died in 1991 at the age of forty-eight after a long illness. Opal was essential over the years helping Carruthers bring in money through their numerous grant proposals. Losing Opal and Heckathorn, among others, limited Carruthers's proposal production in the latter part of the decade, a fact that reduced his effectiveness and prompted division managers to question his future.

## ADOPTING THE CCD AND MANAGING TECHNOLOGICAL OBSOLESCENCE

A 1988 National Research Council report by the blue ribbon Naval Studies Board Panel on Research Opportunities in Astronomy and Astrophysics highlighted the growing problem for UV astronomy overall, especially in the wake of *Challenger*:

> With little or no access to space available for the next several years, the emphasis of the NRL's ultraviolet astronomy program should now be focused on the development of improved instruments for the next decade. The group's work on the enhancement of the ultraviolet sensitivity of charged coupled devices (CCDs) through ion implantation is a good example of the sort of programs that could lead to significant advances in ultraviolet astronomy in the future. This effort should be continued and expanded.[13]

This view certainly called into question the future of Carruthers's electronographic detectors, but he was still developing them and, sporadically, they were flying. And by 1991, there were larger issues: continuing support had become critical. Another National Research Council report, "ONR Research Opportunities in Astronomy and Astrophysics," pointed out bluntly:

> The aftermath of the *Challenger* disaster has led to the cancellation of the SPARTAN program, with the result that the NRL's ultraviolet astronomy program has been severely curtailed. Other small instruments, with modest astronomical capabilities, have also not yet been flown aboard the shuttle, though the launch of a pair of small, low-resolution, ultraviolet cameras employing film as the recording medium is imminent.

And overall:

> The lack of an effective route to space for NRL instruments in recent years has been a major obstacle to the success of the ultraviolet program at the NRL.[14]

The NRC panel, consisting of academic physicists and astronomers from universities, NSF, and NASA research institutes, concluded that NRL should find ways to hire younger staff to increase its effectiveness in establishing collaborative and supportable missions with university-based research groups, using new technologies. The report, in fact, singled out UV astronomy as the major problem area. The X-ray, gamma ray, infrared, and the Naval Observatory's Radio/Optical groups seemed to be maintaining vital and attractive programs. And beyond acknowledging the importance of detector development in the infrared, UV, X-ray, and gamma ray regions explicitly, it also advocated pursuing solid-state arrays like matrixed panels of CCDs for wide-angle surveys for both scientific and military application. It placed NRL "at the forefront" of a revolution for studying the infrared:

> A revolution is occurring in the development and application of these devices, with rapid advances not only in the number of detector elements (now exceeding 60,000 elements per array), but also in the sensitivity of each detector. Such developments significantly enhance the rate of acquisition and the quality of infrared observations and will have major impact on the Navy's mission for target surveillance, detection, and identification and for guidance.[15]

But no such comment was made for the ultraviolet, other than to suggest that NRL collaborate with other groups to develop and use solid-state arrays.[16]

This was a priority that Gursky heavily supported. Even when he was at The Smithsonian Astrophysical Observatory (SAO) in the 1970s, as astronomer David Latham remembered, "Herb Gursky was insistent that we had to move to CCD technology because he recognized it as the future. Nobody really argued with him, that's for sure. . . . He was also always pushing to apply new technologies, like computers and complex software systems."[17] And as historians have recounted, CCDs became the detector of choice over the vacuum tube vidicon for the Large Space Telescope's (later called Hubble) cameras in 1977 even though there were many technical hurdles to overcome, like the spectral sensitivity range, the resolution, and the effective field of view. Soon, even the most active developer of image tube technologies for astronomy in the United States, the Carnegie Image Tube Committee, had effectively ceased by then, for astronomical applications at least.[18]

Despite their limitations in the mid-1980s, including small collecting area, low resolution, and low UV sensitivity, CCDs showed great promise. So as Carruthers and his staff experimented with ways to improve his electronographic detectors, they included CCD technology as a recording medium to replace the film in his electronographic cameras. As early as August 1977, they submitted a proposal to NASA titled, generically, "Far Ultraviolet Imaging and Photometry" that would survey both natural and artificial sources of emission phenomena in "near-Earth space," which again included reconnaissance of rocket exhaust plumes, explicitly stating that Air Force P72-1 flight measurements "first indicated the potential for surveillance applications." To facilitate the latter, they also proposed to include "television readout versions by incorporating charge-coupled array devices."[19]

His electronic amplification system remained intact, however, as the central ingredient of the design. But CCDs were not Carruthers's forte. How he learned to adapt them, but not replace his own technology, contrasts with the reactions of most astronomers who made the transition completely. As Samantha Thompson has shown, how astronomers made the transition depended upon their professional priorities: problem-based or instrument-based.[20] As we know, Carruthers was most certainly in the latter category, but characteristically, he was not reluctant to promote the new technology, especially to those he felt were the future of the profession: his students.[21] Following his path, not to transition but to accommodation, will give us a deeper appreciation of the complexities of an important historical theme: how old technologies are replaced or adapted. A familiar story in both history and fiction.[22]

To use CCDs more effectively, and in fact to become comfortable applying them, Carruthers sought the advice of John L. Lowrance, a Princeton detector specialist. Lowrance had also been a vacuum tube devotee; he supplied SEC vidicons for Lyman Spitzer's failed bid to build *Hubble's* Wide-Field Planetary Camera. So recognizing the obsolescence of his technology, Lowrance was now exploring ways to improve "electron-bombardment CCD arrays and opaque alkali-halide photocathodes" for vacuum UV applications. These detectors promised 100 percent efficiency, with single photoelectron detection. With Lawrence's advice, Carruthers, with Opal's and Heckathorn's assistance, experimented with replacing the film cassette with CCD detectors, proposing to use them with his electronic amplifier on robotic missions where physical retrieval was not possible.[23]

Carruthers continued to explore other options to improve, as his colleagues well knew, his first love, vacuum tube technologies. But he also looked for ways to test the powers of the new technology compared to his own. The convincing test was serendipitous. Carruthers had asked Heckathorn to team up with Opal, then on sabbatical at the University of Texas, to have a look at another detector design that its inventor at McDonald Observatory called "electrographic."[24] Since the early 1970s, Paul J. Griboval, a "special research associate" in the university's department of astronomy, had been experimenting with visual electrographic detector systems for McDonald and by 1979 built a "high resolution and sensitivity detector for astronomy, yet reliable and easy and fast to operate" with "a 5 cm useful field diameter."[25] So to get familiar with, and of course to evaluate, the Griboval camera's capabilities, at Carruthers's direction, Heckathorn and Opal requested observing time with McDonald's 84-inch telescope. Heckathorn was already familiar with the big telescope there, as he and Guido Chincarini had been using it with a Kron electrographic camera to image extended objects like galaxies and nebulae. They justified the new effort with the Griboval to continue their search (as we reviewed earlier) for faint "very very blue" visual counterparts to UV objects in the S-201 fields that needed confirmation and correlation.[26]

So in 1981, Heckathorn and Opal organized an observing expedition with astronomers at McDonald, taking the Griboval camera to the best observing site in the southern hemisphere, the Cerro Tololo Inter-American Observatory in Chile. Their observing sessions at the observatory were once again justified by the continuing need to confirm the nature of "uncatalogued stellar objects" that were detected with the S-201 camera.[27] They used the Griboval on a 40-inch reflector (for its wider field) to find the objects, and then a spectrograph with a vidicon detector on the giant Cerro Tololo 160-inch telescope (now known as the Victor M. Blanco Telescope) to examine them more closely.[28] They did identify more objects from the S-201 catalogue, but their effort also led to a serendipitous test, pitting the Griboval against the capabilities of the CCD, which brings us back to that thread.

At the observatory, they also collaborated with staff astronomer Patrick Seitzer on the Mayall telescope scrutinizing the S-201 objects. Seitzer, as it turned out, was using an RCA CCD camera, so Heckathorn and Opal suggested they compare the performance of CCDs on the Blanco with the Griboval on the 40-inch. They observed the same objects simultaneously to see which detector performed better. They chose a field of stars near the bright globular cluster Ω (Omega) Centauri.

Compensating for the different capabilities of the Blanco and the 40-inch, they concluded that the CCDs had two to three times the sensitivity, but Griboval's electrographic camera had far greater resolution and field of view. They felt that "the overall figure of merit for the GC [Griboval Camera] may be better than the CCD if the areal advantage is sufficiently important." But the significant advantage of the CCD was "that it is a photometric device; that is, once calibrated it is stable and reproducible and can be used to make absolute measurements."[29] In sympathy with the 1965 predication by Gordon Moore, cofounder of Intel, that the powers of solid-state microchip technologies would double every two years (known as Moore's Law), they knew that CCDs would become larger and of higher resolution in the near future. And by the 1980s, they knew too that there was good progress improving their UV sensitivity.[30] But they hoped that film could be improved too, so (possibly in deference to Carruthers's investment in his camera and aware of Griboval's hopes to enlarge his camera field) they predicted that "in the foreseeable future it will always be easier to build a larger electrographic camera than it will be to build a larger CCD chip, so the areal advantage of the electrographic camera is likely to remain."[31]

CCDs did get larger and larger, and there were ingenious efforts to connect them electronically in a matrix (as in the *Hubble Space Telescope's* Wide Field/Planetary Camera), but only when their electronic support systems could be miniaturized did it become feasible to build contiguous arrays of CCDs to increase field sizes comparable to, and exceeding, photography.[32] So the equivocal conclusions reached by Heckathorn et al. were not surprising. CCDs were emerging, but to devotees of vacuum tube technologies like Carruthers, there was still life for the electronographic camera, given his investment in the art and craft of applied plasma technologies.[33]

As it turns out, curiously, Heckathorn and Opal never attempted to compare Carruthers's and Griboval's cameras because they were designed for completely different wavelength regions, and thus, their cathodes were very different. Like the Kron and Lallemand, Griboval's cathodes were transmissive.[34] Carruthers, however, concerned and not complacent, continually looked at other options to keep his cameras competitive. He and Heckathorn, in collaboration with the Marshall Space Flight Center, experimented with ways to intensify electrographic film negatives after they were processed to improve signal-to-noise ratio.[35] To increase sensitivity, Carruthers also added microchannel plates to his proposal for the AMPS payload on the shuttle in the mid-1970s.

There were just so many options, and the different technologies demanded new talents and training. But Heckathorn felt that the writing was on the wall. The increase in resolution and size of the chips even then posed "the death knell for the electrographic camera."[36] By the mid-1980s, Carruthers reluctantly agreed but, as we noted and based upon John Lowrance's advice, continued to propose ways to combine its powers with his detector to further enhance sensitivity. His conversion was slow, and ultimately incomplete, but he was always looking for new talent in the hope of finding a path for a combined technology.

When Tim Seeley came on board in 1984 as a half-time co-op student from the University of Maryland, he was assigned to Carruthers's group to gain experience with detector technologies. Seeley found Carruthers "just a very pleasant person, he just seemed very eager to share his time and his knowledge."[37] Then a student of electrical engineering at Maryland, he worked with Harry Merchant to become familiar with electrical fabrication and with David King on the mechanics for payloads, especially a blockhouse firing panel for two Aerobee flights of Carruthers's spectrograph to observe Comet Halley in 1986.[38]

Seeley's varied work history before NRL, which included a wide range of sometimes rancorous bosses, as well as considerable physical danger, made him especially sensitive to the work environment.[39] He appreciated how smoothly Carruthers's team operated and Carruthers's calm nature: "They just knew what each other needed and wanted. . . . George would give general guidelines, and Harry and Dave would just pick it up and build it." Seeley found Carruthers's collaborative work and management style perfect for his goals, and he continued there throughout his co-op years: "I was in heaven working there." In 1986, upon graduation, Seeley was hired full time and worked on SPARTAN, developing ground support and flight electronics. This was also his first exposure to the possibility of applying the CCD as a detector in NRL's latest attempt at gaining a spot on a SPARTAN payload. Although the CCD was originally considered in development for what would be *SPARTAN-281*, by the time Seeley joined the team, Carruthers had decided to stick for the moment with their tried-and-true electronographic technique, amplified by a microchannel array. But future designs, he predicted, would be using "electron-bombarded CCD arrays in place of electrographic film recording."[40]

Working on SPARTAN required many meetings at the nearby Goddard Space Flight Center in Greenbelt, Maryland. Seeley remembered some harrowing trips to Goddard in Carruthers's "old beater of a car."[41] Instead of an ignition key, George used a screwdriver, and there were various other patches that gave it character and made it "nerve-wracking at times." As his brother Gerald recalled, George "had cars that were just pieces of junk, because he never bothered to get them fixed." And apparently George did not have a driver's license, or a car, until he moved to Washington.[42] These meetings at Goddard provided Seeley valuable exposure to both Carruthers and his interactions with the NASA style.[43] Over time, as Meier observed, Seeley, Merchant, and King all became "very devoted to George, almost protective of him." Informally they called him Murph, after Murphy's law, because George continually invoked it: "If something can go wrong, it will go wrong."[44]

After SPARTAN, in line with Carruthers's stated intentions, Seeley sensed that he would be of greatest value if he learned about CCDs. Carruthers had given him a popular article from *Sky & Telescope* as a gentle introduction to what was still new technology, and one Carruthers himself was not completely comfortable with. He, like Lyman Spitzer and others, were wedded to vacuum tube technologies, and the CCDs seemed just not to be ready yet. But they were the future, and Carruthers had to find a way to adapt, as we have noted.[45]

The now-classic 1987 *Sky & Telescope* article, titled "Sky on a Chip: The Fabulous CCD," was written by James Janesick, who headed an "advanced imaging sensors group" at the Jet Propulsion Laboratory. It was coauthored by Morley Blouke, well known as a CCD pioneer at Tektronix. They acknowledged that the largest CCDs then available to science, in the unclassified world, were 2,048 pixels on a side for a total of 4 million pixels (or picture elements) but were barely two and a half inches on a side. They could not compete with the resolution available from 35-mm film (effectively 25 million pixels or better). But they had other advantages in quantum efficiency and dynamic range. And quite soon, they would also have broad spectral response, though the authors spared readers the details. They also made it quite clear that "a CCD, like a racing car, requires a full-time engineering support team if it is to work at peak performance."[46]

That was exactly what Carruthers was doing, encouraged by Gursky, by directing Seeley to assist Heckathorn and Opal to concentrate on CCDs. Carruthers gave Seeley CCDs and associated electronics from Texas Instruments that were apparently just sitting around in a jumble: "It was all very hodgepodge. And I was given pretty much a free hand to design a new set of electronics." With Chet Opal lending guidance and expertise, Seeley was able "to eventually pull together a set of CCD readout electronics that we then subsequently used to test the back-thinned, electron-bombarded CCDs from Texas Instruments."[47] These were intended to replace the film in Carruthers's electronographic cameras. Although it is not known how Carruthers felt about weakening support for his designs, he never hesitated to modify them to take any advantage that it might bring.

### A NEW REGION TO EXPLORE: THE MAHRSI MISSION

Another good example of Carruthers's transition, not only to a new technology but to a new participatory role, was his assignment to become part of a new mission advocated by Gursky: a problem area he was not familiar with. What came to be known as the "Middle Atmosphere High Resolution Spectrograph Investigation" (MAHRSI) eventually flew on a payload called *ASTRO-SPAS* (Shuttle Pallet Satellite) in November 1994. As another NRL colleague, Robert Conway, put it, "[Carruthers] was very enthusiastic about MAHRSI because it was new. It was an observation that hadn't been done in that way before," and it promised significant improvements in understanding the dynamics of the middle atmosphere.[48] But Carruthers, now with his *Apollo* and *Skylab* experience, knew that there were new technological hurdles to overcome, such as the efficient and accurate transmission of data from the middle atmospheric layers. Accordingly, he took a more conservative engineering approach.

MAHRSI, planned for shuttle flights, would analyze data on airglow in the middle atmosphere, consisting of the mesosphere and the stratosphere. The temperature increases with altitude in the stratosphere, and since warmer air is lighter than colder air, it is generally a stable "stratified" region. But in the mesosphere, temperatures drop with altitude, leading to instability. It is the highest atmospheric layer where gases are still mixed. Establishing the nature of the mixture and its change with altitude would be most helpful in gauging the dynamics of the region, long neglected because NRL's legacy, from its sounding rocket work in the 1950s to Carruthers's images of the geocorona from *Apollo 16*, as well as the continued work of Meier and Mange, dealt with the ionosphere.[49] Sensitive to this limitation in the NRL tradition, the middle atmosphere became one of Gursky's enthusiasms.

What eventually became MAHRSI went through many phases. With Heckathorn gone, Gursky initially gave the lead to George Mount, a National Oceanic and Atmospheric Administration geophysicist and solar astronomer trained at the University of Colorado's Laboratory for Astrophysics and Space Science (LASP). As a recent PhD, Mount already had a good track record in mesospheric studies specializing in the measurement of trace gases in the Earth's atmosphere. He also developed microchannel electronic amplifier arrays within the large LASP group. Gursky may also have tapped him as a means for gaining multi-institutional support from the National Oceanic and Atmospheric Administration for what was initially a balloon-borne test, called the Middle Atmosphere High Resolution Spectrograph (MAHRS) to link it to MAHRSI, which would hopefully lead to a flight on shuttle.[50]

This arrangement did not work out, and so Gursky transferred the lead to airglow specialist Robert Conway. Not to be confused with William Conway, Robert Conway came to NRL in the early 1980s after finishing a PhD thesis at the University of Colorado, titled "Spectroscopy of the Mars Airglow from Mariner 9 Ultraviolet Spectrometer Data." Conway had concentrated on analyzing molecular spectroscopy data and was especially sensitive to the need to understand the instruments that produced the data in order to adequately interpret the results. He came to NRL mainly to do data analysis and synthesis of molecular spectroscopy, "and more than anything, I wanted to learn how to do radiative transfer from Robert Meier and Don Anderson, because I'd learned it in graduate school, but I really wanted to know how to do it for molecular radiation, because I was very interested in molecular spectroscopy."[51]

Once he had the lead, Robert Conway turned to Meier and others for advice but was personally responsible for raising the funds. Charles Brown remembers Conway "always going around with his begging cup." Brown, another member of the group, came to NRL as an NRC postdoc from the University of Maryland in the mid-1970s. He worked first on the calibration of high-resolution spectroscopic data from Richard Tousey's spectrograph on the ATM.[52]

When Brown became part of Conway's team for MAHRSI, he sensed considerable resistance from proposal reviewers at Goddard, "who'd tried to build a spectrometer to measure hydroxyl in the middle atmosphere, and they were always defeated by scattered light. And the dogma was, 'Can't be done.'"[53] Thus, lacking NASA support initially, NRL had established a relationship with a German group that was also proposing to NASA a pallet-launched payload called CRISTA-SPAS, a supercooled infrared spectrometer for atmospheric studies (Cryogenic Infrared Spectrometers & Telescopes for the Atmosphere), and eventually, after some setbacks (it was canceled by the incoming NASA administrator Dan Goldin's staff), Gursky stepped in and used discretionary funds from ONR he had at hand. According to Conway, Gursky advised, "Don't worry about it. Let's just fly it. Let's fly MAHRSI." Conway felt this reflected Gursky's desperation "to bring in new life into the division."[54] This was only one of several examples of Gursky's reaction to the dire observations about UV astronomy at NRL leveled by the Burke NRC panel report. As Meier observed, Gursky was always "willing to go way out on the limb to get something flown."[55] But where was Carruthers in all this?

When Gursky gave him the assignment, Robert Conway knew he had support from Meier for analysis, but he needed advice on designing and building the instrument. So he walked across the hall to seek advice from Carruthers, announcing, "Hey, I want to look at the airglow in the middle atmosphere [and] all of a sudden, here we were, trying to do something in the middle atmosphere, and he was totally open and supportive and excited for it." Conway appreciated Carruthers's help providing the author with additional insights into Carruthers's work habits:

> As a new colleague for George, I wasn't building instruments, but I was there trying to understand the spectroscopy of the airglow and understand what I was seeing in the instruments. No matter how late I worked, what day of the week it was, I could always walk around the corner, and George was in his lab. And I could walk into the lab and say, "George, I have a question." And no matter what he was doing, he would stop and listen, and think. We would talk about it. He would make suggestions. And then I would ask him questions about the instrument, and we would talk about instruments, and he would show me the instruments he was working on.
>
> I went to George so I could understand my theoretical analysis of doing radiative transfer and comparing it to instrument data. And I was trying to say, "George, I don't really understand what the instrument's doing." So, he talked to me about instruments, and that became more and more exciting and interesting. And ultimately, it is what drew me into the process of building. I didn't design instruments, particularly. I worked with people to design them,

> but I built them and flew them in space on the Space Shuttle and on satellites deployed by the Space Shuttle.[56]

Robert Conway approached Carruthers because he needed to understand the instrumentation that would be producing the data that he, Meier, and others would analyze. There was no one better for the task. And Gursky agreed, facilitating the relationship. Conway sensed that "Carruthers's detector program didn't have any money to do their work, so Dr. Gursky, who was our division head at the time, said, 'I'll solve two problems. George, you make a detector for MARHSI.'"[57]

Carruthers, understanding the problem from the technical standpoint, advised that they bring in Seeley because the observations would require long integration times and high sensitivity. By then, Seeley was the go-to for CCDs—according to Conway, "an expert in the problems of building intensified CCD detectors." Their challenge was to combine CCDs with electronic image intensifiers, such as Carruthers's cameras. As Conway sensed, "George was very engaged in that problem, and he committed himself in that, and Tim Seeley, to building the detector for the MAHRSI experiment. And that was a great success. And that's a very important contribution that George made."[58] So they worked together on a design to combine the powers of a microchannel plate-intensified CCD with a Carruthers electronic image intensifier all integrated into the MAHRSI design. Since photography was no longer part of the process, it was no longer an "electronographic" instrument. It was a sophisticated evolution of Carruthers's original design.[59] It was also an evolution from Carruthers as lead, to Carruthers as technical support and mentor, which he did without hesitation.

MAHRSI first flew aboard the Shuttle *Atlantis* (*STS-66*) in November 1994 on the ASTRO-SPAS pallet and observed the distribution of nitric oxide from some 47 to 87 miles above the Earth's surface. Overall, "the data provided a seven-hour snapshot of lower thermospheric and mesospheric nitric oxide from sunrise near 48° S to sunset near 61° N latitude following a period of low solar and high geomagnetic activity."[60] MAHRSI flew again in 1997 for eight days, mounted with the Cryogenic Infrared Spectrometers and Telescopes for the Atmosphere (CRISTA) on the German Shuttle Palette Satellite (SPAS), which was co-manifested with the third flight of the Atmospheric Laboratory for Applications and

Science (ATLAS-3). MAHRSI was designed to measure the dayglow in the 1,900–3,200 Angstrom region and the high distribution of concentrations of hydroxyls and nitric oxide in the mesosphere and thermosphere (some 20 to 90 miles altitude) to an accuracy of about 1.25 miles, this time tuned to a UV hydroxyl band to complement the infrared observations of CRISTA. Together, they examined how the shuttle exhaust gases interacted with the mesospheric medium by creating polar mesospheric clouds that contained known quantities of those molecules, providing insight into the formation of the clouds. This was a notable feat.

It should be evident by now, given the many acronyms and institutions involved, that the complexity of doing science with the shuttle was fostering more integrative, multi-institutional approaches to space science, and these examples provide insight into how Carruthers's world was changing. Creating a new MAHRSI was a new departure for a space spectrograph. It was a sizable instrument, 5 feet in length, and heavy, especially with Carruthers's camera. So Conway looked for a better solution to continue the monitoring program more efficiently, reducing the size of the complex optical system. Coincidentally, a physicist from Wisconsin, Fred Roesler, was visiting Robert Meier, who brought up the issue of modifying MAHRSI. Roesler had been working on an innovative technique called spatial heterodyne spectroscopy and told Meier, "I can build you a MAHRSI-type instrument that can do the same thing as MAHRSI can do, but it can fit in a shoebox." So Meier took Roesler down the hall to talk to Conway. It was a design perfect for flight; as NRL physicist Christoph Englert has explained, "It's an interferometer with a beam splitter, field-widening prisms, and gratings germinating the two interferometer arms, and it's all glued together so it's one chunk of glass at the end."[61] This was still an experimental design in the 1990s, but Robert Conway was determined to make it work, building a prototype that was tuned to detect hydroxyls. When Englert came to NRL on a postdoc in 1999 from the Deutsches Zentrum für Luft-und Raumfahrt in Oberpfaffenhofen, "Bob basically said to me, 'There's this new optical technique called spatial heterodyne spectroscopy. It's very powerful. It's not the solution to all spectroscopy problems, obviously, but it's a very powerful technique. You find out what you can use it for,' not just for OH, but beyond that."[62] Englert was attracted to NRL because of Robert Conway, having had contact with him and other NRL staffers visiting

Germany. He was part of a German group working with the CRISTA team at the time and so was quite familiar with MAHRSI, flying a 2.5-terahertz spectrometer to make hydroxyl measurements from aircraft and analyzing the spectroscopic data.

The amplifying portion of Carruthers's camera was still part of the design, but Carruthers was no longer the lead—more a participant and advisor. He accepted this new role in various MAHRSI and SPARTAN payloads starting in the middle 1980s. In the same manner, he took on both a leading role and a participatory role instrumenting what was called the Global Imaging Monitor of the Ionosphere (GIMI), a payload that would eventually fly in the 1990s on the ambitious multi-payload DOD Space Test Program's Advanced Research and Global Observation Satellite (ARGOS).[63] The "G" in GIMI was also meant to stimulate future monitoring from high geosynchronous orbit, inspired by the global images returned from *Apollo 16*.[64] Carruthers appreciated this legacy but sent it along with the team to let the demands of other experiments on *ARGOS* dictate that it first be in polar orbit.

## STUDENT ASSISTANCE -2-

Given his multitude of projects, and his dwindling funds, Carruthers turned once again to his SEAP (Science and Engineering Apprenticeship Program) and co-op students and his interns, such as Garland Dixon (see appendix A), to help design and test his payloads. This was a "win-win" situation. Not only did this give his students valuable insights into how they could play a part, but they were "cheap labor" paid out of SEAP funding. Dixon came first as a high school student from the summer engineering apprenticeship program at George Washington University. In the summer of 1988, Dixon designed a mounting plate for one of Carruthers's spectrographs that astronauts could operate, including pointing it at the shuttle's fuselage through the back window of the crew cabin to collect local environmental data on the shuttle's passage through the atmosphere.[65]

Dixon returned in the early 1990s and through the decade worked as a co-op student, undergraduate student, and then graduate student at the University of Maryland, teaming up with others, like Melody Finch, an undergraduate from Maryland. Carruthers gave them ever more complex design challenges—not as orders, but as advice—helping them understand how what they were doing fit into the larger picture. Dixon's design for the mounting of what became the High Resolution [Space] Shuttle Glow Spectrograph A (HRSGS-A) finally flew on *Discovery* (*STS-51*) in September 1993, producing images of the glow around the shuttle's surface and tail. Dixon also created a new gimbal-mounted system for FUVIS, the same project that William Glascoe had contributed to in the summer of 1985. Glascoe had mentored Dixon in his 1987 summer stint, so he was up to the task. FUVIS was a dual camera for what was originally hoped to be a *SPARTAN-202* Shuttle payload, but those plans were shelved by the time Dixon joined the project.

Knowing his interest in computer-aided design, Carruthers had Dixon redesign the mounting system several times as possible flight configurations changed. Carruthers was constantly supportive and approving, searching out tasks that fit the talents and interests of his students. Dixon proudly recalled him frequently saying, "Oh! This is pretty good. This solves our problems."[66] And then Carruthers would suggest something larger and more challenging.

Dixon found that Carruthers cared about the interests and capabilities of his students: "Whatever major you had he would try to gear your work towards that." As he worked at NRL in the co-op program and continued at Maryland, moving into aerospace engineering, Dixon preferred flight dynamics to design. Carruthers accordingly put him to the task of developing computer routines to determine the aspect and position of his payloads within orbiting satellites.

Dixon continued working in the Maryland co-op program through graduate school. Among the many reasons he chose Maryland, in fact, was because he could continue to work with Carruthers. By the early 1990s, he worked on various projects, typically providing data processing assistance. When FUVIS finally flew on *SPARTAN-281* in the late 1990s, Dixon and other students, including Melody Finch, helped reduce the data. It required a whole new mode of data reduction because, once again, Carruthers had turned to CCD technology to make the data retrievable. As with all projects of this kind, it was years in the making, and during that time, his students observed Carruthers constantly revising his instrument designs, incorporating solid-state sensors in various ways to make his vacuum tube cameras more competitive. Carruthers mentored informally, and by example.[67]

## MORE PROJECTS IN THE 1990S, BUT LESS COMPETITIVENESS

In the 1990s, together with Tim Seeley, Kenneth Dymond, and his students, Carruthers continued to refine how he adapted his Schmidt cameras to use electron-sensitive CCDs prepared by Seeley. They planned to continuously monitor UV phenomena in the upper atmosphere as part of GIMI.[68] The goal of this effort, routine for scientific and military applications by the late 1990s, was to "characterize local perturbations of the ionosphere due to natural and artificial events . . . and to obtain all-sky surveys of celestial point and diffuse sources."[69] Their instrument was one of nine in the GIMI array, and while it awaited launch (which slipped), Carruthers and another team, which once again included Garland Dixon and Melody Finch from the University of Maryland (see figure 12.1), prepared another payload instrument along with GIMI also to be flown on *ARGOS*. Called the "High-Resolution Airglow and Aurora Spectroscopy Experiment" (HIRAAS), it included three far-ultraviolet spectrographs that would be on the Earth-side of the spacecraft, trained on the nightside of the Earth's limb. These instruments, observing simultaneously, along with GIMI and others, would provide a synoptic correlative view of the upper atmosphere and ionosphere.[70]

*ARGOS* was finally placed into a Sun-synchronous orbit in February 1999 on a *Delta II* launch vehicle. But after launch and checkout, there was a most interesting surprise. During the 1999 Leonid meteor shower that November, Carruthers's GIMI cameras, such as FUVIS, recorded both the image and spectrum of a meteor on November 18 when ARGOS was at 517 miles altitude, providing a highly improved far-ultraviolet record of a meteor's interaction with the upper atmosphere as well as an image of the meteor that gained wide attention. Other satellite observations were made of meteor spectra since 1997, but not reaching Lyman-alpha.

Carruthers had also been helping to devise a wide-angle system for recording meteor spectra, and this chance observation was most stimulating. It was later recalled as an "amazing observation" by colleagues he had teamed with in the early 2000s to extend reach and sensitivity in order to capture the far-ultraviolet spectra of meteors in a mission dubbed "*NEOCAM*" (Near Earth Object Chemical Analysis Mission). As Goddard Space Flight Center's Joseph A. Nuth III explained it, these observations might better capture the chemical diversity of meteors, both in swarms and

Figure 12.1
University of Maryland students (l to r) Melody Finch (undergraduate) and Garland Dixon (graduate student) with the HIRAAS payload scheduled to fly on *ARGOS*.
*Source:* Courtesy of Naval Research Laboratory.

sporadic, and thereby improve our understanding of the chemical nature of our primordial Solar System; in effect, "these bodies represent grab bag samples of the solar nebula."[71] *ARGOS* and Carruthers's GIMI instrument flew again in the early 2000s with improved calibration prior to flight and post flight, mainly to obtain better quantitative information. But there were no recorded interactions with meteors.

NRL, NASA, and Princeton all supported Carruthers's role in *NEOCAM*. The primary instrument was another departure from Carruthers's basic design. A slitless UV spectrometer consisting of a 12-inch aperture wide-field Schmidt camera and a reflective grating dispersing element all fed a transmissive cathode, which then sent an accelerated electron beam to a small CCD. Colleagues had encouraged Carruthers to make this significant change, to transmission from reflecting photocathodes, and to replace film with CCDs, making it competitive for untended satellite platform applications. Effectively, with the rapid advance of CCD capabilities in sensitivity, collecting area, and resolution, there was less need for Carruthers's electronographic amplification system, no matter its design. This was a watershed moment in Carruthers's career.

One hint of this transition was when Seeley left Carruthers's group in 1997 because he sensed that his upward mobility was limited, given the lack of funds. He moved to the Naval Center for Space Technology but returned for *ARGOS* because funding was available. But he knew the writing was on the wall, as did others on Carruthers's staff who were being let go or were leaving for more active programs. Those I interviewed tell the same story but always emphasized how willing and able George Carruthers was to advise, lend a hand, and provide a tool or a needed part. Meier recalled that there were "two or three satellite and rocket projects that I proposed as PI. In each case, I asked Carruthers to be a PI and he willingly jumped in and helped where he could."[72] Although his colleagues appreciated that Carruthers was open to using the latest detector technologies, they knew that he would always try to use them in conjunction with his image intensifiers. MAHRSI was a poignant example, since his colleagues, like Charles Brown, felt that his camera was no longer the best solution.[73]

As the collecting area, spectral range, sensitivity, and reliability of solid-state detectors grew, Carruthers's cameras became less competitive. He also became less prominent in the resulting publications of the results. Although he actively supported MAHRSI, he was not one of the authors on the resulting papers. Still, he was sought out as a resource. Brown described him as "the strong magnet guy," which put Carruthers in an outdated generation who had struggled with shielding problems that we have covered here. But when one needed a strong magnetic field, Carruthers was the one with the expertise. When asked for help assembling a magnet, he would head to his lab to assemble one from "a whole bunch of bars, made in a circle." Even so, despite Carruthers's helpful nature, others outside the section, like Russell A. Howard in Richard Tousey's section, sensed a slow change in the air toward Carruthers after Friedman's retirement and as his detectors became less critical. By the late 1980s, "I think the attitude regarding George, or the respect for George, declined, after some point."[74] Others within his own division, however, respected Carruthers as much as ever but sensed that he "had less of a dominating presence."[75]

Appreciation for his science never faltered, however. In the 2010s, after many people began to realize, partly through the MAHRSI observations, "that the ionosphere is not only driven by the Sun, but also by lower atmospheric weather,"[76] NASA approved the *Ionospheric Connection Explorer (ICON)*, a collaboration that included UC Berkeley, U. Texas, Utah State, Goddard, and NRL to explore how the Earth's lower atmosphere interacts with the ionosphere to alter its structure and properties. Everyone knew that this new recognition of the complexities of ionospheric behavior was "built on the fundamental work and first observations that were done by George's camera."[77] So as they prepared their proposal for *ICON*, David Siskind at NRL suggested that the most iconic image to place on its cover was Carruthers's *Apollo 16* image of the Earth's geocorona. *ICON* was eventually launched in 2019 aboard an airborne *Pegasus XL* rocket carrying four instruments including NRL's successful "Michelson Interferometer for Global High-Resolution Thermospheric Imaging."[78]

## WORKING SPACE BECOMES A PREMIUM

By the mid-to-late 1990s, there was a new administrative problem at NRL. The space division was growing quickly again, partly encouraged by restored funding from several sources and by Gursky's facility with redirecting funds and establishing multiple sources of support. Physical laboratory space was becoming a premium in Building 209 at NRL, which housed the space sciences. Free space was dwindling as different branches competed for physical space, and space was an expensive overhead cost.[79]

From the start, Carruthers never tired of designing and constructing variations of his detectors and cameras. Since he rarely, if ever, threw anything away, his growing collection of instruments and apparatus overflowed his laboratory space, requiring that he acquire more territory, initially supported by NRL through his other projects assigned by the Navy. For example, through this time, he had been applying his cameras to the analysis of rocket plumes, as Heckathorn noted. His classified reports from the mid-1970s through the 1990s analyzed rocket plumes from a wide variety of vehicles, including the Minuteman and Peacekeeper missiles, examining their motor exhaust in test firings in vacuum chambers, at launches, and from space on the Shuttle, in order to "assess the feasibility of far-UV surveillance and tracking."[80] It was this latter application that, Seeley recalled, "certainly . . . brought in money for his projects."[81] Carruthers was still valuable in the Navy's eyes, but it was a stop-gap measure.

In 1999, Gursky directed his associate superintendent, Frank J. Giovane, to solve the work space problem. Giovane, a University of Pennsylvania PhD, had worked on *Skylab* data and had written a thesis on Comet Kohoutek. After positions at NSF, NASA Headquarters, and then at the University of Florida, he came to NRL in 1999 with interests in planetary systems science, comets, and minor planets. But Gursky also assigned him managerial duties as his assistant. Giovane shared Gursky's concerns about space allocation and knew he had to provide younger staff with adequate laboratory facilities, working within the limits of Building 209. As did others, Meier recalled that Giovane was given a tough task. "He had to find more floor space quickly and zeroed in on Carruthers's labs."[82]

Indeed, this was the impression that other NRL staff had at the time. Over the years, Carruthers, working on so many different projects and rarely dropping any of them or the inventory they amassed, had accumulated "an enormous string of labs on the south side of Bldg. 209."[83] But the matter was far more complex, given the relative status of branch heads and senior staff. As Giovane observed:

> If there were reallocations of space it was a Branch and Branch Chief activity. The Branch requested and if needed paid for space. So if the Branch felt they needed more space than their allotment. . . . they paid for it, if not they gave up space and paid less. . . . George was not in any Branch and therefore space and payment for space above a reasonable allocation was his responsibility.[84]

When he retired in 1980, Friedman made Carruthers head of the Ultraviolet Branch. But quite soon, Gursky and others realized that this managerial responsibility was a mistake. Meier agreed, "as it took him away from his research and he was not enough of an extrovert."[85] He was relieved of the position in 1982, and Meier became head of the branch. Carruthers was more than happy to be relieved from management, but then Gursky, urged by NRL director Timothy Coffey, promoted him again into the senior ranks of federal service. Carruthers now outranked Meier, and so, according to Civil Service rules, he could not report officially to him. The branch remained very much alive under Meier's direction and in close collaboration with Carruthers. But Carruthers now was essentially a "free agent," a branch unto himself, reporting to Gursky. But he was alone.

Giovane noted in retrospect that "Gursky and the lab administration saw value in having George promoted to the highest scientific level. Their continued funding of him, covering at least his salary, was a vote of confidence extended to only a very small minority of the Space Science staff."[86] For the Navy, his "senior astrophysicist" rank placed him among "internationally recognized experts in their field, senior scientists, proven leaders, and world-renowned researchers dedicated to public service at the highest levels."[87]

Carruthers's rank was expensive, and in the 1980s, it had given him considerable leeway to pursue many projects simultaneously, time and again. As we have often seen, when he was on the verge of flying one package, he was already proposing for others. All these projects required equipment, staff, and laboratory space. As he continued to develop new instrumentation and the facilities to build and test them, as we noted, he held on to his older equipment, which added to his space requirements, especially by the late 1980s. He also had office space just across the hallway from his seven contiguous labs on the second floor.

His labs and office were notoriously messy, and he reveled in it, as people remember him saying sarcastically, "A clean desk is the sign of a sick mind." Throughout the 1980s and 1990s, when he had high school students and college interns working in his spaces, as Heckathorn colorfully recalled, "You had to be careful walking into his office. You might tip something over," adding,

> He had a very interesting way of storing documents. He had a file system that was all flat. It was one piece of paper on top of another, on top of another, on top of another. And whenever you walked in, you were worried that you were going to tip something over, and the whole pile would come down. I don't know [how] he ever found anything, but he had a system. And the same was true of his laboratory. . . . The only negative thing I ever heard was the condition of his laboratory because it was always a mess. But George knew where everything was, always. If there was an Allen wrench that was missing, he knew I took it [laughs] and I didn't put it back. So, he knew where everything was, and he ran a fairly large lab. He had a lot of lab space that he commandeered.[88]

Carruthers seemed so permanent, as Heckathorn added, "I don't think they could ever move him out." Meier felt that "his reputation as a genius in the laboratory preceded him." Since Carruthers typically stuck close to his work and saw it all through, some worked better with Carruthers than others. Meier recalled that "George would do everything," but when Chet Opal was hired, "working with George was right up his alley. He was one of the few people who could really get in there, elbow to elbow."[89]

His colleagues knew that Carruthers was in his laboratory day and night, even on the weekends. He was a constant and giving resource, an iconic presence in his lab. Anyone who needed anything knew that Carruthers would probably make it available; it "was a great gift."[90] Meier explained that Carruthers's accumulation of instruments and tools and the space to use them was because Carruthers "found it easier and faster to do it himself than to work with somebody else." He would work "all day and half the night. Nobody knew when he ever slept." Often his team members would arrive the next day ready to do something, and Carruthers had already done it. There were times, of course, when Carruthers was elsewhere, and people would sneak in to borrow a tool they knew he had. Some like Meier got caught in the act, but Carruthers did not seem to mind, offering to help if needed. But overall, as Meier added more than once, "His lab was something to behold, I mean there was just stuff everywhere."

In the late 1990s, every so often, someone would find Carruthers sitting on a stool fast asleep. That worried Meier because the lab was filled with exposed high-voltage switches, vacuum evaporation systems, and sharp objects.[91] There had been instances that caused worry among his NRL colleagues. Ken Dymond recalled,

> In his laboratories, he worked with 30 kV power supplies to his detectors. The cables would inevitably break down under the high internal electric fields and the wires would charge up and then spread apart and with a few crackles they discharged and returned to their normal positions. To prevent this, George would carefully "wipe" the accumulated static charge off onto his hand. Then he'd go to the sink and discharge himself to the grounded water supply lines through a high-valued resistor that he carried in his pocket. One day, one of his technicians got a little too close when George was heading to the sink after charge wiping. Zap, the tech got a jolt. Not sure if [it] was better than caffeine, but he was definitely more alert afterward; stung but smiling.[92]

The piles of papers increased in height over the years, and the passageways between them got narrower and narrower. It became something of an office joke, but also a worry. Chet Opal was a man of "ample girth" and, according to Meier, more than once knocked over one of the piles. One staff member wondered if Chet had brushed against the pile, or, as Meier recalled Carruthers saying, "it was caused by the gravitational attraction between the nearest pile and the mass of Chet's tummy. A volume vs mass effect."[93]

Everyone knew that "George's heart was in that laboratory."[94] Brown and others also knew that "with his personality—his shyness, he was not a salesman." But Friedman's records show that up to 1980, Carruthers had prepared and submitted at least three dozen proposals. About a dozen were accepted—a typical success rate. Many were re-proposals for extensions to a project, and some were team efforts or smaller requests for data reduction. He was most successful in the mid-to-late 1970s but less so in the 1980s. By the late 1980s, he and his colleagues darkly knew "the money wasn't flowing."[95] Indeed, though Carruthers kept busy, mostly refining prototypes for his proposals, he was completing fewer of them due to staff attrition and to his growing attention to outreach. And it was getting worse.

NASA funding for new space science programs was also at a low ebb, causing rejection rates to increase in the 1990s.[96] In November 1998, in an endorsement for a Carruthers educational proposal, "Astronomy and Space Science Outreach Programs for DC Public Schools Students and Teachers," Gursky was concerned that Carruthers led no active NASA mission grants. He was still working on FUVIS that flew on NASA's *SPARTAN-204* and had support from DOD for several smaller projects, but his main activities now were educational. As Gursky testified, he has "established partnerships with several educational organizations," listing five activities. Although of definite worth and promise, something NRL could be proud of, it did not bring in much overhead.[97]

Everyone knew that there were funding issues and competition among the branches for space and facilities. This led to what was a painful situation for the staff, as Conway poignantly recalled,

> There were lots of young people that had been hired, and so the idea was to get new blood into the division. But what happened in that process is these young people came in, and they needed to create their own territory. And so, there was a lot of competition. And George's laboratory was in the eyes of a number of those people. And if George wasn't getting supported, then they were more than happy to do whatever they could to take over his spaces. And to me, it's a sad story. Especially, as I mentioned, it makes me very, very sad, knowing what ultimately happened, and the fact that George was moved out of not only his laboratory, but his adjoining office space and ultimately put into the trailer. That's a tragic story, in my opinion. George deserved far, far more than that, because of what he gave. He was a powerful influence in the history of the division. He was a legend. He was an iconic character for the division. And for that to happen is a tragedy. So, I am very sad about that.[98]

The cold fact was, office and laboratory space were overhead expenses, and they were paid for by revenue from outside grants. Senior-grade scientists especially were expected to win grants that would provide major funding for the research infrastructure. As former NRL historian Angelina Callahan described the situation, each staff member had the "latitude to build relationships with sponsors, but each of the researchers has to pay for square footage and office space."[99]

As Conway recalled, Carruthers's laboratory and office ultimately ended up in a forty-foot mobile home–type trailer connected directly to the building in an area that served as the loading dock of Building 209. It was visible at ground level and accessible from the outside. Herb Gursky, advised by Giovane, had given Carruthers a choice: either he reduce his space on the second floor by at least two-thirds, or he move his laboratory into the trailer. Carruthers decided on the latter, since it offered enough space for him to continue his mentoring and his projects.[100] But the optics of this decision deeply concerned the staff, as Conway eloquently expressed it. They may not have known that Gursky gave Carruthers a choice, and indeed, that he was not the only one who lost space. Some of the staff were angered, one of them saying, "It was unforgivable. . . . [It was a] disgraceful period in our building."[101]

Emotions were running hot. Was Carruthers being eased out? Would the trailer body be connected to a cab some night and wheels attached and be hauled out of the building? Or would Carruthers just give up in frustration? As Callahan observed during an oral history interview with Russell Howard, "The fact that he maintain[ed] a footprint at all here is significant." Howard had an answer for that: "I believe it was due to lab management." Management told Gursky, "You're not going to fire him. . . . The lab considered him to be too much of a public figure to take any action against him."[102]

Indeed, Giovane well knew how important Carruthers was to the image of NRL as a place that supported a famous and visible Black scientist. But there were just too many pressures for laboratory space, and that space had to be supported by outside revenue. By the late 1990s, although Carruthers was still working on several instruments and missions such as MAHRSI, he was the lead for only a few of the instruments, such as FUVIS and GIMI, the latter funded by NRL.

One more factor must be considered, tracing Carruthers's move to the basement port. Christoph Englert worked mainly with Robert Conway after he arrived in 1999, but encountered Carruthers from time to time. Like others, he found Carruthers engaged in his work, yet always approachable. But he soon sensed a concern among his colleagues for Carruthers's safety when he came upstairs to work in the machine shops, which he frequently did, to the continuing general annoyance of the technicians there who felt it was their territory. Beyond the instances where he was found asleep amidst active experimentation, his frequent use of shop space, for a man in his sixties, according to Englert, "became a problem later when his health deteriorated. . . . We had to make sure that he wouldn't have access to the machine shop [or] be there alone, for example."[103] From the beginning, Carruthers's penchant for doing everything himself, including much of the shop work, was his hallmark. The technicians were annoyed at first, as we have noted, but now they worried.

Carruthers apparently never complained about the move. He could still use laboratory facilities upstairs when they were available. Still, Meier, Conway, and others expressed deep sadness with the move, but Meier observed that "George didn't seem to mind it because he liked working on his own and nobody bothered him there."[104] Another staffer recalled that he took it in stride, telling them, "In life, you have to adapt. . . . My science will continue on. . . . This won't stop me."[105] And Frank Giovane took the management view: "He could do anything he'd want with it; no one paid much attention. . . . Nobody was going to question what he was doing. He was basically a free agent, operating in a space that no one else wanted, and that he did."[106] Overall, the move to the trailer gave Carruthers enough room and freedom to be at one with his instruments and projects, and it also made him even more accessible to his growing number of students. And in no way was he banned from special instruments in the laboratory facilities on the upper floors when his work, and that of his students, required access.

One of his interns, Jessye Bemley (see appendix A) recalled that the basement location was actually more convenient than the laboratories upstairs. She still had to enter the building but went directly into the laboratory in the trailer as if it were just another big room. It also facilitated outdoor solar observing and videotaping sessions when she and Carruthers with the other students and interns started taking a small telescope out onto the grass next to the building to produce STEM-themed (Science, Technology, Engineering, and Mathematics) instructional videos for classes.[107]

His close colleagues were sad about the move, but the general feeling was that it was necessary, though Meier and others protested to Gursky. Some of his newer colleagues believed that it was this forced move that stimulated Carruthers to become more active in outreach. But well before then, as we have seen, he was an originator of SEAP and hosted a constant stream of summer students and yearlong co-op students. And his outreach became even more visible in the 2000s.

# 13

# REACHING OUT AND GIVING BACK

INTERVIEWER: *"And what do you think your legacy is, Dr. Carruthers?"*

CARRUTHERS: *"Basically, the students who have followed me, worked with me, or heard about me over the years."*[1]

Motivational and instructional outreach was not new for Carruthers. As we have seen in previous chapters, he had long been concerned about the lack of interest and access that minority children had to the worlds of science and engineering. As early as his junior year in college, at a "Launching an Aerospace Career" student conference at the University of Illinois in April 1960, just twenty years old at the time, he presented a long autobiographical profile to help students appreciate that these worlds were accessible. We have provided excerpts from this address earlier, but here we take a larger look at its overall purpose.

He began by telling colorful stories about how his childhood shaped his career. But his focus now was to impress his audience with the great opportunities of the space age. There were now, he began, "vast opportunities that are open for careers in astronautics, or the science of spaceflight and exploration." He predicted that "in a year or so, man will set forth into the vastest unexplored frontier of them all—the space beyond the sky." And even more audacious, "within the next five to ten years, man will set foot on the Moon. . . . The exploration of the Moon will be far more challenging than was the exploration of the Americas." There would be unimagined benefits to society and mankind. Astronomy would be revolutionized. There would be revolutions as well in weather prediction and worldwide communications, to mention only a few. He then posed the challenge: "Sound interesting? Well, you too can concentrate on the conquest of space if you plan now. You may even be that first man on the Moon if you try hard enough. In astronautics now there is a great shortage of qualified scientists and engineers."[2]

Just as he would with Sandra years later, he suggested core courses in science, mathematics, and English that would prepare them for college. And beyond formal courses, just as he did and was still doing with the Chicago Rocket Society, he encouraged that "as a bare minimum—participate in school science clubs and science fairs." Possibilities were endless; young people could still get active in astronautics even if they were not strong in the more abstruse sciences: "After all, someone has to service the rocket engines, develop those pictures of the unknown far side [of the Moon], and write newspaper articles about the latest satellite. Therefore regardless of your field of interest, there is some opening for you in astronautics, so that you, too, can contribute to the conquest of space."[3]

Carruthers felt it was important for scientists to gain public attention and trust. But to do so, they had to know how to speak in terms the world could accept. At first, he felt he was not ready but was willing. In September 1970, his uncle Benjamin wrote asking if he would like to write a popular essay on science for *Tuesday Magazine*, a magazine published in Chicago and distributed nationally as a newspaper supplement. Carruthers replied, "It would be a challenge for me to write an article of this type which would be understandable to the man on the street, since I am used to writing articles to be read by other scientists in my own and related fields."[4] But then he added, expressing a need that is still felt today, "However, I feel that one of the reasons for declining public support for the space program, and scientific research in general, is that scientists do not put enough effort into explaining to the taxpayer the significance and eventual benefits of their research. Therefore, I would be very happy to have the opportunity to prepare an article such as you suggested."[5]

Carruthers's evangelism for science awareness grew in the early 1970s. His appearances were sporadic and usually local at first, but after his discoveries and achievements in 1970 and his 1972 smash hit with *Apollo 16*, as we have shown, people like Roscoe Monroe and Francis Redhead started inviting Carruthers to attend and address science motivational programs in Brooklyn and Harlem. His Caribbean tour in June 1972 with a team of NASA staffers and two astronauts drew wide public notice, and many of these visits were highlighted by honors from the United Nations and a public awarding of NASA's Exceptional Scientific Achievement Medal.[6] Over the years, he made six rounds of visits and tours in the New York area that captured headlines in the Black press.

Carruthers's motivational efforts synchronized with the times. By the 1970s, numerous local and national organizations were calling for support to promote better opportunities and incentives for minorities and women. And African American studies programs were growing in number and visibility. There was also more exposure to the achievements of Black scientists in popular books and magazines. Among the many calls for social action, noting the irony, was the famous spoken-word poem by poet Gil Scott Heron: "I can't pay no doctor bills, but Whitey's on the Moon."[7] Carruthers vicariously was the exception.

NASA and NRL responded to these appeals, and their support in no small way was critical in propelling Carruthers to stardom. He openly parlayed the notoriety he received in his first decade at NRL and earlier, with the notoriety he gained from his rocketry experiments in his teens through college. *Apollo 16* made him a legend among minority activists drawing the attention of the press. The NRL public relations office strongly highlighted his achievements. The press emphasized that he was a Black man, implying that his achievements were even more profound, but NRL stuck to the significance of his science and his technical achievement.

## NRL SUPPORT

NRL supported Carruthers's evangelism because it had for years been sensitive to the importance of community relations. Since the 1950s, the *NRL Bulletin* gave notice of home and family health matters and of youth activities available to staff. The staff organized annual Toys for Christmas drives for local community services. And by the early 1960s, the staff held informational teachers' seminars for local classroom teachers and their students. One attendee from Laurel, Maryland, wrote to the NRL director in February 1962, stating, "There is nothing like it in this area."[8] The Toys for Christmas drive, organized by the NRL fire department, grew to an annual Christmas party for children at the base.

Sensitive to the need to expose young people to its more successful staff as role models, by the mid-1960s NRL encouraged selected staff to function as tutors in local schools and continued this for decades. In 1986, its *Labstracts* newsletter observed that "these activities are intended to improve the employability of members of under-represented groups."[9] Starting in 1971, coordinated by the NRL Community Action Council,

NRL supported a "Summer Aid Program," including tutoring for local secondary schools like Kramer Junior High and Ballou Senior High School. Through the 1970s, responding to the government-wide campaign "Summer Employment for Needy Youth," NRL brought in about one hundred students each summer, in addition to using other federally sponsored program funds to host dozens of college interns.[10] There were also racially inclusive programs like SEAP that Carruthers fostered and various university co-op programs that attracted a wide range of students. Carruthers, enabled by this supportive atmosphere, along with others on the staff, mentored a wide range of students, from high school through postdoctoral, in their laboratories.[11]

Beyond SEAP, NRL supported many programs in the Washington, DC, area, including the District of Columbia Space Grant Consortium, NASA's IDEAS program (Initiative to Develop Education through Astronomy and Space Science), S.M.A.R.T. Inc. (Science, Mathematics, Aerospace, Research and Technology), and the National Technical Association (NTA). The latter two focused on minorities, but the others were general. As NRL physical scientist and STEM coordinator Henry Pickard noted, "In 1987 an NRL EEO specialist attested to a writer for the *US Black Engineer*, 'We target schools (universities and colleges) that have a high population of minorities.'"[12]

Pickard worked on a wide range of projects at NRL, from problems in celestial mechanics to missile and satellite star trackers to multispectral imaging. In the late 1980s, he became active in mentoring programs "within the framework of existing student programs: STEP, SCEP, SEAP, NREIP, and summer hires."[13] He gained the impression that although these programs did not focus on minorities and were "racially blind," there were individual mentors like Carruthers who did target minorities but still had students with a range of backgrounds and ethnicities. Students applied and their applications were reviewed by prospective mentors. Pickard chose students based upon "academic merit, stated scientific discipline of interest, and teacher recommendations." Many staff members chose several students, and some were mentored by several staff. Most of the outreach activities were student tours of the laboratories and visits to schools, but the most effective were students engaging in projects. Carruthers was active in all of them.

Over time, NRL did target minorities, especially reaching out to local schools. Directives from the Navy "explicitly gave the lab broad latitude to do outreach in its immediate neighborhood." For NRL, that meant the Anacostia neighborhood, which was predominantly Black. Two years after Pickard retired in 2011, he returned to NRL as its STEM coordinator, finding that "there were initiatives at all levels of government to promote STEM, and race and gender had most definitely become considerations." He also found that "terminology had also evolved to reflect a broader perspective, so that the term 'Minorities' was typically replaced by 'Populations underrepresented in Science & Technology' or similar terms. Diversity became the new buzz word."[14]

As Pickard's perspective indicates, today we regard such activities as normative and as an integral part of modern social and cultural practice. But in Carruthers's day, starting in the 1970s, they were not universally accepted. Certainly, hiring practices were not. Scanning pages of NRL's *Labstracts*, the employee newsletter in the 1960s through the 1980s, the overwhelming majority of Black "new hires" were working aids, helpers, cleaners, and clerk typists, as well as some machinists, a few in personnel management, the Technical Information Division, and administrators in the Supply Division. Carruthers was among the few Black people hired as an engineer, mathematician, chemist, or physicist. But he was not alone.

Notable was Alvin Goins, who by the mid-1960s was a psychologist in the Engineering Psychology Branch, Applications Research Division. He had joined NRL in 1946 and continued his education, receiving a PhD in 1967 after being detailed to the White House as a member of the President's Commission on Crime in the District of Columbia.[15] And there was Carl Rouse, who by 1965 was working at NRL as a Hulburt fellow concentrating on mathematical methods to interpret observations of the solar atmosphere, chromosphere, and corona by theoretical modeling and simulations. He dedicated himself to understanding how observations of the composition of the solar atmosphere could reveal details about the nuclear processes in the Sun. Rouse had a 1956 PhD in physics from Caltech and had worked at what is now the Lawrence Berkeley National Laboratory for several years, researching and refining Meghnad Saha's classic theory of ionization equilibrium to better understand shock wave phenomena using exploding wire observations of metallic elements—definitely of practical as well as theoretical interest to better understand stellar pulsations and supernovae.[16] A strong theoretical background made him useful to various observational programs in the Solar Division. While at NRL, he held a joint appointment with the Space Sciences Laboratory at Berkeley but by the early 1970s left for a position at the General Atomic Company in San Diego.

The rarity of finding persons of color in leading roles at NRL is poignantly illustrated by a moment during a job interview Carruthers was conducting to find a new office assistant. Carruthers as usual was dressed casually, and, as he was African American, the candidate mistook him for a janitor. There was a hush in the room, but Carruthers looked straight at them saying, "I like you," adding "You're an honest little one." Regarding being mistaken for a janitor, and responding in a way that reveals not only his self-awareness but also his composure, he replied, "That's okay. I dress like one, so everybody usually thinks I am." Carruthers added that his wife often teased him about his dress habits. The staffer recalled, "I figured I didn't get the job . . . but within 10 minutes, he was telling me I was going to be working there with him. So, he was wonderful."[17]

Carruthers and the few scientists of color mentioned above were, indeed, rarities at NRL. These few but significant examples were a preamble. But it would take some time, at least a generation, to identify, attract, and train that broader racial range of talent, turning it from a potential asset into a competitive resource. NRL had responded quickly to John F. Kennedy's establishment of a Committee on Equal Employment Opportunity and his call for a desegregated federal work force, and therefore avoided the public criticism directed toward other agencies, like NASA, which were more visible and far less consistent in their efforts. Indeed, it was not until March 1972 that NASA held its first full-scale public EEO conference, where NASA administrator James C. Fletcher admitted that there were "problems of equal opportunity for minority groups and women in NASA" and promised that "'steps' are being taken to avoid 'serious' discrimination."[18] This was, of course, not a surprise, but at least a formal admission by groups in civilian, commercial, and military circles across the land that the world had to change.

Nevertheless, as Ruth Joy Calvino points out in her 2020 thesis, "'A Phenomenon to Monitor': Racial Discrimination at NASA, 1974–1985," NASA consistently fumbled EEO matters. During the 1960s, she notes, "The intensity of the crash Apollo program created a culture where NASA ignored everything but their technocratic goals."[19] Richard Paul and Steven Moss note that NASA administrator James Webb tried to leverage the 1964 Civil Rights Act to achieve reform in NASA. But "parsing the agency's apparent schizophrenia on race relations" given the fact that the geographically scattered NASA centers had their own distinctive cultures made it virtually impossible.[20] In the early 1970s, with funding struggles in the post-Apollo era and increased pressures due partly to Nixon's re-enforcement of the Civil Rights Act to apply to government agencies, NASA had to respond.[21] One of the ways it chose to do so was to feature Black faces in science, like Carruthers.

Administrator Fletcher had formed an EEO office in 1971, hiring Ruth Bates Harris to direct the effort within NASA's directorate for Organization and Management.[22] Harris, a seasoned civil rights activist with an MBA from New York University in personnel and industrial relations, was a member of the Caribbean tour—organized by Francis Redhead—along with Carruthers. She was also active in local organizational efforts in DC and Maryland.

After the tour, Harris directed Roscoe Monroe to engage Carruthers in NASA outreach programming.[23] Monroe, who had also orchestrated a campaign featuring *Star Trek* actor Nichelle Nichols to recruit a new and more diverse astronaut corps, had already heard about Carruthers a bit before Harris's contact with him, just before the flight of *Apollo 16*. One morning, Monroe happened to be sitting next to a woman from NRL on a commuter train from Baltimore when she asked him if he knew about their "Moon man." Indeed, he did not, but sometime after that encounter, in the excitement after *Apollo 16*, and urged on by Harris, he included Carruthers whenever he could when he "was showcasing NASA employees in STEM careers," even though Carruthers was not a NASA employee.[24] After she interviewed Monroe in 2011, Valerie L. Thomas, a former associate chief of the Space Science Data Operations Office at NASA, former president of the NTA, and a founder of S.M.A.R.T. Inc., observed that at the present time, "since NASA did not have any African American astronauts, Monroe considered Carruthers as NASA's surrogate African American astronaut."[25]

General accounts of "Blacks in space," as in the July 18, 1997, issue of the *Philadelphia Tribune*, gave NASA "credit for recognizing courageous Blacks in space," including some nine Black astronauts and Robert Shurney, a NASA engineer trained at the Historically Black Tennessee State University, who was given credit for developing the Lunar Rover's tires. George Carruthers was also prominently honored, without stated affiliation, making it appear he was a NASA employee.[26] Although this could be considered a form of exploitation, Carruthers embraced it.[27]

## BEYOND THE CLASSROOM

Carruthers's outreach efforts, starting at NRL and continuing at local high schools and then at Howard University, were marked by his distinct perspective on racial attitudes toward science. When asked about his observations as of the early 1990s, he felt there was nothing intrinsically different between the races in nature.

> There may be some difference in the degree because of environment and background. Certainly I don't think there is anything inherent to race that's involved. It's just that most of the African American students come from less well-to-do backgrounds, inner city backgrounds, and aren't exposed to science and technology to the degree that some of the other students are.[28]

His primary concern then was how to rectify the situation. Lecturing did not seem to be effective. He could reach a lot of students that way, and leave a visible mark in the record, but the students required something more personal, more intimate:

> So what we have been trying to do is give them hands-on activities, use videos and demonstrations that get across information in a way that's more like entertainment, because certainly students are interested in seeing science fiction movies on television, they like to see "Star Trek" and "Star Wars" and "Battlestar Galactica." So what we're trying to do is cast real science in a way that's as attractive to them as science fiction is. . . . . It's something that students can relate to better than someone writing some equations on the blackboard or just giving them a lecture without anything other than word vu-graphs.[29]

As his visibility and demand for his appearances enlarged, Carruthers never abandoned the classroom, at the grade school or college levels. He established a wide pattern combining classrooms with field trips. And he always emphasized hands-on experiences, often to fit a need or an event. He teamed up with several organizations that provided infrastructure and exposure. But typically, the activists, like Harris, Monroe, and Thomas, found him first.

## THE NATIONAL TECHNICAL ASSOCIATION

Among his many efforts to increase Carruthers's involvement in outreach and advocacy movements, Monroe encouraged him to attend a local meeting of the NTA. Both a regional and national voice stimulating and facilitating the inclusion of Black people in science, engineering, and technology, the NTA was founded in the mid-1920s by Black technical, scientific, and professional engineers in Chicago, partly to improve inner-city functional architecture for businesses and housing. It soon became a nationwide network promoting professional development in the technical professions. The NTA's leading founder was Charles Sumner Duke, son of an Arkansas newspaper publisher and the first Black to graduate from Harvard. By the 1920s, he was a prominent architect and engineer in Chicago, devoted to improving the living conditions of the Black community.[30]

By the 1980s, the NTA had many regional chapters across the United States. But early in the decade, it could claim a membership of only eight hundred active members. It was hungry for growth, hoping to increase revenue and staff and indeed grew to at least some thirteen thousand members, reaching five hundred thousand readers in five years.

After attending a few meetings, speaking at one of them and receiving their Samuel Cheevers Award in 1976 at a meeting in Philadelphia, Carruthers joined the NTA, participating in meetings and classroom visits. He joined the staff of the NTA *Newsletter* in 1981. By the mid-1980s, encouraged by Valerie L. Thomas, he became editor of the NTA *Journal*, which issued three volumes a year. It was then about to reduce to two regular newsy volumes, replacing the third with a "technical papers supplement" volume, which suited Carruthers's goal to attract and vet technical papers from scientists and engineers that could reach students, as well as essays by students writing for their peers.

Carruthers spoke at these meetings and later gave papers and organized sessions, including some featuring his co-op students and interns who had worked in his laboratory. He also joined local community-based efforts facilitated by the NTA to reach out and mentor students. He impressed those he met as "always very focused, mild mannered, shy, quite responsible and dependable." Hattie Carwell, then a rising activist in the NTA, later to become its president succeeding Valerie L. Thomas, attested that "I never saw him excited, anger[ed], worried or afraid," even though, adding for emphasis, "everyone in the NTA treated him like a rock star."[31] Carwell was a senior health physicist and program manager with the Department of Energy in its San Francisco Operations Office, dedicated to protecting people from radiation hazards. During her career, she engaged in a wide range of educational outreach activities, taking leading roles in many of them.[32]

Both Carwell and Thomas encouraged Carruthers to get more involved in the operational side of the NTA. By 1986, as noted, he was editor of the association's primary publication, the *NTA Journal* and also agreed to edit its newsletter for several years and edit the proceedings of its annual meetings. He organized career booklets, served as secretary of the "Development Fund for Black Students in Science and Technology" and "participated in countless school visitations." As secretary, he prepared and distributed the meeting minutes and worked on scholarship committees, and he was the registered agent for handling the funds accrued from membership fees, investment income, and donations. He also devoted a decade to being president of the local chapter, and, as Carwell asserted in a tribute to Carruthers, by August 2011, Carruthers had devoted more than twenty thousand hours of volunteer service to the NTA.[33]

From the some twenty cubic feet of surviving NTA records recovered from a storage area at NRL, now at Howard's Moorland Spingarn Research Center, it also seems like he served as a de facto archivist for the NTA, from gathering and storing records of NTA activities from minutes of the directors meetings, to third-party correspondence that did not relate to his own responsibilities, to student applications for scholarships and their evaluation, to the financial and procedural records of planning for the annual meetings around the country, and to acting as primary contact with the publishers of the NTA *Journal* and the *Black Collegian* magazine. Clearly, he had assumed administrative as well as content-related responsibilities.

How to explain this amazing devotion in the 1980s, a time when Carruthers was still deeply involved in his scientific activities and proposing for missions on the Space Shuttle? As noted below, NRL supported this activity, but it also demonstrates that he was not blind or indifferent to racial matters.[34] He saw the NTA as an essential organization to work toward racial equality in science and engineering. But he also realized, once he got involved, that the NTA needed assistance. It still lacked a physical central office and staff, and coordination and management problems persisted throughout the organization, from publishing the journals to maintaining support.

Carruthers faced these problems in the convoluted editorial process of producing NTA publications. As he gained editorial responsibilities, encouraged by Thomas, Carruthers looked for ways to improve the editorial process. In August 1984, Thomas specifically requested his views on "Guidelines and Policy Regarding NTA Publications." Carruthers agreed with Thomas that there was a "lack of clearly specified division of responsibilities between the Executive Director and the Editorial Review Committee."[35] He made detailed suggestions for avoiding delays and failures in the complex communication chain, such as ways to improve communications and define responsibilities, and improving how the journal recruited and then supported authors. In marked distinction to his extreme focus at NRL, as one would surmise from his outwardly passive manner and his avoidance of administrative duties, he was actively involved in the health of the NTA, citing and addressing problems with stable funding so that the journal could be published on a regular schedule.

He outwardly praised the staff of the contract publisher, the *NTA Journal* staff at Black Collegiate Services, but he was frustrated by the number of major editorial errors he was seeing in the published editions: figures appeared in the wrong articles, articles were dropped, articles were inserted without his knowledge, many editorial corrections were needed—all leading him to suggest "that in the future, to minimize the possibility of error, I should be sent proof copies of both the final edited manuscripts and their associated figures." A necessary request that in the end increased Carruthers's workload.[36]

The publisher's representative, Sonya Stinson, advised, supported, and sometimes directed Carruthers's editorial work, helping him keep deadlines straight and often suggesting themes and essays. Occasionally, there were sensitive editorial issues relating to gender and race, and she helped him manage those situations. By then, Carruthers's main goal was to make the journal more accessible to students by adding technical articles they authored based upon laboratory or analytical studies they were doing in school or in conjunction with him and other mentors. "The big problem, of course, is to get the students to write up their parts of the project in a manner suitable for combining into [a] coherent paper which meets NTA Journal editorial standards!"[37]

His first recruiting effort to get students to publish targeted high school students from Ballou Senior High School, about a mile from NRL, who had worked in his laboratory and were supported by one of NTA's mentoring programs. He frequently appealed to readers for papers, saying "publication of technical papers is a very important means of communicating the results of your work, and to aid your development and advancement as a student or professional, in all fields of science and technology." In the spring 1993 issue, he described the NTA *Journal* as unique "in that all major fields are represented in the readership and authorship. Send us your papers!"[38]

During his years with the NTA, Carruthers wrote many essays and editorials, including long essays on technical subjects for the journal, keeping its members and broader audience—mainly educators and students—updated on opportunities in aerospace. He prepared these to be highly accessible and informative and to help all parts of the community, especially small businesses that needed to become familiar with what could be gained from improved contact. In 1985, for example, he reported on low-cost approaches to space shuttle experimentation and prospects for minority participation in the commercial use of space. For the winter 1994 issue, he described emerging space applications, highlighting that now, with the end of the Cold War and increased access to space by the shuttle, "there has been increased emphasis by the US Government on technology transfer from military and space programs to private industry." Providing copious references for further reading, bringing knowledge of journals like *Aviation Week and Space Technology* and NASA's *Tech Briefs* to his readers, Carruthers highlighted the promise of growing access to global positioning satellite systems like *NAVSTAR*. The last time the NTA *Journal* had paid attention to GPS, in 1990, it was expensive and complex. Thus, Carruthers added:

> However, in the last few years, electronic miniaturization has brought GPS positioning equipment into the size and cost range that private individuals can use it in their automobiles, boats or even backpacks to determine their positions to within a few meters! The practical benefits of this capability are many and obvious. . . . The development and sale of ground-based GPS equipment for commercial and private use is certainly a lucrative opportunity for small businesses.[39]

Carruthers went on to describe other opportunities for small businesses and individuals, urging them to take advantage of this heretofore exclusive technology. He also recruited other authors to give detailed descriptions of the internet, how technology transfer could be used by small businesses, and how to apply new technologies to their best advantage.

Carruthers constantly reminded readers of the NTA's objectives: "1) to increase the percentage of African Americans in scientific and technical occupations (and to enhance their status within such occupations), and 2) to improve the science, mathematics, and technology education of our youth," even among those who do not pursue those fields. Carruthers took a positive view, saying that much progress has been made and that it was good for the country as well as for the African American community. But he left no doubt that there was still a long road to equity.[40]

While his editorial essays were promotional and evangelical, his articles were research-based, where he identified areas that offered particularly exciting opportunities. Far from parochial, he embraced new technologies that threatened to replace his own. In 1986, he published a long paper, "Charge-Coupled Device (CCD) Arrays for Electronic Imaging," helping his readers understand how they worked: how, in the future, miniaturization in electronics would allow them to work in arrays to increase field of view and how CCDs offered new opportunities in many scientific and technical areas. He emphasized their application to astronomy, providing "significant improvements in low light level imaging, either alone or combined with electronic amplification," and acknowledged that the Wide-Field Planetary Camera on the recently renamed Hubble Space Telescope would employ CCDs, and that Princeton efforts, led by John Lowrance, had now used them in their sounding rocket flights. Although he was aware of technical problems scientists at the Jet Propulsion Laboratory (JPL) and elsewhere were having with the chips, he spoke only of their promise.[41]

Throughout these years, Carruthers organized NTA volumes with themes highlighting computer science and technology, aerospace science and technology, and, after interviewing a NASA assistant administrator, prospects for minority participation in space commercialization. By the 1990s, his frequent column, "State of the Art Update: Aerospace Science and Technology," highlighted profiles of Black people who were prominent names at NASA, like astronauts Charles Bolden and Frederick Gregory.[42]

But even Carruthers had limits. By then, his involvement was so deep in NTA that he asked to be relieved of the editorship of the newsletter to ease the burden. It had less technical material and was more focused on the immediate workings of the association. He was now serving on the NTA's "Science and Technology Advisory Board" to "keep the Journal abreast of current and cutting-edge technologies," and in 1990, Valerie L. Thomas even nominated him to be "Vice President for Public Relations," a request that he declined. But in the same year, he accepted the post of secretary to the "Development Fund for Black Students in Science and Technology," a program Carwell led. Over the next two decades he reviewed some "40,000 pages of applications for review and selection and not once delegating this daunting task."[43]

Throughout his years with the NTA, many of Carruthers's articles reflected his current interests, but he also reviewed in some detail the work of his NRL colleagues like Meier. In 1992, he spoke on "Ultraviolet Remote Sensing of the Middle and Upper Atmosphere" at an NTA conference in Ohio, at the same time contributing an address titled "Stepping into Tomorrow . . . Building on Our Technical Tradition." The point he wanted to make was that there was a lot going on that could welcome new talent. He reviewed "current NRL missions" in the early 1990s, including cameras and spectrographs on the Air Force Test Program *Shuttle mission STS-39*; an "Ultraviolet Limb Imaging Experiment (UVLIM) Air Force Space Test program STP-1 Hitchhiker"; and a project collaboration between NRL and the Aerospace Corporation for composition studies, to be flown also on NOAA's *TIROS-J*, maybe in 1994. He frequently wrote about programs he took part in, like MAHRSI, which would "measure the concentrations and altitude distributions of the trace constituents OH and NO in the mesosphere and upper stratosphere." Other missions he highlighted included "Cryogenic Infrared Spectrometers and Telescopes for the Atmosphere/Shuttle Pallet Satellite (CRISTA/SPAS)." And finally, he described "High-Resolution Airglow and Aurora Spectrograph (HIRAAS)" and GIMI, for a flight on ARGOS in 1995, and then GIMOL, the "Global Imaging Monitor of the Ozone Layer." Indeed, Carruthers demonstrated that there was a lot going on, highlighting the work of his NRL colleagues, describing their work in accessible terms that would make it attractive as exciting opportunities for his readers, inviting them to someday get involved.[44]

Although he provided enough details for his readers to appreciate how the payload experiments were differentiated, his primary goal was to emphasize why these sorts of studies were important for life on Earth. For instance, the atmospheric gases these missions studied had serious effects on ozone in the upper atmosphere. As he sternly warned, "There is now increasingly widespread awareness and concern about the effects of man-made activities on the atmospheric environment: for example global warming, ozone layer depletion, and acid rain."[45]

Viewing these challenges as opportunities, he reminded everyone that "there will clearly exist a variety of career opportunities in the development of new or improved space transportation systems." These systems required new expertise in aerospace and the mechanical and chemical engineering disciplines, as well as electrical engineering, engineering physics, nuclear engineering, and computer technology "to an increasing extent."[46]

Despite its broadly optimistic plans for expansion in the early 1980s, by 1988, the NTA had still not established a permanent national office. All they claimed to want was an address, possibly an office, so they mounted a $50,000 campaign to secure it because, though left unsaid, they still needed staff, including someone responsible for archiving the records.[47] There were definitely continuing financial problems, so severe by 1984 that after a meeting of the board on the "State of the NTA," the NTA president, Gilbert A. Haynes, proclaimed that they needed $100,000 to get out of debt, imploring the board, "WE MUST SURVIVE."[48]

Beyond debt, there were also problems coordinating finances, tax records, and recruitment, and, of course, managing the journal. Well into the 1990s, problems continued with the publications. The lack of staff made it difficult to coordinate both annual and regional meetings, especially the editorial work required after the meetings to produce the all-important proceedings. Carruthers, people felt, was doing all he could and evidently was aware of the situation. So after a meeting at the NASA Langley Research Center at Hampton Roads, Virginia, in 1997, the Langley lead wrote to Carruthers about the problem:

> Part of the problem is that no one takes responsibility for producing the final product. You do a great job at assembling, editing, and compiling the information. The ball is dropped when the conference host chapter does not make someone responsible for taking your output, adding the required cover, foreword, table of contents, acknowledgements, etc. and printing the final product (if you make it "camera-ready" then their job is less involved). Also no one takes responsibility for seeking underwriters for the printing. Since we have now outsourced most of our printing here at Langley, I'm not sure we can do it from here this year.[49]

This stark criticism, coming from NASA, was not taken lightly. Carruthers did what he could to compensate, but the records available do not illuminate the situation further. Even within deliberations over problems in the early 2000s that reported on meeting planning, meeting reports, NTA administration costs, tax filings, and budgets, there were issues raised as to the importance of establishing a formal records office. This had not been done yet. But on the bright side, by that time, the NTA *Journal* circulation was reported to be twenty thousand.[50]

The success of the journal and Carruthers's devotion to NTA infrastructure were no doubt some of the reasons the NTA frequently showered Carruthers with honors over the years. In 1990, the NTA cheered the "Black Achievers in Science" exhibit at the American Museum of Natural History in New York, sponsored by Citibank, which also ran an ad featuring Carruthers: "He Didn't Just Wish on the Stars, He Made a Career of Them." In 1993, Sonya Stinson, the publisher's representative to the NTA, told Carruthers that several supporters, like Citibank, wanted to "profile you as an outstanding member of NTA" (see figure 13.1).[51]

Carruthers was frequently highlighted because he exemplified a successful achiever in science whom promoters wanted to use to support their missions. But from what the accessible record reveals, his devotion and energies were not directed to winning recognition or prizes. He wanted to reach out to young people and find an effective mechanism to do so. While some members may have viewed the NTA meetings as a gathering of peers with all the expected social amenities, Carruthers saw them as an opportunity to reach students. As he reported after the NTA's 57th annual conference in Houston in 1985, which was filled with social functions, dinners, and tours, "For one or two days students were courted and wooed."[52]

Carruthers never ceased to encourage students he was mentoring to prepare the work they had done for publication in the NTA *Journal* as well as give talks at NTA conferences. He also wrote to science department heads and professors at universities looking for good candidates, spreading the word as far as possible. This was his goal.

He took special pride when his own students provided papers that reflected his own work and how they were involved. One might say he agreed with biologist Blake Riggs's view that "we must trust our research students because they are extensions of ourselves in the laboratory."[53] For example, at the 71st Annual Conference in Washington, DC, in November 1999, there were four tiers of technical sessions drawing upon professional speakers, and graduate, undergraduate, and precollege students. Sessions were held in computer science and information systems, aerospace, environmental and mechanical science, and technology, as well as

UPPER EAST SIDE VIDEO 215 East 86th St. TICKETRON

**HE DIDN'T JUST WISH ON THE STARS, HE MADE A CAREER OF THEM.**

**COME SEE THE BLACK ACHIEVERS IN SCIENCE EXHIBIT AT THE AMERICAN MUSEUM OF NATURAL HISTORY.**

George Carruthers set his sights far. All the way to the Big Dipper and beyond. Now he's a world-renowned astrophysicist.

Learn about more than 100 fascinating scientists at the Black Achievers in Science Exhibit. See how their curiosity and determination changed the world we live in. With great achievements in engineering, mathematics, computers, astronomy and more.

It all takes place at the American Museum of Natural History, March 16th through June 10th. Monday, Tuesday, Thursday and Sunday—10am–5:45pm. And Wednesday, Friday and Saturday—10am–9pm.

The Black Achievers in Science Exhibit is made possible by a grant from Citibank.

So come discover the Black Achievers in Science. A fascinating exhibit that can inspire our children to aim for the stars.

American Museum of Natural History

SPONSORED BY

CITIBANK

Figure 13.1
Citibank advertisement, *New York Times*, March 25, 1990. *Source:* Courtesy of American Museum of Natural History and Citibank.

in education and outreach in the physical sciences. Most of the speakers were local, but among the professionals speaking were Hattie Carwell, who described the opening of the Museum of African American Technology (MAAT) Science Village in Oakland, California.[54] Carruthers spoke in the aerospace session describing the objectives, mission operations, and expected data return from GIMI. But his goal here was to introduce his students. Garland Dixon, by then at the University of Maryland but still affiliated with NRL, described how GIMI was controlled to observe specific portions of the sky. In the undergraduate sessions, Melody Finch, also at Maryland and working for Carruthers at NRL, described orbital computations for command and control of GIMI and how the data would be processed through various Air Force facilities. Jessye Bemley, then a high school student at St. Francis Xavier School, reported on the sophisticated Hopfield network method of pattern recognition, a portion of Dixon's work. These were all solidly connected to Carruthers's interests, more so than the presentations between other mentors and their mentees.[55]

## S.M.A.R.T. INC. (SCIENCE, MATHEMATICS, AEROSPACE, RESEARCH AND TECHNOLOGY)

Carruthers remained active in the NTA until 2013, frequently citing the sequence "science, mathematics and technology" in his editorials and writings. By the mid-1980s, he had also become involved with S.M.A.R.T. Inc. S.M.A.R.T. was incorporated as a private nonprofit group in Washington in the mid-1980s. Hardly a unique acronym, it was the brainchild of then-congressman Mervyn Dymally, who chaired the Congressional Science and Technology Subcommittee. In 1985, Dymally "suggested the formation of S.M.A.R.T. as a group to advise on science and technology issues of importance to the black community."[56] One of NTA's brochures described S.M.A.R.T.'s formative history:

> It began as the Science and Technology Interface Group, with the National Black Leadership Roundtable (NBLR). It was initiated . . . to identify the most important science/technology related issues affecting African-Americans. S.M.A.R.T. was formed with the purpose of bringing together representatives of the scientific and professional communities to address the under-representation of African-Americans in science and technology in preparation for the 21st Century.[57]

S.M.A.R.T. soon evolved from high-level policy efforts to direct contact with students. In 1986, Valerie L. Thomas, who was the first female president of the NTA before Carwell, was asked to lead S.M.A.R.T., and Carruthers joined Thomas in that effort. Thomas's background therefore provides a useful perspective on Carruthers's own path.

Thomas took physics at Morgan State (College) University in Baltimore and in 1964 was hired by the National Space Science Data Center at the Goddard Space Flight Center as a mathematician and data analyst and then as an image processing manager on a series of NASA missions. Simultaneously with her work, she received a master's degree in engineering administration from George Washington University, which conveniently held classes at Goddard. She then rose rapidly in the ranks, becoming the computer facility manager for the National Space Science Data Center. She had extensive experience managing data flow for Landsat, first known as ERTS, or the Earth Resources Technology Satellite, heading a large team that developed predictive models for crop production and other applications. In 1985, as National Space Science Data Center computer facility manager, she managed a major upgrade/merger of two previously independent computer facilities, making it far more effective for international access and rapid calculations.[58]

Thomas's activism was ignited in college. She remembered being shocked on the first day of her physics class at Morgan State when she found that she had to be able to handle the calculus prerequisite. Her high school did not emphasize its centrality to the physical sciences: "Knowing the impact on me of not taking trig and pre-calculus classes in high school, I decided to reach out to the young people and let them know the value and importance of taking those high-level math classes."[59]

As DC chapter president, Thomas was always on the lookout for help, and George always complied. "George was always one of the ones to be involved and help make things happen. . . . He was of the quiet type, but he would step up and do work, especially education outreach work. He liked that, too."[60] Both Thomas and Carruthers were attracted to S.M.A.R.T.'s promised plan of action: it was local, was less complex than NTA, involved hands-on activities, involved parents, and was "an informally managed organization with people who had the same ideals about reaching out to our young people, and then informally interacting with each other, the people skills came sort of organically."[61] Most attractive, however, was S.M.A.R.T.'s plan to improve communications with other like organizations across the country, while remaining local, and to assess areas of the city that needed more attention.

Carruthers joined S.M.A.R.T. when the group was planning their first national conference, which took place in February 1989. S.M.A.R.T. was still trying to "decide exactly on a plan of action for the next ten years, our so-called ten-year plan."[62] That plan would change S.M.A.R.T.'s scope considerably. As it gained its own local focus and while it promoted hands-on experiences nationally, it became Carruthers's favorite, even more than his membership in the prestigious National Society of Black Physicists, based locally at Morgan State. S.M.A.R.T.'s efforts at coordination on a national scale also fit his sense of need. As he explained:

> The unique feature of S.M.A.R.T. is not that its goals and objectives are different from these other groups, but it seeks to coordinate the activities of these other groups, and the ten-year plan which was developed in part was an attempt to coordinate the activities of these separate groups that had similar goals. The 1989 conference . . . had as its objective to get people to speak about the programs of these individual organizations so that we could identify gaps where no one was doing anything and also to identify areas where we could have different groups collaborate so as to achieve the goals more efficiently.[63]

Carruthers, however, felt that although the conference was stimulating, its aftermath was a bit disappointing because it was difficult to maintain such a large network, especially for an all-volunteer organization. He was encouraged however that subsequent conferences would be smaller and more effective. And most important, by 1992, much to his liking,

> in addition to having speakers, we have workshops specifically directed to students and their teachers, and we have essentially three-day sessions. We have the first day for elementary school students, the second day for junior and senior high school students and teachers, in which we have hands-on workshops both of those days. And then only the third day, which is on a Saturday, do we actually have a formal set of lectures like at a normal professional conference, because we feel that one of the things that we can do that really is important is to get students interested in science at an early age.[64]

This was Carruthers's mission. The NTA meetings and publications were no doubt effective venues for already committed and accomplished students to demonstrate their work. But S.M.A.R.T. continued to offer less formal and therefore more effective and intimate venues to attract and then showcase students early in their careers. This philosophy aligned with Carruthers's interpersonal nature.

## S.M.A.R.T. SATURDAY WORKSHOPS AT THE NATIONAL AIR AND SPACE MUSEUM

S.M.A.R.T. sponsored myriad activities in the 1990s and 2000s to encourage students to experience astronomical events like solar eclipses, a spectacular evening when all five visible planets were in a line, a rare transit of Venus across the Sun, and a total lunar eclipse. Teaming up with NASA, the DC Public Schools, Howard University, civic associations, and school districts, S.M.A.R.T. provided formal and informal experiences, webcast services, and science and technology fairs, all under the umbrella of what is called STEM today.

One of the most ambitious programs was a six-year series of lectures, demonstrations, and tours cosponsored by NASA and the Smithsonian's National Air and Space Museum (NASM). Between July 1992 and June 1998, S.M.A.R.T. organized and directed over fifty Saturday morning lectures and workshops as a hybrid experience linking classroom and hands-on activities. Held in the Einstein Spacearium (the museum's planetarium) with hands-on activities in its "Briefing Room" before public hours in the morning, the students would later spread around the museum sharing their work with the public. Carruthers led the inaugural session with the talk, "Viewing the Universe and the Earth from Space." It was a general talk outlining the power and promise of the space program and the many areas of science it served. The talk was followed by a workshop that engaged students in the "basic principles of remote sensing" with experiments and demonstrations in optics, spectroscopy, and electronic imaging. The students used diffraction gratings to view tungsten and mercury lamps. They also watched and even manipulated image intensifiers, including night vision goggles and CCD cameras, to appreciate how they enhanced visibility. During these morning programs, NASM typically provided free access to planetarium shows and the IMAX theater, as well as a complimentary lunch for all participants.[65]

Interspersed with these weekend events, S.M.A.R.T. also sponsored special "S.M.A.R.T. Day" mornings. One of them, on January 23, 1993, attracted some eight hundred students, teachers, and parents. The student demonstrations were available to all, including casual visitors, who also saw Carruthers make at least five special presentations to school groups and organizations like the NTA, the Young Technocrats Laboratory School at Ballou Senior High School, and the Joint Educational Facilities program organized by Dr. Jesse Bemley, Jessye Bemley Talley's father. Bemley founded Joint Educational Facilities to advocate computer literacy among marginalized students and was well known for his efforts in past years to bring thousands of students together for coding sessions called "hack-a-thons."[66] The "Young Techs" presented a particularly imaginative award-winning exhibit depicting a Mars-based space city, dubbed "Timbuktu 2071." NASM exhibited it on the floor for another week after the workshop, and during that time, Carruthers and his students visited several times, demonstrating a compressed cold-gas model rocket and displaying more high-powered model rockets. The NTA's 3-T Program (Technologists, Teachers and Targeted Students) had sent four students and a teacher from Ballou Senior High School to work with Carruthers to mount and discuss their own Mars display: a vision of colonizing Mars, centering on producing a breathable atmosphere. There were also environmental themes, like the importance of switching from petroleum to electricity as a means to propel cars, and there was a semi-autonomous roving robot built by Bryce Rowe, a seventh grader from Decatur Middle School High, who sent his creation around the museum, surprising and delighting visitors.[67]

In the following months and years, speakers included community education specialists, astronauts Charles Bolden and Frederick Gregory, an NRL speaker who introduced techniques of Earth observation from the Space Shuttle, speakers from NASA and elsewhere on how to commercialize NASA technology, project engineers from Langley and Goddard, and a former Tuskegee airman, who described careers for Blacks as pilots, astronauts and future prospects. These presentations were followed typically by a tour of the museum's *Black Wings* exhibit, which had been open since 1982. There were also presentations and workshops by Howard predoctoral students in mechanical engineering and by a middle school aerospace teacher on African Americans in space, as well as a series of sessions that were devoted to hands-on space science activities led by S.M.A.R.T. and the affiliated Young Technocrats.

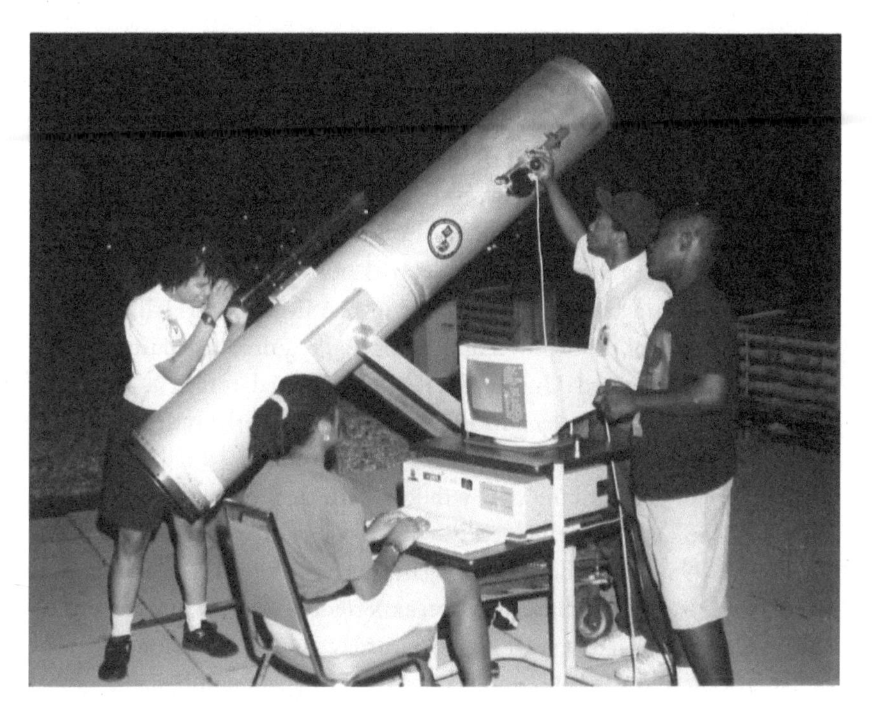

Figure 13.2
Viewing the Shoemaker-Levy comet impact into Jupiter's atmosphere from the roof of the Laboratory School, July 1994, Howard University, Washington, DC. Carruthers photograph. George Carruthers, "Outreach Programs for African-American Students in Washington, D.C.," *Mercury* 24, no. 3 (May–June 1995): 30. *Source:* Courtesy of Valerie L. Thomas.

Howard University became another site for Carruthers's effort. In 1994, Carruthers helped students in the Young Technocrats Laboratory School at Howard renovate a 12.5-inch reflecting telescope, equipping it with a CCD for a special Saturday evening session on July 16, 1994, titled "Jupiter Watch" (see figure 13.2).[68] (They wanted to detect the aftermath of the impact of the first of some twenty-two fragments of Comet Shoemaker-Levy that had entered the Jupiter atmosphere at about 4:00 p.m. Eastern Daylight Time on that day. Jupiter was in the southwest at the first viewing time of 7:00 p.m. as the Sun was setting. Carruthers and the student group continued to follow Jupiter over the next few nights. As the *Washington Post* reported a few days later, the accumulating impacts produced large Earth-sized "holes" in Jupiter's atmosphere that could be seen by "skilled amateurs using relatively small telescopes."[69]

In April 1995, Carruthers organized a session led by students from Ballou's Young Technocrats Laboratory School. They discussed and demonstrated techniques of UV spectrum analysis, solar-electric rocket propulsion, the influence of carbon dioxide on plant growth, and a recap of the Shoemaker-Levy comet impact. Then in May, Carruthers teamed up with teachers from Ballou and Howard to present and demonstrate the principles and safety practices of model rocketry. They demonstrated low-power vinegar and baking soda rockets in NASM's Space Hall and then showed examples of more powerful models that required safe outdoor sites.

NASM and Howard University staff also joined forces at times. In June 1995, in the Saturday morning program, recently retired Howard University astronomer Benjamin Peery gave an illustrated talk of his personal experience, describing his "winding path that led me to astronomy, and why I think it was the luckiest discovery of my life."[70] Following Peery in August 1995, Carruthers spoke again on observing the Earth's atmosphere from space, providing hands-on experiences for how remote sensing works. Carruthers led several other Saturday morning sessions on topics like hot air and gas ballooning; on ARGOS, again emphasizing global resource mapping; and on the Lunar Prospector Mission, known as project "Moonlink." Toward the end of the series, students from area schools like the Anacostia High School gave presentations of their projects, from observing a comet with a CCD camera to exploring the technology of rocketry. And finally, toward the end of the series, in June 1997, there was a video premiere of part of a series titled "Earth and Space Science" that featured students from Anacostia, Ballou, and Arundel High Schools, as well as Howard and the University of Maryland. This form of presentation, where students performed the experiments on screen, became one of Carruthers's favorite strategies: students as "model" scientists (see appendix A).

Over the years, half the presenters in the series were women—teachers, practicing scientists, administrators, and students. There were many voices of diversity offering insights into careers in science, engineering, aviation, and spaceflight technology—the topics were diverse, and the series gave exposure to many local and national organizations ready to further engage students.[71]

While the series was in progress, Carruthers also proposed collaborative programs for funding by NASA and by the NASA-funded DC Space Grant Consortium, which was a group consisting of five local colleges and universities, the NASA Alumni League, Raytheon, and S.M.A.R.T.[72] In 1997, Carruthers and Thomas created a "Proposal for Space Science and Technology Education and Outreach Facilitation in the Washington, DC Area." Carruthers was the principal investigator, and S.M.A.R.T. was the primary agent, allied with the newly formed Young Technocrats, Inc., created out of the Ballou Senior High School group as a "nonprofit hands-on science, mathematics, advanced technology and entrepreneurship training program for children and young adults of African descent, ages 4–21." This partnership, supported by several sources, including a NASA IDEAS grant and the DC Space Consortium, conducted not only more classes and labs, but also, most significantly, produced another video series titled "Pyramids to Planets," shown on Washington, DC, cable television station Channel 28.[73]

Starting in about 2000, Carruthers's involvement deepened and broadened. He and Thomas were now in contact with the Southeast Regional Clearing House, a NASA Education Payload Operations broker and facilitator promoting themes relevant to programs supported by NASA's OSSA and funded by the DC Space Grant Consortium, which also supported S.M.A.R.T. as an affiliate. Specifically, OSSA's goals, then and now, were to improve knowledge of the Sun–Earth connection—in other words, how the Sun influenced life and culture on Earth, in its atmosphere, and in the many types of space applications that NASA and other agencies were developing, such as communications and navigation, that were affected by the connection. The program also covered the exploration of the Solar System, the search for the origin of life, and the study of the structure, evolution, and fate of the universe. But OSSA also wanted deeper and more intimate access to public schools and community centers. That is where S.M.A.R.T. came in.

Carruthers and Thomas reported several times on "space science education and outreach," delineating the many programs S.M.A.R.T. was sponsoring in the Washington, DC, area and elsewhere. In 2002, for instance, they reported on some eight programs, past and present, including the Saturday morning programs at NASM. They also described the video series they were producing where students like Jessye Bemley served as "actors" to increase the "interest and appeal to the targeted students." Other programs provided video and textual material, as well as hardware support for elementary and secondary school teachers. And on several occasions, supported by both NASA and NRL's "Community Outreach Program," they encouraged and facilitated student participation in science fairs and essay contests.[74]

And as Carruthers and Thomas reported at a meeting of the American Geophysical Union in 2005, they had also started developing curricula centered on active learning and hands-on experiences, along with objects and equipment to accomplish the stated goals of onsite interaction in both formal and informal gatherings. In addition to presentations around the DC metropolitan area, they also produced a series of videos as well as digital copies of their presentations for classroom use. "Of particular note," they stressed, "is that students, and their teachers, are active participants in the videos." Anacostia Senior High School and Backus Middle School were two of the initial sites.[75]

Carruthers seemed to most enjoy taking advantage of celestial events, bringing them to the public, and making students feel they were taking a part in exploring them. Thomas facilitated these efforts by establishing associations with sites in the DC area that fostered such events, like "D.C. Link and Learn." In the late 1990s, the US General Services Administration provided some forty thousand square feet of office and warehouse space in the Waterside Mall area in southwest DC to establish "a computer and information technology center that will train youths and adults for the many technology-based jobs available in the Washington, DC area."[76] S.M.A.R.T. utilized facilities like this for many events, such as when the Moon was especially prominent or the planets were visible in the evenings. Carruthers would often bring portable telescopes he built and students he had trained to operate them so multiple lines of visitors could be accommodated. He did much the same at Howard and other sites whenever something was interesting in the sky, like the transit of Venus across the Sun's disk in 2004.[77]

Carruthers liked to tease his students at times, not as a superior adult, but as a peer. He blended in with them in a nonconfrontational, encouraging manner. Thomas recalled that S.M.A.R.T. supported the renovation of the telescope on the Howard campus, which was above what was called the "penthouse" on the fifth floor of one of the buildings, and sponsored numerous days and evenings when people could look through the main telescope as well as small telescopes that Carruthers and his students would bring in. During these events, Carruthers's favorite prank was to push his students into the elevator in the basement with the telescopes, hit the "up" button, and then after the doors closed, run up the stairs to the top floor greeting the elevator as the doors opened, saying, "Hi, you finally got here." As Thomas added, "It was like his little trick on the students."[78] He did this also at NRL with students, always running up the stairs, faster than the elevator.

Even in his sixties, still slim and fit, he continued to amaze his students with tricks like these or wearing wild T-shirts with celestial apparitions. He rarely drove his car, preferring to make colorful entrances on his bike. He kept these antics up when he taught at Howard and in other S.M.A.R.T. ventures. One never knew what surprise would come next. Typically, he would wear a white shirt and his ubiquitous pocket protector filled with pens, pencils, and a small slide rule. But he did not shy away from theatrics. Valerie L. Thomas remembered a Halloween science presentation that Carruthers gave at a local high school. It was an interior room with no windows, and after he finished his formal presentation, he flicked the lights off and ripped open his shirt, revealing a glowing Superman logo. "Some students were so scared they went running out of the door."[79] Most of his many T-shirts carried NASA mission logos or general space themes.

## THE 2000s—EXPANDED OUTREACH AND CONTINUED RECOGNITION

As we have seen, NRL worked with schools to organize minority job fairs to engage "educational partnership and tutoring programs with schools from elementary through high school." There were frequent tours of the facility for school groups and paid summer training programs that managed over one hundred students each year, many engaged directly in research projects.[80]

NRL's support aligned with Carruthers's goals and eventually improved diversity employment. In 1987, NRL could report only some forty Black scientists and engineers on the staff, less than 3 percent of some 1,400 staff in those categories—not inconsequential though definitely still in need of attention. By 2010, however, that fraction had increased, and the *Diversity/Careers in Engineering & Information Technology* magazine, for the fourth year in a row, identified NRL as the "best Diversity Company" from a poll of its readership.[81]

NRL was more than willing to support Carruthers's outreach efforts because he provided positive community exposure for the lab. They had always promoted his achievement in space science, but by the 1980s, they were also recognizing his community efforts. In October 1986, Carruthers, as "senior astrophysicist," was presented with the "1985 Commanding Officer's Award for Achievement in the Non-Supervisory Field of Equal Employment Opportunity (EEO)" for his leadership in the NTA, for his membership on NRL's EEO committees, and "for his role in the recruitment of minorities and women at colleges, universities, and job fairs." He was also cited as "instrumental in implementing and participating in numerous activities for the NRL Community Outreach Program," serving as a tutor, lecturer, mentor, role model, and judge for science fairs, and lauded for his work with the University of the District of Columbia's STARS program (Science, Technology and Research Students) for junior high school students.[82] The same issue of NRL's *Labstracts* announcing the award also had a two-page spread, titled "Reaching Out: To Give and to Receive—The Best for Both," highlighting NRL's history of outreach.

Beyond NRL, Carruthers was frequently cited for his outreach efforts and celebrated as one of a handful of "Racial Stereotype Busters," the title of a 1999 review of a CD-ROM titled "American Science Leaders."[83] The point of the review, however, was to explore just how few Black people were included, some fourteen out of four hundred scientific leaders. Some of the fourteen, in addition, were historical figures: Benjamin Banneker and George Washington Carver, for instance. But there were others listed with brief biographies that, the reviewer contended, warranted greater attention. They included Carruthers, and the astrophysicist Carl Rouse, who, we have noted, was a Hulburt Fellow in the mid-1960s.

Most of Carruthers's early recognition came for his success as a space scientist. In 1970, however, Washington Jaycees awarded Carruthers its Arthur S. Flemming Award for Federal Service. In 1972, he received NASA's Exceptional Scientific Achievement Medal, and a year later, he received the Helen B. Warner Prize for Astronomy in 1973 by the American Astronomical Society, given to exceptional astronomers under the age of thirty-six. The Warner Prize specifically cited his detection of interstellar molecular hydrogen.

Later awards recognized Carruthers as a symbol for the contributions African Americans were making to science and engineering. He was designated "Black Engineer of the Year" in 1987 and cited in numerous compendia, such as the three-volume set *African Americans: Voices of Triumph* from Time Life, Inc. in 1993, and he was inducted into the National Inventors

Hall of Fame in 2003.[84] The awards, prizes, and medals continued, along with news features into the 2000s. One of the highest honors was bestowed on February 1, 2013, when Carruthers received the National Medal of Technology and Innovation for 2011, presented by President Barack Obama at a White House ceremony (see chapter 14).

Most of these awards and recognition in the 1990s, however, were not for his recent efforts but for his *Apollo 16* success and his measurement of the abundance of molecular hydrogen between the stars. In a 2015 four-volume NRL retrospective on the achievements of the Space Science Division, citations to Carruthers's publications diminished in the 1980s and 1990s, although the fourth, for 2000 to 2010, cited him in narratives of his leadership in GIMI, marking him as a "pioneer of far-ultraviolet astronomy.[85]

GIMI, however, was wholly NRL funded, so even with his effectively "free" space in his trailer domain, he still was not bringing in enough funds to pay the bills, which included his substantial salary. However, Carruthers was still highly visible and well liked. The value of his collegiality, his outreach and mentoring, and the positive public exposure he was still accruing for the laboratory could not be ignored.

## YET ANOTHER EVALUATION AND RETIREMENT

In 2000, soon after Carruthers moved to his trailer lab at NRL, he, along with everyone else at NRL, had to be "re-mapped" into a newly revised personnel system.[86] NRL management carefully reworked his position description and the standards he had to meet for acceptable performance, and he did very well. Rated as a "Science and Engineering Professional, Career Level IV," this was second only to the highest level "V," which was the division superintendent, Gursky. Giovane was listed as immediate supervisor and Gursky as higher-level supervisor. In both the 1998 and 2000 versions, the incumbent had to be an authority "within their professional areas" *or* be working at the level of "key program administrator."[87] In addition, Carruthers's draft identified technical consultation as well as "strategic planning & prioritization."[88] The weight for each category, however, could be prioritized; Carruthers's draft personal statement read, "You may or may not perform all of the duties and responsibilities," which, in effect, gave his supervisors considerable latitude.[89]

By 2000, Carruthers was still collaborating on various projects, but not supervising anyone except his students—nor was he engaged in any high-level planning or organization. There is no indication that these criteria factored into his case, but the sad reality was, of course, that he no longer had a full-time support team. Opal, Heckathorn, and William Conway were all gone. And by 2000, Robert Conway had retired and Seeley was part time. GIMI was flying, and Carruthers was still proposing for space missions for studies of the Earth's uppermost atmosphere. But the time had come for a change.

Just as Gursky had negotiated with Carruthers over his space allotment, he also sweetened the pot for considering retirement. Sometime in 2001, Gursky crafted a plan to designate Carruthers a "rehired annuitant" ***if*** he agreed to retire. This would technically provide a retirement income supplement that was available to both military and civilian employees who had worked for more than twenty years.[90] The exact agreement is not known, but Gursky left no doubt that Carruthers would retain his laboratory space and full access to all facilities for his students and his research. Further, and significantly, he promised that NRL funds, those that Gursky held as "discretionary" funds from the overhead that the division's many contracts accrued, could be applied, as needed, to support Carruthers's continued research expenses.[91]

There was evidently another reason for Carruthers to accept retirement. In spite of, or in addition to, his agreement with Gursky, Carruthers made it clear on his resignation form that "I am resigning from my position at NRL for personal reasons, to care of family."[92] We will provide background confirming this statement in chapter 14.

Carruthers retired from being a full-time staff member in 2002 at age sixty-three after thirty-five years of service, but he remained active at NRL for the next decade, professionally identified with his office and status as a "rehired annuitant," eventually receiving forty- and forty-five-year service pins and certificates.[93] He appeared at the lab almost every day, continued to mentor younger staff whenever asked, and continued to use his trailer for his widening student activities.

By the 2000s, although he collaborated on GIMI reports and proposals for more missions, most of his publications and presentations had shifted to education and outreach, as well as to addressing social and cultural issues like global warming. He wrote semitechnical introductions to topics in the space sciences, especially the production of image tubes, since the 1970s for science-based audiences, including "Space Astronomy in the Far Ultraviolet" for *The Science Teacher* (1973) and "Photoelectric Devices in Astronomy and Space Research, Parts 1 and 2" for *The Physics Teacher* (1974), delineating the basic principles of photoelectric point-source and multistage imaging devices. And he continued to provide these informational articles to broad popular astronomical audiences even after he became editor for the NTA *Journal*.

Over these years, he spoke frequently and wrote essays like "Outreach Programs for African-American Students in Washington, D.C." in the popular astronomy journal *Mercury*. He spoke on broader educational topics at meetings of the American Astronomical Society and was very enthusiastic about NASA's IDEAS program cosupported by NRL, Howard University, and the DC Space Grant Consortium. On several occasions, he reported on the many space science videos he was producing with his students; they had a total running time in many segments of some nine hours demonstrating hands-on activities and special astronomical events, like a Venus transit across the Sun's face in 2004 and the crash of Comet Shoemaker-Levy into Jupiter's atmosphere in 1994.

*Mercury* had invited Black astronomers, women, and Native Americans to provide personal commentary. The editors made the situation quite clear: "There are about 15 African-American professional astronomers in the United States, Not 15 percent, *fifteen*."[94] Most of those invited gave personal accounts, but Carruthers focused on the many mentorship programs he was involved in "exposing" young students "to African-American science and technology professionals (who serve as role models and mentors) and by providing them with opportunities for hands on involvement." He gave much the same message in a 1999 summer teacher training course held both on the campus of Howard University and in the labs at NRL, once again singling out how important it was for students to gain practical laboratory experience by assisting in the development of scientific instruments, like spectrographs. He attached one of them to Howard's campus observatory, which was becoming more and more a center of his professional life.[95]

## HOWARD UNIVERSITY PROFESSORSHIP

In 2002, continuing as an annuitant NRL staffer, Carruthers became an adjunct professor at Howard University and was funded by an education initiative sponsored by a NASA Aerospace Workforce Development Grant.

In addition to creating special events at the observatory during the day and night, he taught a two-semester course on the Earth and space sciences two evenings a week, which allowed him to continue his NRL activities and his frequent speaking trips.[96] Carruthers was not the first professor of astronomy at Howard, so a short diversion to describe the program and the astronomers who created it before he entered will help foster appreciation for Carruthers's involvement.

Astronomy at Howard began formally when Harvey Washington Banks arrived in 1969. A 1961 PhD from Georgetown University, in the words of Robert Fikes, Banks "became the first African American to earn the doctorate specifically in astronomy."[97] At Howard, he focused on teaching, first in astronomy and then in physics. Remembered as a "fighter, a champion of students, staff and faculty alike," he built a planetarium on campus and a students' observatory at a site in Beltsville a few miles away from campus. Banks was also active in the National Science Teachers' Association and remained active at Howard until his untimely death in 1979 at the age of fifty-six.[98]

Banks was followed by Benjamin Peery, the second Black American to be awarded a PhD in astronomy. Peery, born in Missouri in 1922, was the son of a railway mail clerk and a teacher. But his mother quit working when he was born. They were one of the few Black families in the area, and in words reminiscent of Carruthers's, Peery felt that growing up in a nonurban environment had advantages:

> It not only provided me opportunities for some sort of contact with nature, which I think is very important if a person is science-bound, but there was this sort of preservation of innocence, which I don't want to make too light of, because I do think it's terribly important.[99]

And, Peery emphasized:

> One of my great depressing feelings about the lack of blacks in science has very largely to do with just this sort of debilitating influence that I think black urban life has on this loss of innocence, for want of a better term. I think there are lots of other reasons why there are so few blacks in science, too, but I think this is one of them.[100]

Given this outlook, Peery, who had obtained his master's degree in astronomy from Fisk University and his PhD from the University of

Michigan in 1962 with a thesis on spectroscopic binary stars, felt he had a social obligation to engage in outreach in astronomy, but his core goal was research and graduate-level teaching.

Peery was a professor for some eighteen years at Indiana University in their strong astronomy department, chalking up a solid record of research and publications and a successful stream of PhD students.[101] In 1977, he decided to leave Indiana to teach at Howard, feeling "a responsibility to increase . . . and to improve educational opportunities generally for the African Americans," even though he realized that this "new move would mean less time for research, and more emphasis on teaching, services, and efforts to bring more African Americans into scientific careers."[102] Peery's goal was "to start a graduate program in astrophysics." It was a challenge he could not deny. As much as he and his family loved Indiana, Washington beckoned.[103]

The prospects of establishing a graduate program seemed exciting, but soon he found it to be anything but straightforward. Interviewed soon after he arrived at Howard, he felt that

> "Howard has enormous problems." They were primarily problems that "any university faces in transition from an undergraduate school of very high quality, devastated by the action of the late sixties and early seventies. . . . Howard perhaps suffered more than any other campus in the country. . . . I thought I could walk in and just have my way, and I finally had to accept the fact that that is not easily done."[104]

Peery fought to establish his graduate program over the next thirteen years until he retired. He hoped for "a real program, a permanent sustained program of utilizing space data."[105] He remained active as emeritus professor in astronomy through the 1990s and also, like Carruthers, began to lecture widely in public venues and at colleges, appearing on the PBS series *The Astronomers* in 1991.[106]

Peery did manage to develop the department. Cidambi K. Kumar joined Peery in 1974 after a few years at the Carnegie Institution of Washington working with astronomer Vera Rubin. In 1988, Prabhakar Misra arrived with a PhD in physics from Ohio State University. Together with Kumar, Peery taught four graduate courses in astronomy: "Radiation and Matter," "The Galaxy," "Basic Astronomical Data," and "Stellar Structure and Evolution." By the 1990s, the department had grown a bit, with Anand P. Batra and Lewis Klein on the staff.[107]

The department diversified in race and specialty. By 1996, it has been expanded into an interdisciplinary "Center for the Study of Terrestrial and Extraterrestrial Atmospheres" within the Department of Physics and Astronomy. In parallel with this effort, a "Graduate Program in Atmospheric Sciences (HUPAS)" emerged, with faculty from the departments of physics and astronomy, chemistry, and mechanical engineering. "HUPAS is the first advanced degree program in the atmospheric sciences, or related fields such as meteorology and Earth system sciences, instituted at a historically black college or university or minority-serving institution."[108]

During these years, as we noted, Carruthers had contact with Howard University, inviting co-op students to his laboratory.[109] He also helped to organize special events at the observatory, where he brought small telescopes to augment what was there. In 1988, he helped to stage a S.M.A.R.T. conference there.

His contact increased in the 1990s when Venable, Misra, and their colleagues invited him to serve as a member of the advisory board for a new "Center for the Study of Terrestrial & Extraterrestrial Atmospheres," a NASA-funded program.[110] He expanded his evening sessions at the Howard University Observatory, engaging his Howard undergrads and grads who returned to help out, doing what he could to restore and complement the instruments in the old observatory on the roof of Locke Hall. His "family nights" delighted teachers, families, and students alike. Misra felt that "these 'family nights' served to inspire numerous students and DC public school teachers and help[ed] them witness some of Mother Nature's astronomical grandeur at work first-hand!"[111]

Starting in 2002, as adjunct professor, Carruthers engaged in more than teaching at Howard. In 2004, when Misra left the campus for a Fulbright sabbatical year, Carruthers filled in, managing Misra's proposals to NASA and elsewhere to support their burgeoning curriculum.[112] He also assisted and endorsed many proposals for funding Space Grant Consortia projects from colleagues like Richard C. Henry of Johns Hopkins. In 2006, he coauthored an extensive report and paper with Prabhakar Misra and Gregory S. Jenkins in the *Journal of Geoscience Education*, where they outlined curricula for physics, the atmospheric sciences, and the earth and space sciences to prepare students, "especially underrepresented minorities," for careers in those broad areas.[113] Further, as Misra recalls, the courses Carruthers created with other faculty "were leveraged by colleagues in the department" to secure large grants from NASA, NOAA, NSF, the Universities Space Research Association (USRA), and the DC Space Grant Consortium: "In no small measure, this in turn helped launch the graduate-level Howard University Program in Atmospheric Sciences (HUPAS) that has been very successful in producing African-American Ph.D.'s in Atmospheric Science."[114]

Misra also recalls that Carruthers was a visible character on campus. As he did at NRL, he rode his bike everywhere, committing the same antics he played out at NRL with his students: "I have fond memories of George riding his bike to the Howard campus and bringing his bike inside the research building where my lab was located and chatting with students in the hallway with his helmet still on his head!"[115]

Beyond Howard, Carruthers reached out to a wider generalist but still science-oriented audience. In the 2000s, he directed a greater fraction of his time to science teachers and elementary grade–level classrooms. Of his eighteen publications and abstracts of talks after 2000 recorded in the Astrophysics Data System, six were educational, but he wrote many more not indexed.[116] With Valerie L. Thomas, he presented the many activities of S.M.A.R.T. With Misra, he reported on the revised Howard programs in astronomy and physics and the atmospheric sciences. And there were articles on upcoming celestial events like solar eclipses and how everyone could participate. In 2006, he published at least three articles generally on science education and outreach.

Clearly, Carruthers knew, as Harry Heckathorn recalled after seeing Carruthers for the first time at a meeting at Northwestern, that even brief and casual contact could help minority youth realize that careers in science were possible for Black people. For example, Frank Padgett (see appendix A) had been interested in astronomy since second grade but did not feel it was something he could do until Valerie L. Thomas introduced him to Carruthers, first in fifth grade and then in middle school at Howard when Carruthers invited the class to tour his laboratory at NRL. This contact propelled Padgett to attend Space Camp in Huntsville and then a summer school astronomy course at Harvard, which led to a master's degree in physics at Fisk University and a budding career as a radiation effects test engineer (see figure 13.3).[117]

Carruthers's attention to minority outreach eventually dominated his life. His last recorded publication in a refereed journal cited by the SAO/NASA Astrophysics Data System appeared in March 2009, based on a talk

Figure 13.3
In 2012, with a scholarship obtained through Valerie L. Thomas, Frank Padgett was able to attend Space Camp at the US Space & Rocket Center in Huntsville, Alabama, living out his dream inspired by Carruthers. *Source:* Courtesy of Frank Padgett.

he gave in a "Dynamics of Diversity in Astrophysics" session at the American Physical Society (APS). The speakers represented many backgrounds, and the session was organized to demonstrate diversity in the APS. Hakkeem Oluseyi, a specialist on solar physics and variable stars from Florida Institute of Technology, addressed the matter, speaking on "The Participation and Research of Astronomers and Astrophysicists of Black African Descent." The three other invited speakers, including Carruthers, served as examples of Black people active in frontier science. Two of them centered on their contributions to understanding the dynamics of the accelerating dark universe. Carruthers, however, chose to speak on what was, and remains, possibly the most pressing of all problems facing humanity: climate change and ways to mitigate it.[118]

In his later years, having concentrated on understanding the middle and uppermost regions of the Earth's atmosphere, Carruthers became evermore concerned about environmental issues like climate change, but he saw this as "no major changes in direction" in his own scientific endeavor.[119] Indeed, his scientific goals varied little, from the Earth's upper atmosphere to the nature of the interstellar medium. But as we have seen here, what changed over time was his increasing attention to mentoring and outreach on subjects ranging from science to the environment.

There is also little question, we have shown here, that his prominence after *Apollo 16* made him attractive to those who were seeking symbols of successful Black scientists to present to diverse audiences, the media, and skeptical reviewers of EEO efforts. And he responded primarily to give back by making himself accessible to countless students, of all backgrounds, seeking inspiration. He never tried to impose himself as a superior being, but as an accommodating, nonintimidating equal, demonstrating the principles and mechanisms of physics and astronomy (as his former mentees attest in appendix A).

Carruthers continued to teach at Howard and be engaged in related community activities for about a decade until his health and energies started failing.

# 14

# FAMILY LIFE AND RECOGNITION IN LATER YEARS

Carruthers was, indeed, "NRL's Not-So Hidden Figure."[1] Throughout his career, he received significant recognition through awards and prizes; literally wherever he spoke, the hosts would provide him with a certificate of appreciation, or a mug, medal, or desk ornament. At home, life remained quiet. Carruthers was always loyal to his mother and siblings and was especially protective of Sandra. There were family reunions, in Chicago and Brooklyn, and these were enjoyable moments for the very quiet couple.

Having risen rapidly in the ranks at NRL, as we noted, becoming a high-level GS-14 in the late 1960s, a GS-15 in 1971 (in the mid-$20,000s, over twice the median salary range), then head of the Ultraviolet Branch in 1980, and a nonsupervisory senior-level astrophysicist two years later, George and Sandra were comfortable financially. Carruthers was generous with his family. His brother Gerald observed that:

> George has always been willing to help anyone in need regardless of how much. He loaned Mom and Mr. Martin [his stepfather] down payment on a house in Barbados [Martin soon passed], and from then on paid the remaining money [to] our Mom! He has helped all of us, whether we repaid or not!! After Mom passed [in 1990], he gave the house to Tony . . . [and] . . . He lent me money when I was out of work, my son when he was in college, Barbara Ann for fixing up her house etc.[2]

After her marriage to Martin, Carruthers's mother Sophia assumed the peripatetic life of a military wife. She and her younger children followed him at least once to a new assignment. But after his retirement, they moved back to Chicago, where Martin served as dean of St. Augustine Cathedral and Sophia became active in Christian social organizations for several years until Martin died. They had already established a vacation home in Barbados, and Sophia visited the family often after Martin's death, continuing her social activism in many organizations both in Barbados and in Chicago, including the Barbados Museum, Meals on Wheels, the Post Office Chorale, and the Peace Corps. She passed away in Chicago in February 1990.[3]

## LIFE AT HOME

George and Sandra's move to O Street turned out to be a blessing for both of them, especially for Sandra. It was nearer to many amenities and much closer to Howard University, George Washington University, and the Smithsonian. George's bike ride commute to work was longer, requiring that he safely negotiate a bridge over the Anacostia River, but by the 1990s, Carruthers's attention was not only on NRL but also on reaching out and giving back, which took him around the DC region to schools and to Howard University.

After their marriage, one of Sandra's cousins reflected, "I [knew] her to be very social especially with her friends—I wondered about their relationship as they were total opposites from my observation. She never complained however." The cousin also recalled darkly that "Sandra wanted children and that was one of the reasons she suffered mental issues."[4] The symptoms, as described by the cousins, included believing she was pregnant and gathering baby paraphernalia and leaving them scattered around the house.

Her mental condition preceded their marriage. An aunt recalled that even in her teens, "she worried about not having kids," since several of her aunts were childless. And in time, she exhibited deeper delusional states. Her father tried to get her medical attention while keeping her mental condition secret from the family as well as Carruthers. To the cousin, "it was being dealt with as something shameful so I do not know if she ever got [help] and every time I asked Uncle Francis [Redhead] about her he would only say that she was ok."[5]

Sandra outwardly was bright, sociable, and cheerful, so her troubles were not apparent to Carruthers during their courtship. But the troubles gradually deepened after their marriage. Beyond the false pregnancies, she suffered from other delusions. Even though he sent her postcards during his many travels for space launches, according to one cousin, Sandra "told me that George was abducted by aliens and was gone for weeks and when he returned it was actually a clone and not her husband." She did travel on her own, visiting family in Brooklyn and Grenada. She also

Figure 14.1
Carruthers enjoys teasing his nieces and nephew at his mother's home. *Source:* Courtesy of Gerald Carruthers.

traveled to Canada, England, and Europe, "or wherever the whim hit." Inevitably, there were mishaps: getting lost, losing her luggage, or missing a connecting flight.[6]

Carruthers always supported Sandra in areas where he perceived need, like her education and her travels to and from family. After she finished college and became frustrated trying to find a job, she worsened. During visits, Uncle Ben definitely noticed her deterioration, her fantasies, and erratic behavior. He became deeply worried and even suggested to Carruthers that he divorce her, which George refused to do.[7] But as her state of mind deteriorated over time, Carruthers became deeply concerned. Her desire for children must have been very worrisome. Carruthers definitely liked children (see figure 14.1) but must have known that if they did have a child, Sandra would not be able to manage. Sandra's mother also started to visit them in Washington, staying for days to help manage Sandra's needs.

By the early 1980s, likely urged by George, Sandra was seeing a therapist weekly and a psychiatrist when she needed medication.

In their courtship correspondence, Carruthers had bluntly described what life would be like living with him. And indeed, during these years, Carruthers focused evermore intensely on his laboratory and his research activities, keeping long hours, working weekends, and traveling. Through the late 1970s, he continued to work with Page and others, analyzing the *Apollo 16* S-201 data and proposing for payloads on Skylab and other missions, and, as we know, by 1976 had applied to be an astronaut.[8] Even during their courtship, beyond NRL, and the pressures of the limelight focused on him by Monroe and Redhead, Carruthers also was a part-time associate professor at Johns Hopkins University for the academic year 1972–1973, teaching a course in introductory astrophysics.[9]

With her college degree from George Washington, Sandra probably sought out various jobs that appealed to her, but no evidence has been uncovered that she established a successful professional career and stable independent life. Little correspondence has been found between them after their courtship. What has been found was addressed only to Carruthers. Although life at home was evidently quiet, as Carruthers promised, they traveled together frequently, taking short vacations to various garden spots like St. Croix in the Virgin Islands, to California, and also to Huntsville, Brooklyn, and Chicago to see family. During these vacations, surviving photographs were mainly scenery centered on Sandra, taken by George, or of George taken by Sandra.[10]

What we do know, however, is that in March 1987, Carruthers secured a lawyer to help him acquire a "Live-out Housekeeper 35 hours/week 9:00 to 4:00" to provide "Care of Mrs. Carruthers who has mental problems, clean the house, grocery shopping and prepare meals." He had found a suitable housekeeper who might bond with Sandra; she was a nonresident, so he needed to acquire a work permit for her through that lawyer.[11] In the next few years, he engaged several people but did not find one fully capable of the task until July 1991, when he hired Christine James.

By then, Sandra was an outpatient at a community-based clinic, the DC Institute of Mental Hygiene on Connecticut Avenue in northwest Washington. Christine would make sure that Sandra met her appointments and took her medications as prescribed, three times a day. Carruthers had prepared a job description for Christine, saying that beyond the medication and doctor appointments, "She [Sandra] also needs someone to encourage and (if needed) assist her in her housework and other activities, and to keep her company during the day." And he cautioned:

> Please be aware that because of Sandra's mental condition, she will not always be cooperative in taking medicine, keeping appointments, and doing her housework. She will often appear to be distracted, talking to nonexistent persons, and it may be hard to get her attention. She will need to be reminded to do things several times, not just once. Therefore, it is necessary to be firm with Sandra.[12]

Christine not only kept the O Street townhome neat and tidy but also saw to it that both George and Sandra were properly cared for, which included shopping, meals, and general housework. Christine made a special effort to have food on the table; she was, according to one of Sandra's cousins, "a very sweet person, she showed me how to make bread at my visit." It was evident to the cousins during their visits in the 1990s that Christine's company and comfort lessened Sandra's loneliness as George still left for work early in the morning and returned late at night.[13]

In the 1990s, George and Sandra were also supporting her divorced mother, who was living in St. George's, Grenada. Francis Redhead, no longer ambassador to the UN, was living in Brooklyn and owned a travel agency. By 1997, with Carruthers's support, Sandra's mother frequently visited them in Washington, providing companionship for Sandra and taking care of her needs.[14]

From all those interviewed at NRL, there is no evidence that Carruthers ever spoke about Sandra's condition. One close neighbor in Riverpark rarely saw them and recalls no house visits. She would sometimes see Carruthers on the sidewalk, alone and preoccupied, but always friendly when hailed.[15] Family life for George and Sandra, made manageable by Christine, carried on quietly through the early 2000s. Carruthers remained devoted to the NTA and S.M.A.R.T., continued to work as a rehired annuitant at NRL, taught at Howard, and was frequently called upon to speak or be present at numerous events around Washington and the nation. All this was now facilitated by Christine.

Two weeks before Christmas 2009, Carruthers returned home as usual late in the evening but found Sandra injured on the floor. Sometime after Christine had left that day, she had fallen and seemed to be unconscious.

Carruthers called an ambulance, and she was transported to the George Washington University Hospital, where she was pronounced dead of a stroke.[16] Carruthers sought help from Sandra's family in Brooklyn. They quickly took charge, had her cremated, and then sent her home to Granada. Carruthers, some recall, was definitely incapacitated and in shock. And this must have been a terribly difficult time for Sandra's family as well because Sandra's father, Francis Redhead, had just died the previous August, well into his nineties.

Carruthers, however, possibly in denial, did not inform his own family about Sandra's passing for some time. When his brother Gerald finally learned the sad fact in March 2010, almost four months later, he also sensed that Carruthers was somehow different. They had no physical contact at this time, so Carruthers's mental state remains unclear, but there was a change that Gerald attributed to deep mourning.

After Sandra's death, Carruthers was alone, save for Christine and his family, who were distant in Huntsville, Chicago, and in-laws in New York. He still had his colleagues at NRL and Howard, and his personal contacts through his outreach efforts. But since no record or public obituary of Sandra's passing has yet been found, one can only assume that life for Carruthers went on as before, including his service for the NTA and to S.M.A.R.T., courses at Howard, and continuing activities as an annuitant at NRL. But his brother Gerald's recollection is that Sandra's death affected him deeply, weakening his cognitive powers.[17]

Sensitive to Carruthers's weakening state, in 2010, NTA and S.M.A.R.T. leaders Hattie Carwell and Valerie L. Thomas organized a two-pronged campaign to recognize Carruthers's legacy, highlighting his participation in their dual missions. One was a private "Tribute . . . An Evening of Honoring an American Living Legend, A Stellar Astrophysicist and Inventor" held at Howard University in September 2011. Speakers celebrated Carruthers's science, his mentoring, and the organizations he served. Personal remarks by representatives from NRL, NASA, the National Air and Space Museum (presented by the author of this book), and the National Society of Black Physicists were followed by loving and eloquent remarks from the organizers, speaking for the NTA and S.M.A.R.T.

The author was deeply impressed by the warmth of the occasion. As usual, Carruthers remained reserved and seemed somewhat detached, a bit more than my previous contact with him. But when he took the podium, he blossomed, thanking everyone who was part of his life and career and, typically, saying nothing about his personal life. It was characteristic of many celebrations in his honor in his last decade.[18] And by 2011, he was no longer alone.

## MARRIAGE TO DEBORAH THOMAS

Carruthers worked on outreach activities with many people, including Deborah Irene Thomas, known informally as Debbie or Debra. They met at the National Air and Space Museum during a Tuskegee Black Aviation Program in late February 2004. Carruthers was one of the stars, of course, and they stayed in contact through many mutual STEM outreach activities with the NTA and S.M.A.R.T.

Debra, who has described herself as a proud and lively "Sea Island Gullah/Geechie,"[19] came from an educated family in Charleston, South Carolina, where she was born on October 21, 1958. Her mother, Pamela, was a licensed practical nurse and her father, William Gregory Thomas, was a senior chief petty officer in the US Coast Guard. His duty stations varied widely, including a three-year posting in Greece in the late 1960s when the Coast Guard deployed a detachment to help train Greek pilots.[20] Debra became fluent in Greek and regards that as a formative period in her life.

Debra first read about Carruthers in her preteen years, thinking he was "cute" when she saw him on television after the *Apollo 16* flight. She was planning to watch a Jackson 5 concert, her "crush" at the time, but somehow ended up watching the Apollo news conference. She recalled that seeing a Black man in that role "blew my mind."[21] By then, her family had moved to the Washington, DC, area. Debra had entered Prince George's County public school system and honed her already wide-ranging interests, from sewing to sports. She also became, in her words, a "teenage feminist," writing papers about the economic disparity of women in society.

Attracted to mathematics and computers early on, she first studied computer management at Prince George's Community College and then systems management at the University of Maryland. In the 1980s and 1990s, she worked in various technical positions as a computer programmer, including stints at NASA Headquarters and the Smithsonian, where she also became a volunteer. An ardent activist in STEM education, joining programs supported by the National Society of Black Engineers, she volunteered to be a surrogate parent volunteer at Ballou Senior High School, tutoring in algebra and helping in youth computer science programs organized by groups in the area.

Debra sought out Carruthers at the NASM event, and they quickly struck up a conversation about the lack of math literacy among the young and how they both wanted to reverse the condition. They shared their interests and experiences, and Carruthers told her about an upcoming math contest organized by his NTA chapter and invited her to attend. From that point on, they became colleagues.

Debra became an active partner and organizer in both NTA and S.M.A.R.T. programming. At the June 2004 Transit of Venus, just a few months after they met, she set up the computer that captured a real-time image of Venus crossing the solar disk that Jessye Bemley and others, using a small telescope, operated for their video shoot (see figure 14.2). Debra remembers that the Venus event was where she and Carruthers became "buddies," describing herself as his "side-kick."[22]

They teamed up again in September 2004 to help a contract writer prepare historical materials for a volume on the NTA and, over time, worked on many projects and events together. During these encounters, Debra came to sense a real compatibility. They were both nonconformists, and though she was more outgoing, she respected and was comfortable with his privacy. Her friendship with Carruthers also helped her become more assertive. On one occasion, she recalled, when she complained about how she felt mistreated and disrespected by some of the older men at Howard, Carruthers responded, "Put on your high-heeled steel-tipped shoes and kick them in their butts."[23]

Debra never met Sandra and knew nothing about her condition. But when Sandra died, Carruthers confided in Debra before others, even his family. Carruthers still met all his responsibilities but evidently was in deep mourning. Debra felt that George needed help. She soon was managing his finances, taking care of the bills, and keeping his calendar up to date, by then equally split between NTA and S.M.A.R.T. activities and an increasing frequency of medical appointments for him at Howard Hospital.[24] Their friendship deepened as Carruthers continued to grieve, which only redoubled her efforts to support him in every way possible. On occasion, she took the lead planning new programs. In early 2011, she created a session for the "Discover Engineering Day" at the National Building Museum that presented what it would take "Returning to the Moon." Writing back

Figure 14.2
Jessye Bemley with the small telescope that both produced a projected image of the Sun visible in the box and also recorded it on computer. Venus was barely visible in this image. *Source:* Courtesy of Valerie L. Thomas.

and forth (she called him "Dear Georgie" by then), Carruthers helped her revise her proposal drafts.[25]

They married in a ceremony at Howard University's School of Divinity on April 19, 2012, followed by a small reception at the National Air and Space Museum (see figure 14.3).

Debra soon sensed that George needed physical and mental support and that she was the one to give it. She has a poignant memory of a female minister telling her sternly that "when you get married you are going to treat your husband like a king." At seventeen years of age, she rejected it fully, but now in her fifties, it became her mantra. She felt needed and filled that need. After their marriage, she was soon fixing up the O Street address, having the plumbing repaired and the carpets cleaned, and organizing the myriad plaques, posters, awards, and professional files that were scattered around the house. Debra kept the books and made sure that support was equitably distributed to the family, anticipating Carruthers's wishes and needs.

Figure 14.3
Debra and George Carruthers at the National Air and Space Museum in the *Apollo to the Moon* gallery. Scene from an informal wedding party at NASM, April 19, 2012, after their wedding at Howard University. George Carruthers Papers, Smithsonian National Air and Space Museum (NASM 9A20347). *Source:* Courtesy of Deborah Carruthers.

She also made sure that George was properly dressed when he was called to attend a formal event and was at his side to make sure he stayed for receptions and was sociable after he won a prize or award. They enjoyed movies and television series, mainly space westerns like *Star Trek* and *Star Wars*, shopped together, and traveled a bit, mainly seeing family in Chicago.

Debra certainly lifted his spirits, as his companion, coworker, and muse. She happily did the cooking, taking care to meet his needs and desires. He always liked spaghetti and meatballs and was especially delighted when she prepared "hummingbird cupcakes" made of cream cheese, crushed pineapple, bananas, and coconut.

In one of his last known formal interviews, in August 2012 with the HistoryMakers, it was clear that at seventy-two years of age, Carruthers was confused by the interviewer's pointed questioning. He had trouble

Figure 14.4
George and Debra visiting the von Braun diorama at the Alabama Space and Rocket Center in Huntsville, July 2011. *Source:* Courtesy of Gerald Carruthers.

saying his name and spelling his middle name. He could not state the day's date, and, being asked when he and Debra were married, made only vague and incorrect guesses.[26] And this was only three months after their marriage. Asked about this session, Debra replied strongly that the interviewer made them both feel uncomfortable, which confused and upset Carruthers.

By then, most of his contact with the NTA, Howard, and NRL had ceased. Debra did all she could to keep his spirits up, and it was clear from those who saw them at various award ceremonies and community functions that she was most attentive, holding his hand and directing his attention.

One special occasion was a visit to Gerald's family in Huntsville, for the marriage of Gerald's son. Of course, they also toured the Alabama Space and Rocket Center, where Carruthers once again met Wernher von Braun, figuratively, as a full-scale von Braun mannequin greeted them from a diorama of his Huntsville office, bringing Carruthers's life dreams, one might say, full circle (figure 14.4).

## THE PRESIDENT'S NATIONAL MEDAL OF TECHNOLOGY AND INNOVATION

Among Carruthers's countless awards and prizes, one stands out. Hattie Carwell initiated the proposal for the President's National Medal of Technology and Innovation by recruiting many of Carruthers's former colleagues and students and notable names in his field. From NRL, she received testimonials from Robert Meier (then retired from NRL and a professor at George Mason University), along with about a dozen former colleagues from NRL, Johns Hopkins University, Goddard Space Flight Center, the University of Nebraska, and Princeton University. Victor McCrary, a physical chemist and president of the National Organization for the Professional Advancement of Black Chemists and Chemical Engineers, lent a wider perspective, and Carwell added praises that Herbert Friedman, who passed in 2000, had written on many occasions.

Carwell, using Friedman's words, cited Carruthers's work on interstellar molecular hydrogen as "one of the most important findings in astrochemistry at the time."[27] An astrochemist who supported the nomination for the National Medal was more specific: Carruthers's observation was "the first step in building up of larger molecules in the interstellar medium that are the building blocks of planets and ultimately [of] life itself."[28]

Carwell also highlighted Carruthers's persistent attention to detail, quoting Kenneth Dymond: "That one aspect of Carruthers's work that is often overlooked because of his innovations is the smaller but very important task of testing and calibrating the instruments he developed." Indeed, this hit the bullseye describing Carruthers's genius in the laboratory and on the launchpad.[29]

Dymond also provided some vivid examples. He used his experience with the HIRAAS payload on sounding rocket test flights in the late 1990s to illustrate Carruthers's "can do" attitude ("Which I try to emulate," he noted). Case in point: facing a launch deadline, they were having difficulties reducing electronic noise in the camera. Two days before the deadline, after a grueling day,

> I left for the evening after working a long day tired and dejected but in search of a solution to the problem. I returned the next morning . . . ready to tear the electrographic camera apart to try and isolate the problem. I stopped by George's office to talk about a solution and George told me that the camera was

ready for calibration. He had stayed all night in the lab and fixed the noise problem. He had taken the camera apart and smoothed off the burrs [he had found] that were the source of the noise. Then he made some of his renowned black epoxy paint and painted over the smoothed surface. He then dried the paint in the oven in his lab and afterward put the painted part into a vacuum chamber where he finished curing the paint while removing any of the gases that would be produced by the drying paint. After that, he reassembled the camera so I could calibrate it.[30]

This episode made an indelible impression on Dymond, especially because he knew that Carruthers was also rushing to meet "his own deadline to deliver flight instruments to fly on the Space Shuttle!" He was also impressed by how willing Carruthers was to speak to students, no matter how busy he was. Accordingly, Dymond arranged for his daughter to interview Carruthers for her high school engineering class, and when she presented the interview before the class, "the minority students were astounded that a minority man had become a world-renowned scientist who had designed and built an instrument that not only flew in space, but is still on the Moon!"[31]

Victor R. McCrary, the national president of the Organization for the Professional Advancement of Black Chemists and Chemical Engineers and also the business executive for the Johns Hopkins Applied Physics Laboratory, recalled vividly not only how Carruthers's research informed his and his thesis advisor's research at Howard but also how Carruthers early on supported Howard University chemistry graduate students to attend national conferences. H. Warren Moos summed up the feelings of the supporters saying simply that Carruthers's persistent efforts "must be considered heroic."[32] Indeed, Moos's observation reflected statistical studies at the time showing slow but steady increases in the percentage of Black undergraduate and master's-level science and engineering majors, which one reviewer "characterized as positive but incremental" since 1991.[33]

Among the ten medal recipients, two were men of color—Carruthers and Rangaswamy Srinivasan, from the IBM Thomas J. Watson Center. There was one woman, Frances H. Arnold from Caltech. All the medalists except Carruthers worked in applied technology areas, from pollution mitigation, drug release systems, biomaterials, and other areas of direct application to medical, industrial, and social needs. Carruthers's brief citation was

Figure 14.5
(L–R) Debra and George Carruthers, President Obama, cousins Cassandra and Caroline Dunlap, and Gerald Carruthers at the awards ceremony for the 2011 President's National Medal of Technology and Innovation, held at the White House on February 21, 2013. Photo by Lawrence Jackson, The White House, February 13, 2011. *Source:* Courtesy of Barack Obama Presidential Library.

"For invention of the Far UV Electrographic Camera, which significantly improved our understanding of space and Earth science" (see figure 14.5).[34]

So in effect, he won it for the camera, not for the science done with it, which indeed is the purview of the medal. His proposers were certainly aware of this fact. As Meier stated at the outset of his supporting letter, the camera was "Carruthers' most important invention," and its applications went far beyond science to defense: "He is one of the giants of the field who succeeded where others failed."[35] Of course, there was considerable overlap between the Medals of Science and the Medals of Technology and Innovation; that year, a number of science medalists were recognized for new tools facilitating discovery and commerce.[36]

By the time he received the President's Medal, those he was still in contact with at NRL and Howard sensed that Carruthers's physical condition was continually weakening, and his mental alertness was lessening, though he was always friendly and responded to questions. He had stopped riding his bicycle by then; there was some talk at NRL that he had suffered an accident that left him with a brain aneurism. His widow Debra confirms that he was injured and suffered but said it was from being mugged one day walking home from shopping. One way or another, Carruthers was on the wane.

### LAST YEARS

The last time I visited George and Debra at their home on O Street SW in December 2019, just before the COVID-19 pandemic hit, George was sitting on the couch picking at the remains of his dinner on a TV tray, focused on NASA TV. Debra invited me to their home to identify and acquire some of his many awards, honors, certificates, and medals for the collection at the National Air and Space Museum. There were dozens, with photo albums and testimonials to his scientific career and educational outreach. As I worked with Debra making selections, George remained seated, fixed on the TV. Whenever we asked him something directly, he would not speak, but he would brighten; a glint of friendly recognition came to his eyes. He seemed at peace.

In his final year, Carruthers's continuing deterioration from a heart condition sent him to area hospitals and nursing homes. Debra valiantly kept the household together, which included helping members of Carruthers's family in Chicago manage their financial affairs. The inevitable finally came on Saturday, December 26, 2020, when Carruthers passed quietly of heart failure at the George Washington University Hospital in Washington, DC.

News of Carruthers's passing spread instantly. Debra's family rushed in to help her. Valerie L. Thomas quickly organized a team to aid Debra with "George's home going/memorial service," a combination virtual streaming and live physical event, given the continuing horror of COVID-19. Thomas also recruited Carwell, brother Gerald, NASA dignitaries, former students, and the author. Carwell also helped Debra prepare the family response. Many others pitched in to hold the service a month later at the Solid Rock Full Gospel Baptist Church.

Among the many speakers and friends, the family was joined by leaders from NRL, S.M.A.R.T., the National Technical Association, Howard University, NASA, the US Patent and Trademark Office, former medalists representing both the National Inventor's Hall of Fame and the National Medal of Technology and Invention, former students, and the author. Charles F. Bolden, astronaut and former NASA administrator delivered the eulogy, remembering Carruthers as "a friend, mentor, colleague," adding,

> During my time at the astronaut office at the Johnson Spaceflight Center in Houston . . . I was fascinated by the brilliance of this young Black man in what then, and in many ways remains today, a white man's world. He persisted against all odds to build and place what was the first, and still is the only, telescope on another celestial body. Carruthers never sought credit or recognition.
>
> Dr. Carruthers was a drum major for good, devoting his life in service to others. He tried to live to serve humanity. Perhaps most impressive to me, throughout my experience with Dr. Carruthers was his ability to reach young students in his very humble quiet way. He was able to take the very complex principles of science and give them practical meaning for youngsters as well as older folks, such as me. . . . He always, *always*, knew how important it was to be able to talk with and understand our fellow human beings.[37]

Bolden's eulogy offers a fitting tribute to Carruthers's life and broad contributions to society. From NRL's perspective, Commanding Officer Captain Ricardo Vigil cited how Carruthers's "stunning images" of the Earth from the Moon "helped to confirm emerging understanding of upper atmosphere morphology . . . fundamental to DOD systems' orbital dynamics, communications, and remote sensing." Carruthers was, moreover, an "outstanding example of how successive generations of scientists build communities around diverse skills and evolving research problems."[38] In his appreciation, Captain Vigil traced Carruthers's legacy that led to many contemporary missions focused on geospheric and heliospheric questions, including characteristically employing the acronyms "ICON, GOLD, SWARM and other DOD work."[39]

And of equal importance, an even more lasting legacy was the growing "cloud" of students who benefited from his mentoring, and they could now "be found at diverse institutions including AFCRL [Air Force Cambridge Research Laboratory], the Aerospace Corporation, Fisk University, Morgan State University," within a growing array of aerospace businesses.[40]

Captain Vigil's description of Carruthers's scientific legacy highlights what is sometimes underappreciated in both academic history and the public mind. Carruthers's contributions helped guide the goals of many later missions to explore and monitor the Earth's upper atmosphere. This demonstrates the lasting impact of his work, from understanding the nature of the interstellar medium to the dynamics of the Earth's upper atmosphere and its interaction with solar radiation. A characteristic of the distribution of citations to his most cited papers is that they have two peaks: one in the first few years after publication, and the other in the 2000s.[41] As late as 2018, he was cited in the scientific literature as the first to design a reflecting-cathode electronographic system that could work in the vacuum of space.[42]

Over the next weeks and months, notices and obituaries in newspapers and magazines remembered him in grand fashion. *The New York Times*, *The Washington Post*, *The Philadelphia Tribune, Nature,* and *Sky & Telescope* among many other periodicals recalled his life and accomplishments, celebrating his legacy on Earth and what is still, hopefully, standing on the Moon. On December 2, 2022, NASA remembered him in grand fashion by naming a new heliophysics mission in his honor, the "George Carruthers Geocorona Observatory." Someday, when we return to the Moon and create memorials at the Apollo landing sites to the astronauts, it would be appropriate to place a tribute next to the camera that George Carruthers designed, built, and, indeed, sent from his laboratory to the Moon.

# APPENDIX A

## TESTIMONY FROM MENTEES

There is no telling how many lives Carruthers touched. Here we provide brief profiles of just a few of them who generously gave over some time to share their impressions of Carruthers and how their time with him changed their lives. Their voices provide poignant portraits of Carruthers, as mentor and scientist.

### JESSYE BEMLEY TALLEY

Jessye Bemley (now Talley), at this writing an assistant professor in the Department of Industrial and Systems Engineering at Morgan State University, was born in 1986 in Washington, DC. When she was a high school student at Archbishop Carroll High School in northeast Washington, DC, her father, Dr. Jesse Bemley, who as we noted in the text was a well-known promoter of computer literacy and a professor at Bowie State, introduced her to Carruthers at an educational event in the SEAP program. Her family had long been devoted to marginalized youth-serving activities, promoting STEM. Talley felt "that sort of jump started us." She had already acquired an interest in space and astronomy, so her father introduced her to Carruthers, and by 2001, as a tenth-grade high schooler, she was enrolled in NRL's summer internship program.[1]

Talley explored Carruthers's laboratory in the trailer in the Building 209 loading dock. Then they went upstairs to the laboratories. Her impression was that, as they toured the building, Carruthers took care to sense what attracted her the most and then created a curriculum to suit. He directed her to two activities: examining the images from his cameras, supplemented by reading to become better familiar with the science he was doing, and also preparing her for video production projects aimed at school audiences: "What I helped him with the most was to create videos" illustrating and demonstrating concepts and techniques. Carruthers provided a host of visual aids supplemented by readings to help his interns "put into practice" the themes and concepts they would present in the videos. She particularly remembered his guidance as they built a small telescope that helped them appreciate how telescopes work.

Carruthers's mentoring style exposed students like Talley directly to original data, making them part of the research. He shared this with many students across these later years. As Carruthers's colleague Tim Seeley sensed, in addition to his continuing designs for new detectors and more flights, he "was mostly looking at what he'd done" guiding his students and "having students help him sort out what he had done."[2]

Talley observed that Carruthers had a light sense of humor. When something did not go right in the laboratory or on a video shoot, he would giddily exclaim, as if to draw his student's active attention, "Oh my goodness" or "This always happens to me" and even "I don't know what's going on" as he deftly corrected the problem, helping his students experience the problem and the process of correction.

He used humor to lower whatever nervous tension existed in the videos or in the lab experimentation. Talley poignantly recalled one instance when, preparing for a video scene, Carruthers turned to her saying, "You have to put your model potential, your model look, on for the demonstration." By that she understood that she was the "model for the demonstration, like have that type of look."[3] Presumably, Carruthers wanted her to authoritatively project her voice, make it accessible, and behave as if she were a scientist. In any event, she felt the process was informal and not in any way intimidating. Carruthers would place a video camera on a tripod either in his laboratory or on an outdoor grassy area and just let it run as he stood by out of the scene, informally handwaving suggestions. "But it would get funny sometimes, when he was telling me what to do." She greatly enjoyed the video work, especially the telescope. The video started in the lab—building the telescope, introducing its components—and then continued out on the grass, using it to view the Sun.

Carruthers took care that his students were not overworked and did not feel obligated to be there at all hours like he was. He watched after them and gauged their hours, promptly telling them, "Okay, it's time for you to go home." He was also not overly directive of their interests, letting them explore the lab and the topics to their satisfaction. As Talley testified,

> Just having that experience and the fact that I was able to do other programs as well . . . helped me to be a little more well-rounded and to see all of the different types of the STEM careers that I could have . . . I ended up [in] engineering . . . because I like to design things. So, working with him that summer, just seeing how to put things together and do stuff like that, that kind of showed me all of the possibilities of what you can do.[4]

At the end of her internship, Carruthers and her father encouraged her to write up her experiences and present them at a meeting of the NTA sometime in 2003. Her father continued to tutor her, and she matriculated at the North Carolina Agricultural and Technical State University in Greensboro in industrial engineering, right through from undergrad to PhD. Without doubt, her father was her constant guide and advisor, but, when he put her in contact with Carruthers, it became "the biggest thing . . . having that experience at the lab . . . it was another avenue for me to see more about space science, but it also just kind of opened me up more to STEM to see what was possible."[5]

### FRANK (TRÉ) PADGETT III

Born in Washington, DC, in 1998, Padgett was in graduate school at Fisk University, majoring in physics with an astronomy track, when the author interviewed him in 2020. Padgett's father Frank was a consultant in construction and project management, and his mother Sherine was active in community outreach. They encouraged both of their children by seeking out stimulating educational opportunities for them. Padgett gained an interest in astronomy in the second grade after learning that there were other planets.

Padgett met Carruthers at a science fair when he was ten years old, was in fifth grade, and, with his parents' support, was about to enter Howard's special Middle School of Mathematics and Science. "I spoke with him for about three to five minutes, he told me that he was an astrophysicist, and ever since then, the term stuck with me. So from that point on, I told myself I wanted to be an astrophysicist."[6] Meeting Carruthers made all the difference:

> Well, the first thing that shocked me, I guess, was that he was an African-American. . . . I always wanted to go into the space industry, but because I never

> knew any other African-Americans who did such a thing, I didn't really know how that would be possible. But just seeing him, being an African-American and being someone who was pretty easy to talk to, it just made my goal that much more relatable, and he made it seem like it could be that much more realistic, that there are Black people who work in the space industry. It's actually an attainable and realistic goal.[7]

Padgett also recalled that every so often, Carruthers visited his class at the Howard University middle school.[8] And on occasion, he invited students from the middle school to his lab at NRL: "He showed us some of the equipment, or . . . some of the simulations, maybe some videos that could explain things to middle school students." If there was an astronomical event to experience, like a solar eclipse, as Thomas, Padgett, and other students and colleagues remember, Carruthers would drag a small reflector either to the roof or lawn or use the student observatory that he had now extensively renovated and expanded. Access to the campus laboratory provided the hands-on experience that was so valuable: "But for my life and my personal path that I decided to take, I think it was the most impactful, just being able to meet Dr. Carruthers and see Dr. Carruthers, to have that motivation that there are other Black space scientists out there, and if he could do it, then I could do it, too. So I always held onto that and the experiences that I witnessed, and just kept pushing forward, and it got me to where I am today."[9]

Building on that experience, Thomas introduced Padgett, then in high school, to space camp in Huntsville, Alabama, which he felt was also "one of my really life-changing experiences."[10] Padgett also attended a summer school at Harvard University, where he took classes in astrophysics and astrobiology and interned under Dr. Susan DeStefano at the Center for Astrophysics. There he learned how to develop a database, coding it to look for exoplanets made detectable from tiny flickers in starlight caused by the gravitational influence of an orbiting planet, called gravitational microlensing. He also interned at George Washington University, becoming familiar with gamma ray studies.

Padgett felt that contact with Carruthers "caught me a break." He added, "Dr. Carruthers has inspired me to become the next African-American Astrophysicist of my time." His path to that role, however, was not without trial. Accepted by the University of Rochester, he was the only Black student in the department and experienced a "huge cultural and academic shock." He felt "completely isolated" and invisible. However, buttressed by his positive experiences made possible by Carruthers, Thomas, and DeStefano, he had the strength to persist: "I protected my identity by owning it."[11] He graduated from Fisk University with his master of science degree in physics and is now (December 2022) a radiation effects test engineer at BAE Systems (a subsidiary of British Aerospace), testing electronics destined for spaceflight against harsh radiation environments, like those found in outer space.[12]

In a memorial address, Padgett, representing S.M.A.R.T., remembered the impact Carruthers had on him at first contact: "Just from hearing his story alone, at that time I had already made up my mind that I wanted to be the next George Carruthers."[13]

### WILLIAM O. GLASCOE III

William Glascoe grew up in the Randle Highlands neighborhood of southeast Washington, DC. His father was a US Postal Service police officer and his mother a beautician. He attended Maryland's "Boys State," a participatory program where students experienced what it would be like working in local, county, and state government jobs. He also worked as a summer science aide at Ballou Senior High School during his sophomore summer. One day at Ballou, he saw a poster advertising the competitive SEAP program and applied for the summer of 1985. He hoped that he would be accepted by NRL because it was close to his home. He previously had been picked to attend a series of Saturday classes at the Space Telescope Science Institute at Johns Hopkins in Baltimore, but it was too distant from his home.[14]

Glascoe was nervous on his first visit to NRL. But his first glimpse of Carruthers's office put him into a state of amazement, which he described in retrospect: "Just the amount of stuff that was in his lab . . . journal articles [stacked in piles], a drafting board, a computer, as well as HP calculators and even a slide rule." It was a world Glascoe had never encountered. Glascoe's unfamiliarity with slide rules "tickled" Carruthers, which warmed up the encounter considerably. With Carruthers's guidance, he became acquainted with vacuum chambers and chemical and physical tools. "He's got chemical hoods in order to do some of the disposition of chemicals on the photodiodes," Glascoe remembered. "He had just so much stuff."[15]

Carruthers demonstrated how to use a darkroom and how to behave safely in a cleanroom. He also introduced him to the laboratory testing facilities, like vibrating "shake tables" in another part of the building that tested the mechanical integrity of his instruments. And he always advised, "Don't mess with stuff," which Glascoe took seriously. Carruthers "never had any anxiety that I would mess things up or break expensive equipment, because I was all ears. You know?"[16]

Glascoe's job was to perform "preflight calibration" for instruments that hopefully were bound for space on SPARTAN and other DOD missions. "That kept me in the lab right there on the second floor with hydrogen or helium lamps, and just running the experiment and collecting data over a range of either voltages or gains on the photomultipliers in the sensors themselves, to detect their quantum efficiency."[17]

At the end of Glascoe's second summer with Carruthers, he graduated with high recommendations, and, having served in Boys State, he won an appointment to enter the US Air Force Academy. He was able to pay the $1,000 entrance fee with a $2,000 award from the National Space Club's Olin E. Teague Scholarship, support the family deeply appreciated.

During his years in college, whenever he visited his family, he also visited Carruthers. Before Glascoe graduated in 1991 with a physics degree, he was able to work again with Carruthers on the *STS-39* mission and stayed in touch through a series of Carruthers's projects. Glascoe went on active duty in August 1991, stationed at the Holloman Air Force Base near White Sands, New Mexico, in its "6585th Central Inertial Guidance Test Squadron" that evaluated airborne Global Positioning System (GPS) receivers and inertial navigation systems.

At an early point, he had "ambitions of being a mission specialist or a payload specialist" as a means of "going into space," so he specialized in astronautical engineering as a commissioned officer. When he learned of the International Space University Summer Session Program in 1994, he wanted to go, but finances and work prevented attending the ten-week program in Stockholm, Sweden. But he had enough support to attend the ISU-SP 99 ten-week program in Khorat, Thailand.

After ten years of commissioned service as an astronautical engineer on active duty, in September 2001, he separated from the US Air Force just after he graduated from the University of Colorado at Boulder with a master of science degree in telecommunications. He then joined the Computer Sciences Corporation, working there until 2013. He continued to serve in the US Air Force Reserve with assignments in many offices where he contributed to space enterprise architecture and data-driven policy changes for pilots. He also was assigned to the National Reconnaissance Office where he became director (acting) for Continuity and Critical Infrastructure Protections. At this writing, he is still heavily engaged in systems engineering enterprises.[18]

Reflecting on his career, Glascoe credited his experience with Carruthers as learning "how people can concentrate and get the work done.":

> Even in a physically cluttered environment . . . you surround yourself either in a cone of silence, or you bring clarity to the outcome that you want. Right? And so, for George, it's like: I want to be in the bay, with a working instrument, waiting for my baby to come home. And I'm going to make sure I put the best baby on the pallet. Period. My stuff is going to be together. You check. You recheck. You know? And you document.[19]

One of Glascoe's most poignant memories of Carruthers's influence was at the Holloman Air Force Base in New Mexico, field testing GPS receivers. At one point, he realized that he was entering data in pencil in his lab book, which was not proper: "I started to erase something in my book, as opposed to a line through it and continue. . . . You want to see the mistakes that you made, so that you don't make them again, or you can share how your thoughts went awry. You know? So, it's those sorts of habits that George helped me be able to draw upon, just having that focus."[20]

Back home in the Maryland suburbs, Glascoe kept in contact with Carruthers, joining and becoming active in the National Technical Association, and joining in the many ceremonies honoring his mentor. As of this writing, he is a retired Air Force Officer, having worked most recently at the Air Force Phillips Laboratory's Technology for Autonomous Operational Survivability, contributing to GPS-related activities.

### GARLAND DIXON

Garland Dixon overlapped with Glascoe at NRL and learned a lot from him. Born in Fort Washington, Maryland, in 1972, his mother Dr. Freddie Mae Dixon was a professor of biology at the University of the District of Columbia. His father, Garland L. Dixon, worked in the construction industry and, as Dixon proudly recalled, was a good auto mechanic. His parents met while his mother was attending Howard University, and so a college education was part of the family. One of Garland's sisters became a medical doctor.

Reminiscent of Carruthers and his space-themed comic books, once Dixon saw the original *Star Wars* in the late 1970s, he decided, "That's what I want to do. I want to get into space. . . . I knew I had to study."[21] He benefited from some pretty tough but fair and encouraging teachers, especially in mathematics and science. He tested well and was accepted into Oxon Hill High School's Science and Technology gifted and talented program, which gave him entrée into George Washington University's science and engineering apprenticeship program. Once in the program, as a sophomore, he learned that there was a "Black astrophysicist at the Naval Research Lab" and applied there, with support from the SEAP program.

On his first day in the summer of 1988, as remarked on by virtually all the students interviewed for this book, Dixon found Carruthers's office and laboratory "very, very, very, very messy, unlike any office I've ever seen." He was charmed by what he saw, even though at the time, he did not understand the significance of Carruthers's contributions. William Glascoe, however, helped him appreciate what he was faced with. Glascoe advised:

> Garland, George is like no other person you're going to meet. He's not going to talk about anything else out of work. No matter how early you get here in the morning, he'll be here. No matter how late you stay, he'll be here. He was like, 'George is a genius.' And you know, . . . don't ask George about the game, [laughs] or anything like that. You know? George is about his work, and he's about his students.[22]

"I had experience in drafting," Dixon added, and so when Carruthers learned of it, he immediately focused on what Dixon might do with his training. After a bit of checking, Carruthers was satisfied that Dixon knew his way around T-squares and other drafting instruments. "So he saw that I had interest in design, and so I did a lot of design work while I was with George."

Dixon worked on the mechanical aspects of ARGOS, such as the GIMI and FUVIS cameras during his years in contact with Carruthers. "And he tasked me, a high school student, with designing . . . so I designed the plate that the camera attached to, to go on the Shuttle, and he actually used it." Throughout his early experiences, he felt that Carruthers "made an effort to touch everyone's interests, and it was a two-way street kind of thing." He never felt like he and his cohorts were "faceless workers for him."

For someone as focused as Carruthers was, Dixon found him "very flexible. He made sure that what you were doing, you had interest in, because he wanted you to enjoy yourself." Of course, the work had to be done, and Carruthers made sure, by example, that everyone was motivated to get it done.

Dixon remained in contact with Carruthers's laboratory from 1989 through 2000, working first as a SEAP and then a co-op student as he graduated from high school and entered the University of Maryland. He continued to visit frequently as a graduate student after receiving his master of science degree in aerospace engineering in August 2000. Dividing his time between NRL and the University of Maryland Neutral Buoyancy Facility's Space Systems Laboratory, he focused at the latter on how launch systems and payload robotic systems behaved in simulated zero gravity conditions, and from there, he got into flight dynamics.

In the early 1990s, Dixon frequently returned to NRL to assist in Carruthers's projects, including the restoration of one of Carruthers's backup *Apollo 16* lunar cameras so that it could be put on display at the National Air and Space Museum (see appendix B). Later in the 1990s, he and undergraduate student Melody Finch worked on ARGOS.

After obtaining his master's degree, Dixon decided he would rather seek employment than continue his studies. "Basically I had been in school so long, it was time to move on." With his extensive practical experience in Carruthers's laboratory and in the Neutral Buoyancy Laboratory, he was well equipped and competitive. His first job was with the Iridium Satellite Corporation to become an orbital systems analyst, but quite soon, a position opened up at the Goddard Space Flight Center, which was closer to home and where his responsibilities were expanded to include attitude dynamics. He worked at NASA for about ten years and then moved on to the National Oceanic and Atmospheric Administration (NOAA) as lead orbit analyst for a series of satellite systems operated jointly by NOAA and NASA. At every stage, he preferred technical assignments to administrative posts: "I'm motivated, but I don't want to be a high-level manager. Let me just put it like that. I like to do things. Just like George, I like to figure stuff out myself, and if I need help with anything, I'll delegate it." To this, he added, "Of all my mentors, he was the most impactful. The main reason was because of George's work ethic, and the fact that he was a Black man that did the things that he did."[23]

When George passed away the day after Christmas, 2020, Dixon was among many who expressed their appreciation for his kindness: "Behind my Mother, Dr. Carruthers was my number one influence in pursuing and achieving my career in Science, Technology, Engineering and Math (STEM). . . . He was the most intelligent and dedicated person I ever met in my life but still had a sense of humor and truly cared for his mentees. He will be missed but never forgotten."[24] Like George, Dixon also devoted his energies to STEM through several national and local organizations, receiving many awards and honors for his work.

As of this writing, Dixon is a senior principal engineer at the Arctic Slope Research Corporation involved with developing polar satellite systems.

# APPENDIX B

## CONSERVING AND DISPLAYING CARRUTHERS'S CAMERA

In 1981, the museum acquired two Carruthers cameras—a backup engineering model and the test model—as part of a large transfer of objects from the Johnson Space Center. NASM curators Gregory Kennedy and Louis Purnell, a former Tuskegee airman, managed the acquisition. He and others put some of the objects on display, but the cameras were stored because they were not in exhibitable shape. Although they were technically part of the Apollo Collection, the author recognized their scientific legacy and asked that they be transferred to the Space Sciences collection.

In the early 1990s, with new museum director and astronomer Martin Harwit in charge, sympathies for exhibiting scientific instrumentation increased, and Carruthers's camera became a point of discussion. The idea was to exhibit the complete test model next to the LM in the east end of the building, reminiscent of the positioning of the real one still on the Moon. In 1992, it was loaned to NRL so that Carruthers and Garland Dixon could restore it prior to putting it on display (see figure B2.1). As part of the restoration, Carruthers attached the original film cassette and transport mechanism that the astronauts brought back from the Moon. During the restoration, he inserted it on the back end of the electronographic camera and added other components.[1]

After the instrument had been disassembled, cleaned, supplemented with missing parts, and then reassembled, the author and other Smithsonian staff, sponsored by the Smithsonian's Lemelson Center, visited his laboratory to video record the end of the process. Typically, Carruthers made sure that Garland Dixon and his other students were front and center, helping to demonstrate the characteristics of the telescope and showing how he used various computer-aided drafting devices in the restoration. The camera went on display in 1993 next to the LM with astronauts climbing out of the flight model of the LM and onto a simulated lunar surface. It was placed

Figure B2.1
Garland Dixon working with Carruthers restoring the backup lunar camera for display at the National Air and Space Museum. 1994 *NRL Review*, 275. Photograph NRL 1994-16077 AD-A279 832. *Source:* Courtesy of Naval Research Laboratory.

so that it was shielded from sunlight in the mornings from the east window wall, a very appropriate reconstruction of the lunar configuration (see figure B2.2). It has now been prominently repositioned in the museum's *Destination Moon* exhibit, complete with the original film cassette returned from the Moon.

When the lander was moved into the Milestones gallery in the center of the museum, the camera was retrieved for further conservation from exposure to the museum atmosphere (denim dust, cafeteria drafts, and emissions from idling busses) and was placed in a sealed protective vitrine in the *Destination Moon* exhibit. Erin Ober, a conservation specialist, inspected the instrument a few years later, a standard procedure, and noticed a peculiar odor coming from the camera. One conservator thought, "It smells like cough drops, the really yucky kind," and another described it as a visit to the dentist.

The cassette was removed, and a full inspection was scheduled to see if it contained deteriorated film.[2] As with many conservation projects, it was a delicate procedure requiring many talents (see figure B2.3). Tim Seeley, John J. Hardgrove, and Angelina Callahan visited the museum's support facility at Dulles to advise the staff, including conservators Ober, Robin O'Hern, Lisa Young, Malcolm Collum and the registrar, Erik Satrum. Curators Jennifer Levasseur and the author anxiously attended. Why would there be film in the cassette, and if there was, what do we do

Figure B2.2
Museum director Martin Harwit with George Carruthers inspecting the camera in the original lunar surface display at the east end of the National Air and Space Museum, 1993. Photo by Michael Savell, NRL *Labstracts* 1993_03_15. *Source:* Courtesy of Naval Research Laboratory.

Figure B2.3 (L–R)
John J. Hardgrove, Tim Seeley, and Robin O'Hern opening the cassette for inspection at the Udvar-Hazy support center. Author photo.

with it? There was corrosion on one of the magnesium alloy film spools, but thankfully (or maybe sadly?) it was empty—the one hundred feet of film having long been removed to be scrutinized by Carruthers, Page, and their colleagues.

# ORAL HISTORY INTERVIEWS

Online versions not paginated. Interviewer is David DeVorkin, unless otherwise noted.

Boggess, Albert, April 20, 1984. STHP/NASM.

Brown, Charles, with DeVorkin and Angelina Callahan, December 27, 2019. AIP/NASM.

Byram, Edward T., and Chubb, Talbot. July 8, 1987. AIP. https://www.aip.org/history-programs/niels-bohr-library/oral-histories/28309.

Byram, Edward T., and Chubb, Talbot. July 20, 1976. Interview with Richard F. Hirsh. AIP. https://www.aip.org/history-programs/niels-bohr-library/oral-histories/31488.

Carruthers, George R., video history interview, September, 1993. Smithsonian Institution Archives, accession 03-079, box 1.

Carruthers, George R., August 18, 1992. Smithsonian Institution Archives, accession 03-079.

Carruthers, George R., with Glen Swanson, March 25, 1999. NASA Johnson Space Center Oral History Project. https://historycollection.jsc.nasa.gov/JSCHistoryPortal/history/oral_histories/CarruthersGR/CarruthersGR_3-25-99.htm.

Carruthers, George R. with Racine Tucker Hamilton, July 27, 2004, The HistoryMakers® video oral history interview, The HistoryMakers® African American Digital Archives. https://www.thehistorymakers.org/biography/george-carruthers-41.

Carruthers, George R., with Larry Crowe, August 27, 2012. The HistoryMakers A2004.112, HistoryMakers Digital Archive.

Carruthers, Gerald, August 3, 2020. AIP/NASM.

Chubb, Talbot, video history interview, 1986. Sloan/SI. https://web.archive.org/web/20090409052654/https://siarchives.si.edu/research/videohistory_catalog9539.html.

Conway, Robert R., August 4, 2020. AIP/NASM.

Dixon, Garland, December 13, 2020. AIP/NASM.

Englert, Christoph, with DeVorkin and Callahan, December 27, 2019. AIP/NASM.

Friedman, Herbert, with Richard Hirsh, August 21, 1980. AIP. www.aip.org/history-programs/niels-bohr-library/oral-histories/4613.

Friedman, Herbert, Talbot Chubb, Edward T. Byram, and Robert Kreplin, November 12, 1986. SIA RU 9539 Collection Division 2, Sloan/SIA, https://www.aip.org/history-programs/niels-bohr-library/oral-histories/31436.

Giovane, Frank, March 14, 2021. AIP/NASM.

Glascoe, William O. III, November 21, 2020. AIP/NASM.

Guhathakurta, Madhulika, with DeVorkin, Katie Boyce-Jacino, and Greg Good. February 28, 2018. AIP.

Heckathorn, Harry, June 11, 2020. AIP/NASM.

Henry, Richard C., April 15, 2015. Johns Hopkins University Oral History Program.

Howard, Russell A., with DeVorkin and Callahan, January 30, 2020. AIP/NASM.

Johnson, Charles, and Julian Holmes, July 30, 1987. SI archives, RU 9539 Collection Division 2: Aeronomy, session 1 and session 2, Sloan/SIA, https://web.archive.org/web/20090320065928/https://siarchives.si.edu/research/videohistory_catalog9539.html.

Johnson, Charles, video history interview, July 8, 1987. Sloan/NASM/SI.

Kowtha, Vijay, with DeVorkin and Callahan, February 1, 2021. AIP/NASM.

Latham, David, October 8, 2005. NASA/AIP/NASM.

Lean, Judith, with DeVorkin, Angel Callahan, and Vincent Femia, August 22, 2016. AIP.

Mange, Phillip, with Ron Doel and Fay Korsmo, March 25, 2003; December 30, 2002. AIP, https://www.aip.org/history-programs/niels-bohr-library/oral-histories/31144-1.

McIntire, Deborah, with DeVorkin and Callahan, February 8, 2021. AIP/NASM. Restricted.

Meier, Robert R., with DeVorkin and Callahan, February 6, 2020. AIP/NASM.

Newell, Homer E., with Richard F. Hirsh, July 17, 1980, and October 20, 1980. AIP. https://www.aip.org/history-programs/niels-bohr-library/oral-histories/4795-1.

Padgett, Frank, III, August 8, 2020. AIP/NASM.

Peery, Benjamin, November 5, 1977. AIP, https://www.aip.org/history-programs/niels-bohr-library/oral-histories/33698.

Phillip, Suzanne with Gwen Ochoa, Zoom and telephone interviews, December 12, 2020, and December 15, 2020. Zoom and telephone interviews.

Roman, Nancy G., August 19, 1980. AIP. www.aip.org/history-programs/niels-bohr-library/oral-histories/4846.

Seeley, Tim. October 15, 2020. AIP/NASM.

Simon, Cynthia, December 7, 2020. Telephone interview.

Spitzer, Lyman, April 8, 1977, and May 10, 1978. AIP, https://www.aip.org/history-programs/niels-bohr-library/oral-histories/4901-1.

Talley, Jessye Bemley, July 27, 2020. AIP/NASM.

Thomas, Valerie L., May 22, 2020. AIP/NASM.

# ARCHIVAL RESOURCES

Bernard Burke Papers, BB/NRAO.

Benjamin Carruthers Collection, Schomberg Center, New York Public Library, BC/SC.

George Carruthers Papers, National Air and Space Museum, GC/NASM.

Herbert Friedman Papers, American Philosophical Society, HF/APS.

Karl Gordon Henize Papers, The Dolph Briscoe Center for American History, University of Texas, KGH/UT.

Nancy Roman Papers, Center for History of Physics, AIP.

National Archives and Records Administration, National Archives at College Park, MD. RG 255, Records of the National Aeronautics and Space Administration, RG 255, NARA.

Niels Bohr Library & Archives, Center for History of Physics Oral History Collections, American Institute of Physics, College Park, MD, AIP.

National Technical Association records, Carruthers Collection from NRL, NTA/Howard.

Thornton L. Page Papers, MS1983-002, Virginia Tech Special Collections, Newman Library, TP/VT.

Thornton Leigh Page Papers, Wesleyan University vertical subject files, 1000-176, Special Collections & Archives, Wesleyan University, Middletown, CT, TP/WU.

Space Astronomy Oral History Project, SAOHP/NASM.

*Skylab 4* papers, Naval Research Laboratory, Skylab/NRL.

Space Telescope History Project, NASM, STHP/NASM.

USSRC—US Space & Rocket Center, USSRC.

William Baum Papers, Putnam Collection Center, Lowell Observatory, WB/PPC.

# GLOSSARY

**Altazimuth** A telescope mounted so that it can swing horizontally and vertically, in altitude and azimuth.

**Angstrom (Å)** A unit of length equal to 4 billionths of an inch. Light is physically characterized by its wavelength, expressed in Angstroms, named for the physicist Knut Ångström. Light visible to the eye ranges approximately from 4,000 to 7,000 Angstroms, expressed in color ranges, from violet to red. The Earth's atmosphere admits this range but blocks shorter wavelengths starting from the ultraviolet from about 10 to 3,000 Angstroms. The ultraviolet range of greatest interest to Carruthers was from 900 to 1,500 Angstroms, called the "far ultraviolet." (See also Spectrum)

**Cassegrain telescope** An all-reflecting telescope where light reflects off a large concave primary mirror that concentrates the light onto a smaller convex secondary mirror that lengthens the beam and sends it back through a hole in the primary mirror to a focus.

**Cathode** See Photocathode.

**Coma** The radial distortion of telescopic images off the optical axis (i.e., not at the center of the image).

**Diffraction grating** An optical element made up of very fine grooves that can break up light into a spectrum like a prism either through reflection or transmission.

**Equatorial** A telescope mounted with one axis pointing directly to the pole of celestial rotation.

**Electronography (also, electrography)** The amplification of an optical image by converting it into an electron pattern, focusing it magnetically and amplifying it electronically, and then recording that pattern with electron sensitive film.

**Exosphere** The topmost part of the Earth's atmosphere.

**Flux (electron, photon)** The flow of matter or energy. Energy transported by electrical (charged atomic particles) or optical (photons) means.

**Focal ratio** The ratio of the focal length of a lens or mirror to its diameter (designated f/##). Lower numbers are optically faster, recording extended objects (images) in less exposure time, and have wider fields of view.

**Gamma rays** The highest energy form of electromagnetic radiation.

**Geocorona** Visible in the ultraviolet, it is the luminous portion of the Earth's outermost atmospheric layer. The extent of the geocorona is still unknown, but its behavior, what causes it to change over time, remains a key element in the study on how solar radiation influences weather, life, and long-range communications systems on Earth.

**Ionization** The loss of one or more electrons from an atom due to collisions, increased temperature or reduced pressure. (See Plasma)

**Ionosphere** A region of the Earth's atmosphere where solar radiation can split molecules and atoms into ions and electrons, rendering the region reflective to radio and microwave radiation and therefore of importance to long-range communications. It is also where aurorae occur when the ionized gases recombine.

**Lyman-alpha line of hydrogen** Spectral line of hydrogen when its electron transitions from the first excited state back to the ground, or lowest energy state ("bright line" emission), or from the ground state to the first excited state ("dark line" absorption). Named after physicist Theodore Lyman. Solar Lyman-alpha was long suspected of influencing the long-range radio reflectivity of the Earth's ionosphere, a key Navy interest.

**Magnetosphere** The region of the Earth's upper atmosphere where magnetic fields most influence the motions of charged particles from space.

**Mesosphere** The region above the stratosphere where temperature decreases with height.

**"Missing mass" problem** Matter not yet detected by any known means but sensed by its gravitational influence to explain the dynamics of clusters of galaxies and the overall dynamics of the universe.

**Observatory** A facility engaged in observing nature, consisting of instruments designed to detect specific phenomena, an infrastructure facilitating the observations, and a protective housing. For astronomy, this usually consists of a telescope and some means of recording what the telescope senses (a camera).

**Photocathode** A material that absorbs or reflects photons (a beam of light) and converts them into electricity (electrons). Typically an alkali-halide formulation. Also "photocell."

**Photon** A particle of light that also is a form of electromagnetic radiation having wavelike properties. The shortest known wavelengths (highest energies) are called the gamma ray and X-ray ranges. Then, with lengthening wavelength in sequence are the extreme ultraviolet, ultraviolet, blue through red, infrared, microwave, and radio.

**Plasma** An electrically conductive gas consisting of electrically charged (ionized) particles caused by heating sources sufficient to tear electrons from atoms. Reducing the pressure in the gas lowers the temperature limit required for releasing electrons. It is known as the fourth state of matter, in addition to solids, liquids, and gases. Lightning, arc welding, neon, and fluorescent tubes are examples of plasmas. The solar wind is a plasma. (See Ionization)

**Schmidt camera** A reflecting telescopic design developed in the 1930s by Bernhard Schmidt developed to capture wide fields of view. Light first passes through a very thin lens, which adjusts the beam so that the primary reflecting surface, a large spherical mirror, focuses the light into a curved plane, where a photosensitive surface records the image. For space applications the entrance window is typically made of lithium fluoride to admit as much ultraviolet light as possible.

**Schwarzschild reflector** An all-reflecting telescopic design developed in 1905 by Karl Schwarzschild. Similar to a Cassegrain, but with aspheric optics, the light falls on a large concave mirror, which concentrates it onto a small convex mirror. The convex mirror then reflects the light, producing a longer effective focal length with greater magnifying power but smaller field of view. Carruthers developed variations on this design, but eventually had the beam directed back toward the large primary mirror and pass through a hole at its center, to come to a focus for recording.

**Solar wind** Charged particles ejected from the Sun during eruptions called solar storms that are propelled into space along the Sun's magnetic field lines.

**Solenoid coil** A helical coil of wire that can produce a uniform magnetic field when an electric current is applied to it.

**Sounding rocket (or "rocketsonde" after the nautical term "depth sounder" for an instrument that senses and measures deep ocean conditions)** Small suborbital rockets that carry sensors to measure the conditions of the Earth's high atmosphere and then, for a few moments at peak altitudes over 60 miles and effectively above the Earth's absorbing atmosphere, glimpse the X-ray and ultraviolet and infrared universe, portions blocked by the atmosphere and therefore invisible from ground-based observatories.

**Spectrograph** An optical instrument that uses prisms or gratings to disperse light into a spectrum and records it using photographic or electronic means.

**Spectrum** When used as an optical term, an optical spectrum is produced when an optical element, like a prism or grating, spreads light of all kinds into a continuum of energy levels. For visible light, it can be thought of as a rainbow of colors, each color deriving from photons of different energies. Dark and bright lines in spectra result from the absorption or emission of photons from atoms and molecules. Their position in a spectrum is determined by the unique structure of the atom or molecule producing them, and hence, mapping these lines is a key step in identifying the chemical identity of the element producing the lines.

**Telescope** An imaging device using either lenses or mirrors that collects and intensifies light (radiation in all wavelengths). When used in conjunction with a recording device, it is also referred to as a "camera," as is done interchangeably here. Carruthers consistently called his instruments "cameras" even when they were parts of telescopes.

**Thermosphere** A region of the atmosphere above the mesosphere and below the exosphere where temperature increases with height.

**Ultraviolet** High-energy light invisible to the eye and mostly (but not completely—hence sunburn damage) blocked by the Earth's atmosphere. To view the ultraviolet universe unobstructed, which was Carruthers's goal, required getting his instruments beyond the atmosphere on rockets or satellites.

**Van Allen radiation belts** Broad curved bands of charged particles around the Earth, captured from the solar winds by the Earth's magnetic field. Discovered in 1958 by physicist James Van Allen from early Explorer flights.

**Wavelength** The distance from crest to crest of an oscillating current in all forms of media. Visible light wavelengths range from 4,000 to 7,000 Angstroms. (See Angstrom)

**X-ray radiation** Photons of energies higher and wavelengths shorter than ultraviolet photons.

# NOTES

## CHAPTER 1

1. Compton, *Where No Man Has Gone Before*, 244–245.

2. Jones, *Apollo 16 Lunar Surface Journal*, "Loading the Rover," 119 hr., 44 min., 11 sec. The lunar excursion module was renamed the "lunar module" after 1967.

3. Harry Heckathorn, interview with DeVorkin, June 11, 2020.

4. Weitekamp, "Space History Matures," 15.

5. Holloway, "NRL's Not-So Hidden Figure." See also Shetterly, *Hidden Figures*.

6. For general histories relevant to this biography, see Wilkerson, *The Warmth of Other Suns*; Sluby, *The Inventive Spirit of African Americans*; and Fouché, *Black Inventors*.

7. Pedrick, "Percy Julian and the False Promise of Exceptionalism," 5. Adapted from discussions in "Innate: How Science Invented the Myth of Race," Science History Institute Museum & Library, accessed May 15, 2023, https://sciencehistory.org/innate.

8. Latour and Woolgar, *Laboratory Life*, 188.

9. Barton C. Hacker, review of DeVorkin, *Science with a Vengeance*, *Technology and Culture* 35, 3, 628–629.

10. On sounding rockets, see Newell, *Sounding Rockets*.

11. See *The Post-Apollo Space Program: Directions for the Future*, September 1969, accessed July 20, 2024, https://www.nasa.gov/history/the-post-apollo-space-program-directions-for-the-future/

12. Carruthers, image converter for detecting electromagnetic radiation.

13. "Background Information," June 1, 1967, in "Astronomy Disciplines Goals and Objectives NASA/OSSA Prospectus 1967," box 1, "Meeting Files, 7/1/1966—6/30/1939, folder "Astronomy Subcommittee No. 2-FY68," Attachment 23, Acc 71A 6314. RG255/NARA.

14. Lubkin, "Charge-Coupled Devices Would be Cheap, Compact," and Smith, *The Space Telescope*, 329–331.

15. Smith and Tatarewicz, "Replacing a Technology."

16. Quote from Miller, "Astronomical Instrumentation Acquires Pedagogical Tools," *Physics Today* 42, 98. See further discussion in Miller, "Optical Instrumentation for Large Telescopes in the Next Decade," *Bulletin of the Astronomical Society*, Vol. 21, No. 4, 1146–1147.

17. On how space technology promoted the development of the CCD, see Morton and Gabriel, *Electronics*, 120–123.

18. Carwell, *Blacks in Science*.

19. Carwell, *Blacks in Science*, 9.

20. Astronaut Thomas Mattingly also conducted visual astronomical photography of the Sun and nebulae from inside the orbiting command module on *Apollo 16*. He used Hasselblad and Nikon cameras with Kodak film. See Mercer, Dunkelman, and Mattingly, "Gum Nebula, Galactic Cluster, and Zodiacal Light Photography"; and MacQueen, Ross, and Mattingly, "Solar Corona Photography," chapter 31. *Apollo 15* also conducted near-ultraviolet photography of the Earth and Moon from a 70-mm camera in the orbiting command module.

## CHAPTER 2

1. Remarks by Barbara Carruthers at the funeral for her mother, February 23, 1990. "Funeral for Sophia Barbara Martin," folder 4, box 1, GC/NASM. Quotation from "History," Chicago State University, accessed November 29, 2022, https://www.csu.edu/collegeofeducation/history.htm.

2. For the statistics, see, among several sources, Christopher Manning, "African Americans," Encyclopedia of Chicago, accessed July 29, 2023, http://www.encyclopedia.chicagohistory.org/pages/27.html; and Thomas D. Snyder, "120 Years of American Education: A Statistical Portrait," Center for Education Statistics, accessed July 29, 2023, https://nces.ed.gov/pubs93/93442.pdf.

3. On educational attainment as a mark of elite status, see McBride and Little, "The Afro-American Elite," 110.

4. Central Tennessee had the largest population of slaves in the state before April 1865 until the legislature ratified the Thirteenth Amendment. Goodstein, "Slavery."

5. Imes, "The Legal Status of Free Negroes and Slaves in Tennessee," 272.

6. Gatewood, "Aristocrats of Color," 10.

7. Quotes from Benjamin F. Carruthers Jr., untitled family history, Carruthers Papers, Schomberg Archives MG433, box 1, folder 1, BC/SA. The Knights of Pythias position comes from, an obituary (see "Benjamin J. Carruthers") and has

not been confirmed. On the Knights, see Bell, "D.C. Knights of Pythias Hope to Eliminate Color Barrier."

8. By the 1940s, some 90 percent of the Black people in Chicago lived in that area. Winder, "Residential Invasion and Racial Antagonism in Chicago," 239.

9. Quotes from untitled family history, Ben Carruthers Papers, Schomberg Archives MG433, box 1, folder 1, BC/SA.

10. Wilkerson, *The Warmth of Other Suns*, 36–37.

11. See the review of Roi Ottley, *The Lonely Warrior* (Chicago: Henry Regenery, 1955) by Blue, "The Black Mr. Hearst," 149–151.

12. Carruthers family history. Benjamin F. Carruthers Jr., untitled family history, Carruthers Papers, Schomberg Archives MG433, box 1, folder 1, BC/SA.

13. Carruthers family history 11. Benjamin F. Carruthers Jr., untitled family history, Carruthers Papers, Schomberg Archives MG433, box 1, folder 1, BC/SA.

14. "Benjamin J. Carruthers," *The Chicago Defender* (November 11, 1939): 10.

15. Wilkerson, *The Warmth of Other Suns*, 269.

16. Ryan A. Ross, "Being Black at Illinois," UIAA, accessed August 3, 2023, https://uiaa.org/2023/04/17/being-black-at-illinois/.

17. See, for instance, "Kappa Alpha Psi Celebrates Its History as Illinoisâ€™ Oldest Black Fraternity," *Daily Illini*, February 8, 2012, https://dailyillini.com/uncategorized/2012/02/08/kappa-alpha-psi-celebrates-its-history-as-illinoisae-oldest-black-fraternity/.(Accessed April 6, 2023). Courtesy Barrett Caldwell.

18. "Young Set Enjoys Gay House Party." She was also a Chi Delta Sigma sorority member in college: "Reunion Dinner Served Chi Delta Sigma Sorority," *Chicago Defender*, March 30, 1935.

19. Benjamin J. Carruthers to "Dear Son", April 15, 1935, B. F. Carruthers Papers, Schomberg Archives, NYPL MG433, box 1, folder 4.

20. 1950 Census.

21. Daniel, *Dispossession*, 15.

22. Gerald Carruthers, interview with David DeVorkin, August 3, 2020.

23. Gerald Carruthers, interview, August 3, 2020.

24. Gerald Carruthers, interview and follow-up phone conversation, August 5, 2023.

25. George Carruthers, interview, 1992.

26. George Carruthers, "Speech, Career Conference," rough draft April 3, 1960, Carruthers Papers, NASM Archives Division, folder 12, box 3, GC/NASM.

27. Carruthers, "Speech, Career Conference."

28. George Carruthers, interview, 1992.

29. George Carruthers, interview, 1992. Gerald Carruthers interview, August 3, 2020.

30. "Clermont County Schools" report cards for 1947–1948 and 1950–1951, folder 2, box 1, GC/NASM.

31. George Carruthers, "Speech: Career Conference." On the influence of Buck Rogers comics, see Weitekamp, *Space Craze*.

32. Carruthers, "Speech, Career Conference."

33. George Carruthers, interview, August 18, 1992. Magazines like *Popular Science* frequently carried these types of advertisements. George may well have been attracted to a January 1946 issue featuring the new 200-inch telescope on Mount Palomar on the cover and found a full-page Edmunds Salvage advertisement featuring a 14-power lens set with a separate tube for $1.55. "Bargains in War Surplus Lenses & Prisms," *Popular Science* 148, no. 1 (1946): 233. George's brother Gerald also recalls that once when George pointed the lens toward the sun, it ignited dry brush that caused a fire.

34. Gerald Carruthers, interview, August 3, 2020.

35. Ryan, "Man Will Conquer Space Soon!" On the wide influence that the series and contemporary writing and films had on future space scientists, see Miller, "The Archaeology of Space Art," 139–143; Corn and Horrigan, *Yesterday's Tomorrows*; and McCurdy, *Space and the American Imagination*. See also *Collier's* Magazine "Man Will Conquer Space," Scribd, accessed November 16, 2022. https://www.scribd.com/doc/118710867/Collier-s-Magazine-Man-Will-Conquer-Space.

36. Roger Launius, commenting in "Sputnik Declassified," PBS, NOVA, November 6, 2007, https://www.pbs.org/wgbh/nova/transcripts/3415_sputnik.html. See also Neufeld, *Von Braun*. On the *Collier's* series specifically, see McCurdy, *Space and the American Imagination*, 37–41.

37. Carruthers, interview, 1992.

38. Carruthers, interview with Swanson.

39. George Archer Carruthers, *Chicago Daily Tribune*, September 27, 1952. ProQuest https://www.proquest.com/docview/178398550/A598C0BA66FE4C9FPQ/4?accountid=46638.

40. Gerald Carruthers, phone conversation with DeVorkin, August 5, 2023.

41. Gerald Carruthers, phone conversation with David DeVorkin, August 9, 2023; "Who Was Emmett Till?" PBS, American Experience, Accessed July 21, 2024. https://www.pbs.org/wgbh/americanexperience/features/biography-emmett-till/. On Till, see, among many sources, Mamie Till-Mobley and Christopher Benson,

*Death of Innocence: The Story of the Hate Crime That Changed America* (New York: Ballantine Books, 2005).

42. See Robert Nemiroff, "Introduction," to Lorraine Hansberry, *A Raisin in the Sun*, 10-11 (New York: Vintage, 1994). Hansberry's "Raisin" portrayed a Black family moving into a white neighborhood as evidence of "striving."

43. "Chicago Englewood High School 'Eagles,'" Illinois High School Glory Days, accessed May 20, 2022, https://illinoishighschoolglorydays.com/2022/03/01/chicago-englewood-high-school-eagles/. On the cultural enrichment, see Manning, "African Americans."

44. Phelps, *They Had a Dream*, chapter 2. See also, in brief, "Robert Henry Lawrence Jr.," Wikimedia Foundation, accessed November 7, 2022, https://en.wikipedia.org/wiki/Robert_Henry_Lawrence_Jr.

45. "Our Students Win Science Awards," *Englewood News* June 1957. Carruthers Papers, NASM Archives.

46. Opened in 1930, the planetarium was funded by Max Adler, a concert violinist who married into the Julius Rosenwald family, the cofounders of Sears, Roebuck & Co. He became vice president of Sears until he retired in 1928 to devote his wealth to philanthropy. Through his wife, Adler also acquired a legendary collection of astronomical instruments from Europe that established the planetarium as the most significant astronomical museum in the United States. Adler Planetarium and Astronomical Museum, *Courses*, 19. Author's collection. On Adler, see Fox, *Adler Planetarium and Astronomical Museum of Chicago*. And on planetariums in the United States, see Jordan D. Marché, *Theaters of Time and Space*.

47. Carruthers, "Speech, Career Conference."

48. Adler Planetarium and Astronomical Museum, *Courses*, 26. Author's collection. Quotations used here are from a paper in press. The shop stimulated careers in optics and astronomy. When the author worked at the Yerkes Observatory, he learned about the Adler's legacy from Richard Monnier, the Yerkes optical shop director, a warm spirit until his untimely death in 1967.

49. George Carruthers, 1992 interview with DeVorkin.

50. Adler Planetarium and Astronomical Museum, *Courses*, 19. Author's collection.

51. See, for instance, Ashdowne Bros. Advertisement, "Deluxe Pyrex Reflecting Telescope Kits," *Sky and Telescope*, October 1955, 513. The cost of 6-inch and 8-inch kits ran from $10 to $17. The largest was 12½ inches at $52.95. Six dollars in 1955 is equivalent to roughly seventy dollars in 2023. See "Inflation Calculator," accessed August 15, 2023, https://www.in2013dollars.com/us/inflation/1955?amount=10.

52. Carruthers, video history interview; Milton Rosen, *Viking Rocket Story*.

53. Carruthers, interview with Swanson.

54. Green and Lomask, *Vanguard, A History*.

55. "George Carruthers Wins Science Contest" and "Public School Star Seniors Are Selected." On Science Service, see Bennet, "Science Service and the Origins of Science Journalism, 1919–1950."

56. Paul J. Copeland (IIT) to George Carruthers, June 3, 1957, folder 20, box 1, GC/NASM.

57. Gerald Carruthers, interview with DeVorkin August 3, 2023. Unfortunately, none of his cartooning or artistic efforts have been found.

58. See https://www.sciencemadness.org/whisper/viewthread.php?tid=157318 for insight into the reaction Carruthers describes. The contemporary version of the song reads,

> O Furlong, our Furlong,
> How many times before
> You've taught us how to faithful be
> You're turned defeat into victory
> O Furlong, our Furlong,
> We'll stand by you today
> Hurrah for the Purple and White
> Hurrah for the Purple and White.
> "Englewood Technical Prep Academy," Wikimedia Foundation, https://en.wikipedia.org/wiki/Englewood_Technical_Prep_Academy.

59. George Carruthers, interview with DeVorkin, August 18, 1992.

60. George Carruthers, interview with y DeVorkin, August 18, 1992.

61. Gerald Carruthers, interview, August 3, 2023. Studies have shown that prejudice was greatest when there was "intermediate contact" between races, and least when it was minor or no mixing. See Winder, "Residential Invasion and Racial Antagonism in Chicago," 240.

62. George Carruthers, conversation with Richard Paul, June 15, 2009, courtesy of Richard Paul and reproduced in Paul and Moss, *We Could Not Fail*, 107–108.

## CHAPTER 3

1. George Carruthers, interview with DeVorkin, August 18, 1992.

2. George Carruthers, interview.

3. George Carruthers, interview.

4. George Carruthers, interview. See also transcript, "Application for Research Associateship or Assistantship in the E.O. Hulburt Center for Space Research," attached to Carruthers to E.O. Hulburt Center for Space Research, July 4, 1963. HF Papers 130016, box 17, APS.

5. Gerald Carruthers to the author, January 14, 2021.

6. George Carruthers, interview by David DeVorkin, August 18, 1992.

7. Early reactions are analyzed by Greenberg, "Post-Sputnik,"1281–1283. For a later perspective, see Launius, An Unintended Consequence of the IGY," See also Launius, Logsdon, and Smith, *Reconsidering Sputnik*.

8. On Project Moonwatch, see McCray, *Keep Watching the Skies*.

9. George Carruthers, interview with DeVorkin, August 18, 1992.

10. George Carruthers, interview.

11. Transcript, "George Robert Carruthers," June 13, 1963, University of Illinois-Office of Admissions and Records, "In Good Standing," folder "EOHC George R. Carruthers Material," box 17, HF/APS.

12. George Carruthers, interview with DeVorkin, August 18, 1992, 35.

13. It was barely a first step. "Student Life at Illinois: 1950–1959," Illinois Library Student Life and Culture Archives, accessed May 8, 2023, https://www.library.illinois.edu/slc/research-education/timeline/1950–1959/.

14. George Carruthers, interview by David DeVorkin, August 18, 1992. On social isolation on white campuses, see Komanduri, Roebuck, and McCamey, "Race and Class Exploitation," 156–173.

15. Diane Jeffers and Harry H. Hilton, "Emeritus Prof. Shee-Mang Yen: In Memoriam," University of Illinois, Aerospace Engineering, published September 7, 2016, https://aerospace.illinois.edu/news/emeritus-prof-shee-mang-yen-memoriam.

16. "Twinkle, Twinkle Little Star, Now We Know Just What You Are," *U. of I. Aerospace Alumni News*, May 1971, 19–20. Copy in folder 1, box 7, GC/NASM.

17. Tolchin, "Teen-Age Rocket Groups Need a Parental Brake."

18. See "Group Builds Experimental Rocket Motor" and "Chicago Rocket Society Juniors Shoot for the Sky."

19. George Carruthers, "Speech, Career Conference," rough draft April 3, 1960, Carruthers Papers, NASM Archives Division, folder 12, box 3, GC/NASM.

20. It is regarded as a founding field in quantum physics, starting with the work of Ernest Rutherford, Joseph John Thomson, and Niels Bohr. See Brown, "A Short History of Gaseous Electronics," 1–18.

21. See "Group Builds Experimental Rocket Motor" and "Chicago Rocket Society Juniors Shoot for the Sky." See also Carruthers, "Speech, Career Conference."

22. Pete De Paolo (Plasmadyne Corporation) to George Carruthers, July 27, 1959, folder 1, box 2, GC/NASM.

23. Carruthers, "Speech, Career Conference."

24. Art Harrison to George Carruthers, February 17, 1959, folder 1, box 2, GC/NASM.

25. "5 Boys Suffer Burns in Two S. Side Blasts." George Carruthers (The HistoryMakers A2004.112), interview by Racine Tucker Hamilton, July 27, 2004, The HistoryMakers Digital Archive, session 1, tape 2, story 3. Gerald incorrectly recalled that George was not involved. See Gerald Carruthers's interview. In November 1960, George's stepfather, Wellington Martin, advised him to provide an affidavit, to be verified by his personal attorney, stating (1) that he was not "manufacturing fireworks for sale in my home," (2) that he was doing experimental work at home because he was an aeronautical engineering student attending summer school at the University of Illinois Branch on Navy Pier "and the experiment was in one with my work there," and (3) that the case was dismissed in August at a hearing in Boy's Court because there had been "no grounds for an arrest." Dad to "Son!" November 28, 1960, "Personal Correspondence," folder 4, box 1, Carruthers Papers, GC/NASM.

26. George Carruthers interview, The HistoryMakers A2004.112, interview by Racine Tucker Hamilton, July 27, 2004, The HistoryMakers Digital Archive, session 1, tape 2, story 3, "George Carruthers Describes His Experiences at Englewood High School in Chicago, Illinois."

27. Kenneth Dymond to David DeVorkin, December 29, 2021.

28. Joyce Bedi to David DeVorkin, March 3, 2021.

29. Transcript, "George Robert Carruthers," June 13, 1963, University of Illinois-Office of Admissions and Records, "In Good Standing," folder "EOHC George R. Carruthers Material," box 17, HF/APS.

30. George Carruthers, interview with David DeVorkin, 1992.

31. George Carruthers to "Dear Mom and Dad," April 15, 1961, folder 4, "Personal Corr," box 1, Carruthers Papers, GC/NASM.

32. Carruthers, "Control of Missiles & Space Vehicles," working draft of equations, sketches, and logic control diagrams of how the system had to be connected, Carruthers Papers, NASM Archives.

33. Plummer, "The Newest South," 77.

34. Gerald Carruthers to George Carruthers, May 21, 1962, Carruthers Papers.

35. George Carruthers, interview, 1992.

36. George Carruthers, interview, 1992.

37. George Carruthers, interview, 1992. As William Whyte has noted, the dominant characteristic of an outstanding scientist is "fierce independence," the antithesis of the "company-oriented" organization man. Whyte, *The Organization Man*, 232.

38. George Carruthers, conversation with Richard Paul, June 15, 2009. Quoted in Paul and Moss, *We Could Not Fail*, 117.

39. Wellington S. Martin ("Dad") to George Carruthers, July 7, 1962, "Personal Correspondence," folder 4, box 1, Carruthers Papers, GC/NASM.

40. On Goldstein and laboratory history, see Joseph T. Verdeyen, "The Gaseous Electronics Laboratory," accessed August 15, 2023, https://web.archive.org/web/20071017121448/http://lope.ece.uiuc.edu/history.htm.

41. See, for instance, Goldstein, Narasinga Rao, and Verdeyen, "Interaction of Microwaves in Gaseous Plasmas," 121.

42. George Carruthers, "A Study of the Recombination of Atomic Nitrogen behind Shock Waves Propagating in Partially Dissociated Nitrogen," courtesy of the University of Illinois Archives, Goldstein Papers, Record Series 11/6/24, box 7, folder U.S. Naval Research Lab, 1966, LG/UI.

43. Booker, "Fifty Years of the Ionosphere," 2113.

44. Carruthers, "Experimental Investigations," 1.

45. Carruthers, "Experimental Investigations," 3.

46. Verdeyen, "The Gaseous Electronics Laboratory."

47. Strong et al., "Glassblowing as a Central Practice."

48. Carruthers, "Experimental Investigations," figure 17. See also page 155. On photomultipliers, see Engstrom, *Photomultiplier Handbook*. On its history, see also DeVorkin, "Photomultiplier," 458–460.

49. Carruthers, "Experimental Investigations," 11–18.

50. Carruthers, "Experimental Investigations," 133, 155.

51. Carruthers, interview, 1992.

52. Carruthers, interview, 1992.

53. John N. Howard, "Edward O. Hulburt: Frederic Ives Medalist, 1955," OPN Optics & Photonics News, https://www.optica-opn.org/opn/media/Images/PDFs/13293_26887_114200.pdf?ext=.pdf. Accessed July 21, 2024.

54. Carruthers to "Dear Sirs," May 13, 1963, folder "George Carruthers Material, 1963–1976," box 17, HF/APS.

55. Carruthers to "Dear Sirs," July 4, 1963, folder "George Carruthers Material, 1963–1976," box 17, HF/APS.

56. Carruthers to "Dear Sirs," referring to a category in an NRL program booklet announcement. "George Carruthers Material, 1963–1976," box 17, HF/APS.

57. S. M. Yen, Professor to E.O. Hulburt Center, September 10, 1963, folder "George Carruthers Material 5517, 1963–76," box 17, HF/APS. Curiously, Carruthers does not acknowledge Yen in his thesis.

58. Herbert Friedman to George Carruthers, October 17, 1963, folder "George Carruthers Material 5517, 1963–76," box 17, HF/APS.

59. George Carruthers to Herbert Friedman, October 25, 1963, folder "George Carruthers Material, 1963–1976," box 17, HF/APS.

60. Edward O. Hulburt to Herbert Friedman, January 21, 1963, folder "Hulburt, Edward 1959 -1981," box 2, HF/APS.

61. Moss, "NASA and Racial Equality in the South, 1961–1968," 48.

62. After completing his PhD at Johns Hopkins in 1939, he applied for positions in academe and industry, but "there was a lot of anti-Semitism in the physical sciences and engineering fields in those days. I found out almost immediately that I couldn't even get an interview at places like Bell Labs or GE and Westinghouse." Posts in academe were equally unlikely, so finally his thesis advisor, Herman Pfund, arranged a civil service position for him at NRL, and Friedman spent the rest of his career there becoming a world leader in X-ray astronomy. Friedman interview with Hirsh, 1979, 155–156; and Hirsh, *Glimpsing an Invisible Universe.*

63. Panel on Astronomical Facilities, *Ground-Based Astronomy A Ten-Year Program*. Washington, DC: National Academy of Sciences, 1964. On the origin and focus of the Decadal Reviews, see DeVorkin, "Who Speaks for Astronomy?," 55–92.

64. Friedman to Gerald F. Mulders, NSF, July 22, 1964, box 17, "George Carruthers Material 5517, 1963–76," Herbert Friedman Papers, APS.

65. See Part III of "Executive Order 10925—Establishing the President's Committee on Equal Employment Opportunity," The American Presidency Project, accessed May 9, 2023, https://www.presidency.ucsb.edu/documents/executive-order-10925-establishing-the-presidents-committee-equal-employment-opportunity.

66. An especially insightful, personal, and vivid account of Birmingham and the Civil Rights Movement is found in McWhorter, *Carry Me Home.*

67. Mumm, "AE Alum to be Presented University's Highest Honor." The first at Illinois was Arthur B. C. Walker Jr. in 1962, who then worked at the Aerospace Corporation and became a professor of physics at Stanford University, mentoring Black PhD students over the years. Like Carruthers, Walker was active in minority outreach, and he was also an alumnus of Englewood High School. See Petrosian, "Arthur B. C. Walker (1936–2001)."

## CHAPTER 4

1. See, for instance, Dupree, *Science in the Federal Government*; and Sapolsky, *Science and the Navy*. See also "NRL History, About Us," US Naval Research Laboratory, accessed on August 14, 2024. https://www.nrl.navy.mil/about-nrl/history/Edison/.

2. Friedman interview with Martin Harwit, June 7, 1983, AIP. And, in general, Bruce William Hevly, "Basic Research within a Military Context," and Allison, *New Eye for the Navy.*

3. DeVorkin, *Science with a Vengeance*.

4. See Hevly, "Basic Research within a Military Context," and Allison, *New Eye for the Navy*.

5. DeYoung, *National Security and the U.S. Naval Research Laboratory*, 8.

6. Hevly, "Basic Research within a Military Context," 1–2.

7. In his 1992 interview with DeVorkin, Carruthers considered all bureaucracy as anathema to creativity.

8. On the Aerobee, see Newell, *Sounding Rockets*; Corliss, *NASA Sounding Rockets, 1958–1968*, 18–21, 46; and DeVorkin, *Science with a Vengeance*, 74–94. On sounding rockets in general, see National Research Council, *Sounding Rockets*, accessed April 26, 2023, https://nap.nationalacademies.org/catalog/12400/sounding-rockets-their-role-in-space-research; and Leak, *Rocket Research at NRL*.

9. Corliss, *NASA Sounding Rockets, 1958–1968*, 40.

10. Data gleaned from "NSSDC Index of International Scientific Rocket Launches Ordered by Sponsoring Country/Agency," National Space Science Data Center, Goddard Space Flight Center, February 1972, NASA Scientific and Technical Information Facility.

11. Oral history interview, "Dr. Homer E. Newell," with Richard F. Hirsh, July 17 and October 20, 1980, 20.

12. Hevly, "Basic Research within a Military Context," 79; Friedman interview with DeVorkin. See also DeVorkin, *Science with a Vengeance*, 237–243.

13. US Naval Research Laboratory, "Collection and Identification of Fission Products of Foreign Origin," date illegible [circa September 21, 1949], Top Secret, National Security Archive, part I; and King, Peter, and Friedman, "Collection and Identification of Fission Products of Foreign Origin," accessed October 17, 2022, https://nsarchive.gwu.edu/document/19586-national-security-archive-doc-15-document-7-u-s. See also Friedman, Lockhart, and Blifford, "Detecting the Soviet Bomb," 38–41.

14. William Rense was the first to determine the structure of the Lyman-alpha line. See DeVorkin, *Science with a Vengeance*, 228.

15. Gursky, "Herbert Friedman," 98.

16. Among those who left for what became the Goddard Space Flight Center, such as Albert Boggess, there was a feeling that they had "jumped ship." Interview with Byram, Chubb, Friedman, and Kreplin, and Boggess interview with DeVorkin.

17. The classified project GRAB, for "Galactic Radiation and Background," was merged with SOLRAD. https://airandspace.si.edu/collection-objects/satellite-electronic-intelligence-galacticradiation-and-background-grab-1/nasm_A20020087000.

18. Gursky, "Herbert Friedman," 100.

19. As late as 2016, sounding rockets were still considered essential for space research and other applications. As an article on the NASA sounding rocket program argued, "in addition to science observations, sounding rockets provide a unique technology test platform and a valuable training ground for scientists and engineers. Most importantly, sounding rockets remain the only way to explore the tenuous regions of the Earth's atmosphere (the upper stratosphere, mesosphere, and lower ionosphere/thermosphere) above balloon altitudes." Christe et al., "Introduction to the Special Issue on Sounding Rockets and Instrumentation."

20. Walsh, "News and Comment: Graduate Education," 788.

21. Walsh, "News and Comment: Graduate Education," 789.

22. Stecher and Milligan, "An Interim Interstellar Radiation Field," 268–270.

23. Hirsh, *Glimpsing an Invisible Universe*, 25.

24. Lean interview with DeVorkin, Callahan, and Femia, 61.

25. Chubb and Byram, "Stellar Brightness Measurement at 1314 and 1427 A," 630.

26. Hallam, "Astronomical Ultraviolet Radiation," 219.

27. Friedman, "N.R.L. Equipment and Plans Report," 24.

28. Logsdon, "Space Research: Directions for the Future," 581.

29. Homer Newell to Harry J. Goett, Director, Goddard Space Flight Center, May 10, 1965, Newell Papers, 4490 NHO.

30. Friedman interview with Hirsh.

31. Friedman to Gerard F. Mulders, June 2, 1964, folder "George Carruthers Material 5517, 1963–76," box 17, HF/APS.

32. Carruthers to Friedman, October 25, 1963, folder "George Carruthers Material 5517, 1963–76," box 17, HF/APS.

33. Carruthers to Friedman, May 23, 1964, folder "George Carruthers Material 5517, 1963–76," box 17, HF/APS.

34. Carruthers to Friedman, July 22, 1964, folder "George Carruthers Material 5517, 1963–76," box 17, HF/APS.

35. Carruthers to Friedman, July 28, 1964, folder "George Carruthers Material 5517, 1963–76," box 17, HF/APS.

36. Talbot Chubb, in Byram, Chubb, Friedman, and Kreplin video interview with DeVorkin, December 12, 1986, SI/SLOAN/AIP.

37. Carruthers to Friedman, August 1964, "A Study of the Far Ultraviolet Spectra of Stars and Nebulae Using a Rocket Spectrograph," folder "George Carruthers Material 5517, 1963–76," box 17, HF/APS.

38. Carruthers interview with DeVorkin, 1992.

39. George Carruthers to Herbert Friedman, August 11, 1964, folder "George Carruthers Material, 5517, 1963–76," box 17, HF/APS.

40. An excellent review of the technology is given in Thompson, "The Carnegie Image Tube Committee." And for a perception of how his technology was regarded in the early 1950s, see Armstrong, "Electron Image-Forming Devices in Astronomy," 147–151.

41. Carruthers oral history, 1992. His Illinois professors worried that his proposal should not be too explicit about the design in his proposal, but Carruthers evidently persisted. The article was most likely Baum, "Electronic Photography of Stars," 81–92.

42. Baum, 92.

43. Thompson, "The Carnegie Image Tube Committee."

44. Hembree, "Summary of British Image Tube Symposium."

45. Carruthers to Friedman, August 11, 1964, folder "George Carruthers Material 5517, 1963–76," box 17, HF/APS.

46. Appendix A in Thompson, "The Carnegie Image Tube Committee."

47. DeVorkin, *Fred Whipple's Empire*, chapter 10.

48. Joel Stebbins, "The Law of Diminishing Returns," 8. *Science* 99, no. 2571 (1944).

49. Wood, *The Present and Future of the Telescope of Moderate Size*, 35.

50. Carruthers, "Electronic Imaging Devices in Astronomy," 347. For the general problem, see Thompson, "The Carnegie Image Tube Committee," 162. One of the greatest drawbacks to this type of detector was that the electron-producing photocathode had to be in a high vacuum, whereas photographic film could not because it would outgas water vapor, which would oxidize and thereby damage the photocathode. The tubes also had to be cooled to liquid nitrogen temperatures. Lallemand had placed the film in a separate glass canister sealed away from the vacuum. Therefore, after an exposure was completed, the glass connection had to be broken to access the film, which exposed the photocathode to the air and rendered it unusable, making the whole observing process extremely inefficient. Other designs avoided this problem, such as a design created by the Lick Observatory astronomer Gerald Kron, which placed the film in a mechanically separable chamber. But still, it was a single-shot device.

51. Carruthers to Friedman, August 26, 1964, folder "George Carruthers Material 5517, 1963–76," box 17, HF/APS.

52. George Carruthers interview, August 18, 1992. Perspective on this tension is provided in Newell, *Beyond the Atmosphere*, 124–125.

53. Heckathorn, "George Robert Carruthers," 64.

54. George Carruthers interview, August 18, 1992.

55. George Carruthers interview, August 18, 1992. Labor relations is noted in Owen, "The Mid-60's at NRL," 109.

56. In a HistoryMakers interview, Carruthers confirmed that Taylor was the only other Black scientist in his division. Interview #1 George Carruthers (The HistoryMakers A2004.112), interview with Racine Tucker Hamilton, July 27, 2004, The HistoryMakers Digital Archive, session 1, tape 1, story 3. The earliest on record is University of Chicago chemistry PhD Warren Elliot Henry who joined NRL's cryogenics division in 1948 after war work at MIT. He worked there as a physicist for over a decade before moving on to industry and academe. Daniel Parry, "Pioneering NRL Physicist had Tuskegee Ties," NRL News and Press Releases, US Naval Research Laboratory, February 23, 2012, https://www.nrl.navy.mil/Media/News/Article/2570036/pioneering-nrl-physicist-had-tuskegee-ties/. See also "William E. Henry, Scientist Born," (accessed August 13, 2024); https://aaregistry.org/story/a-scientific-pioneer-warren-e-henry/ (Accessed August 13, 2024).

57. George Carruthers interview. August 18, 1992.

58. Chubb video history interview.

59. George Carruthers interview, August 18, 1992.

60. "Julian Cliford Holmes," Wikipedia, accessed November 30, 2022, https://en.wikipedia.org/wiki/Julian_Clifford_Holmes.

61. Chubb, video history with DeVorkin, as part of Friedman, Chubb, Byram, and Kreplin interview, 1986, RU 9539 Collection Division 2, SIA.

62. Byram, video history with DeVorkin, 1986.

63. Byram, video history with DeVorkin, 1986.

64. Johnson, video history with DeVorkin, Research Laboratory RU 9539 Collection Division 2: Aeronomy, session 1 and session 2, 1987, SIA.

65. As Bruno Latour once observed, the laboratory scientist "accelerates the frequency of trials, allowing many mistakes to be made and registered." Latour, "Give Me A Laboratory," 165.

66. On the Carnegie project, in particular, see Thompson, "The Carnegie Image Tube Committee."

67. George Carruthers, "Magnetically Focused Image Converters," 7–13. See also Carruthers, "Magnetically Focused Electronographic Image Converters for Space Astronomy Applications," 633–638.

68. Carruthers interview, August 18, 1992.

69. Carruthers, "Electronic Imaging Devices in Astronomy," 335.

70. Carruthers, interview, August 18, 1992.

71. Richard Tousey, "Rocketry," 42; and DeVorkin, *Science with a Vengeance*, 221–230.

72. Gerard F. W. Mulders (NSF) to H. Friedman, September 8, 1965, folder "George Carruthers Material 5517, 1963–76," box 17, HF/APS.

73. Martin Harwit oral history with DeVorkin, session II, Accessed August 18, 2023, https://www.aip.org/history-programs/niels-bohr-library/oral-histories/28169-2.

74. Herbert Friedman, "Reports of Observatories: The E. O. Hulburt Center for Space Research, U. S. Naval Research Laboratory, Washington, D. C.," 765–777.

75. Friedman, 775.

76. Mange interview with Ron Doel and Fay Korsmo. Mange received his PhD from Pennsylvania State University in 1954 and had studied under, among other leaders in atmospheric physics, Marcel Nicolet. He joined NRL in 1959 after several years at the National Academy of Sciences on International Geophysical Year programs. See also Meier, "Philip W. Mange."

77. Kondo and Kupperian, "Interaction of Neutral Hydrogen and Electrons and Protons in the Radiation Belts," 860.

78. This was true especially for later larger payloads. See Brown interview with DeVorkin and Callahan.

79. Wallops Island Missile Range, *Sounding Rocket Program Handbook*, 91.

80. "Operations Requirement," attached to "Solar Ultraviolet Imaging Experiments," submitted by George Carruthers, "Summary of Field Operations," attached to H. Friedman to Commanding Office, U.S. Ordnance Missile Test facility, through the Director, Naval Research Laboratory, May 10, 1966, courtesy of the University of Illinois Archives, Record 11/6/24, box 7, folder U.S. Naval Research Lab, 1966.

81. "Operations Requirement," 6.

82. Hoidale, *Meteorological Data Report*. See also Corliss, *NASA Sounding Rockets, 1958–1968*.

83. George Carruthers, interview, August 18, 1992. Corliss, *NASA Sounding Rockets, 1958–1968*, 122, noted it was partly successful. In an appendix accessed May 5, 2023, https://history.nasa.gov/SP-4401/app-b4.htm. However, it was listed as unsuccessful. 121–125.

84. "Operations Requirement," 11.

85. Carruthers to Goldstein, September 5, 1966, courtesy of the University of Illinois Archives, Record 11/6/24, box 7, folder U.S. Naval Research Lab, 1966, LG/UI.

86. George Carruthers to "Uncle Ben," September 25, 1966, B. F. Carruthers Papers, Schomberg Archives, NYPL MG433, box 1, folder 4. Carruthers Papers, National Air and Space Museum, box 1, BC/SA.

87. Carruthers, "Magnetically Focused Image Converters," 7–13.

88. George Carruthers to "Uncle Ben."

89. Gerald Carruthers interview. See "Dad" (Wellington Martin) to "Dear Son George," October 19, 1961, box 1, folder 4, GC/NASM. Gerald won first place at the Illinois Junior Academy of Science, meeting John Glenn at the awards ceremony in 1963.

90. McWhorter, *Carry Me Home*, 427. See also Launius, *NACA to NASA to Now*, 110–111. Somewhat ironically, a brochure describing the "Rocket City Astronomical Association" reprinted an autobiographical excerpt by Wernher von Braun wherein he stated that the coming of the rocket to Huntsville raised astronomy out of the category of being a "poor" science. Wernher von Braun, "Forward," Rocket City Astronomical Association brochure. Reprinted remarks from von Braun, "Where Are We Going?" *Space Journal* (Summer 1957). Author's collection.

91. United States Army Ordnance Munitions and Electronic Maintenance School. https://web.archive.org/web/20151110135450/https://alu.army.mil/alog/issues/NovDec09/ord_move_ftlee.html.

92. Laney, *German Rocketeers in the Heart of Dixie*, 51.

93. Gerald Carruthers interview, and subsequent informal communications.

94. Certificate, Military Training Cert Reserve Officers Training Corps Junior Division, June 28, 1957, box 1, folder 2, GC/NASM.

95. It became voluntary in 1964. See "Army ROTC—University of Illinois at Urbana-Champaign," University of Illinois Urbana-Champaign, accessed August 5, 2023, https://publish.illinois.edu/army-rotc/history/.

96. S. N. Ross, Captain, USN to Selective Service System (Chicago), June 10, 1971, folder "EOHC, George Carruthers Material," box 17, HF/APS. These were tense times given President Johnson's escalation of the Vietnam War. When George's number came up again in March 1969, Selective Service called him up and did so again in June 1971. Each time, he appealed through his line of authority and each time NRL requested deferments.

## CHAPTER 5

1. Undated notes, clipped to Gerard F W Mulders (NSF) to H. Friedman, September 8, 1965, folder "George Carruthers Material 5517, 1963–76," box 17, HF/APS.

2. In a review of the Hulburt Program in September 1966, Friedman stated that the center "should always be aimed at university cooperation." He identified some twelve universities identifying only Carruthers as being appointed to the research staff at NRL after his postdoctoral years. Herbert Friedman, "Financial Support of E.O. Hulburt Center for Space Research at NRL," folder "EOHC Proposal to NASA 1967–1968," box 17, HF/APS.

3. Carruthers, interview, August 18, 1992.

4. Carruthers, interview, August 18, 1992.

5. Carruthers, "Magnetically focused image converters with internal," 7–13.

6. Carruthers, "Image converter for detecting electromagnetic radiation."

7. Livingston, "Image-Tube Systems," 102. Livingston pointed out the advantages of reflective cathodes. 347–384.

8. See Sluby, *The Inventive Spirit of African Americans*, and Holmes, *Black Inventors*.

9. Carruthers, "Magnetically Focused Electronographic Image Converters," 633. Carruthers's appeal certainly reflected the opinion of many astronomers. A year later, W. C. Livingston of the Kitt Peak National Observatory, in a major review of the promise and the limitations of the new detectors, made clear that "astronomers have long looked forward to the availability of an image tube which would combine the high quantum efficiency and linear response to light of the photomultiplier with the panoramic image properties of the photographic emulsion." But, he concluded, there was still much to be done to improve their performance, such as using reflective photocathodes, but not mentioning Carruthers, though he did so several years later. See Livingston, "Properties and Limitations of Image Intensifiers," 347, 381; and Livingston, "Image-Tube Systems."

10. See "Nuclear Emulsion," Stratopedia, accessed August 19, 2023, https://stratocat.com.ar/stratopedia/427.htm.

11. In the 1930s, Lallemand used transmissive photocathodes at first because the design was optically simpler and "would considerably simplify the optical." At the time, available cathodes were thin, and so "the efficiency is considerably reduced," which made him convert to reflective designs. But by the late 1940s, with thicker new formulations and thicker cathode materials available, he converted back to transmissive photocathodes. My thanks to Dr. Frederic Soulu who kindly provided translated excerpts from Lallemand' s writings. See "Application de l'optiqué électronique à la photographie," 203, 243; and "L'optiqué électronique et les telescopes," 28, accessed April 22, 2023, https://gallica.bnf.fr/ark:/12148/bpt6k3155r/f243.item. The translated quotes are from Selme, "L'optique électronique et les télescopes," 26–33, accessed April 22, 2023, https://gallica.bnf.fr/ark:/12148/bpt6k9679634v/f8.vertical.

12. Thompson, "The Carnegie Image Tube Committee," 33.

13. Carruthers, "Magnetically Focused Electronographic Image Converters," 633.

14. Carruthers, "Magnetically Focused Image Converters," 8.

15. Carruthers, "Magnetically Focused Electronographic Image Converters," 634.

16. Focal variations of the focal plane in the far ultraviolet are far less than in the visible. See Karl Henize, March 22, 1974, Proposal "Deep Sky Survey in Ultraviolet Wavelengths from Space Shuttle," Appendix B by G. Carruthers, p. 2, figure 3, folder "NRL, Geo. Carruthers Mat. 1974," box 19, HF/APS.

17. Carruthers, "Magnetically Focused Electronographic Image Converters," 635, figure 3.

18. Carruthers, "Far-Ultraviolet Spectroscopy," 272–273. Brief descriptions of the stabilization systems are in Bowyer et al., "Observational Results of X-ray Astronomy," 795; and Morton and Spitzer, "Line Spectra of Delta and Pi Scorpii in the Far-Ultraviolet," 2–3.

19. Carruthers, "Far-Ultraviolet Spectroscopy," 283–284.

20. It launched in 1972. Spitzer and Zabriskie, "Interstellar Research with a Spectroscopic Satellite;" 412–420; and Spitzer, "Behavior of Matter in Space," 1–17. Spitzer briefly discusses the conditions for the formation of interstellar molecular hydrogen in Spitzer, *Diffuse Matter in Space*, 125–126. Lyman $\alpha$, named for the discoverer, Theodore Lyman—no relation to Spitzer—is the resonance line of atomic hydrogen, caused by absorption or emission of energy when an electron moves from the ground state of the hydrogen atom to the first excited state.

21. Sullivan, "Ninety per Cent of the Universe Found Missing by Astronomer," 10.

22. Author's personal recollection.

23. Henry, "Possible Detection of a Dense Intergalactic Plasma."

24. Friedman, "Reports of Observatories: The E. O. Hulburt Center for Space Research," 223.

25. Carruthers, "Observation of The Lyman Alpha Interstellar Absorption Line in Theta Orionis," L98.

26. Carruthers, "An Upper Limit," L141–L142.

27. Carruthers, "Far-Ultraviolet Spectroscopy."

28. Undated notes, evidently in late 1967, clipped to Gerard F W Mulders (NSF) to H. Friedman, September 8, 1965, folder "George Carruthers Material 5517, 1963–76," box 17, HF/APS.

29. See Yeager, *Bright Galaxies, Dark Matter, and Beyond*, and Mitton, Mitton, and Bell-Burnell, *Vera Rubin: A Life*.

30. NRL Press Release, "The Invisible Mass Budget of the Galaxy," May 10, 1967, Friedman Papers, APS MS 113 1967.

31. Carruthers, "Atomic and Molecular Hydrogen in Interstellar Space," 481.

32. On the history of pointing control capabilities, see Teeter and Reynolds, *Final Report on the NASA Suborbital Program*, part 2, table 2-1. By 1968, Aerobee payloads could be stabilized in pitch to +/− 10 seconds of arc. Friedman, *Annual Report of the E.O. Hulburt Center for Space Research, Naval Research Laboratory, for Fiscal Year 1969*.

33. Carruthers, "Rocket Observations of Interstellar Molecular Hydrogen," L83.

34. Heiles, "Physical Conditions and Chemical Constitution of Dark Clouds," 298.

35. "Twinkle, Twinkle Little Star, Now We Know Just What You Are," *U. of I. Aerospace Alumni News*, May 1971, 19–20, copy in box 7, folder 1, GC/NASM.

36. Shull and Beckwith, "Interstellar Molecular Hydrogen," 163.

37. Guy Brandenburg and Cal Powell to the author, November 18, 2022. Northern Virginia Astronomy Club, "Eight Decades of Amateur Telescope Making in the DC area," YouTube video, at 8:24, November 15, 2022, https://www.youtube.com/watch?v=RVvqkN9nBfE.

38. Carruthers, "Rocket Observations." The citations are from the "Astrophysics Data System," Harvard University, accessed August 8.2023, https://ui.adsabs.harvard.edu/. As noted, this highly cited paper, according to the ADS compilation, helped to establish Carruthers's visibility. Citations included some 166 classified as astronomy, 35 in physics. They were from many journals, domestic and international. Most of the citations came in the first year, and a second smaller peak was in the period 2016-2020. It is characteristic of Carruthers that he cited it only once, in 1971.

39. Brett A. McGuire, "2021 Census," 5. Quote from Wakelam et al., "H2 Formation on Interstellar Dust Grains," 1–36.

## CHAPTER 6

1. Newell, *Beyond the Atmosphere*, 290.

2. Neufeld, *Von Braun*, 391–392.

3. Neufeld, *Von Braun*, 399. See also Compton and Benson, *Living and Working in Space*, part 1, chapter 1, accessed November 17, 2022, https://history.nasa.gov/SP-4208/ch1.htm.

4. Compton and Benson, *Living and Working in Space*, part 1, chapter 1; Neufeld, *Von Braun*. On intercenter competition, see McCurdy, *Inside NASA*.

5. Carter, "Space Budget," 170.

6. Carter, 171.

7. Carruthers, "Rocket Observations of Interstellar Molecular Hydrogen."

8. A particularly sensitive moment came when von Braun was asked to address the Astronomy Mission Planning Board meeting in Boulder, Colorado, in mid-March 1968. The Board was advisory to NASA's Office of Space Science and Applications and included national leaders in astronomy and observatory directors. Herbert Friedman was the only member not from academe. A briefing paper for von Braun, possibly prepared by his former Peenemünde associate Ernst Stuhlinger, accused the board of being dubious about using Marshall's OWS for astronomy, preferring free flyers like the Orbiting Astronomical Observatory series. They also doubted MSFC's genuine interest in science. It was von Braun's task to assuage the doubters and "profess sincere desire to support astronomy." [Stuhlinger], "Astronomy Mission Planning Board," March 12, 1968, Von Braun Papers, WVB/USSRC 807-7. There were countless differences of opinion. At about the same time, the Astronomy Subcommittee of the Office of Space Science and Applications, reviewing "NASA Goals and Objectives," felt that, faced with the continued fact of manned spaceflight such as the planned Apollo Telescope Mount, they preferred the "shirt-sleeve environment" over EVAs because there would be less preparation and more dexterity. This tension lasted for years. Review of "NASA's Goals and Objectives," Astronomy Subcommittee meeting, date uncertain, circa 1967–1968, 7, KH/BC.

9. See Compton and Benson, *Living and Working in Space*, chapter 4. On the technical demands for AOSO and its state of development in 1963, see Cervenka, "One Approach to the Engineering Design," 31–43. One might speculate that AOSO was cancelled to strengthen the argument for a manned space station, but it is also likely that the state of the art of electronic image recording, in this case "raster scanning," and of pointing accuracy were not able to provide data of the desired quality. NASA was aiming for better than +/– 5 seconds of arc general pointing accuracy and +/– 1 second in the scanning mode, "an order of magnitude improvement over the first generation of OSOs." Another "key" problem was the integration of the Sun tracker with the control subsystem. Further, as of the spring of 1963, the "current spacecraft weight estimates [1000 lbs.] exceed the Thor Agena limit." 43.

10. Donald D. Brousseau (NRL) to Code 7100, January 4, 1966, folder "NASA Carruthers Proposal, 1966–1968," box 41, Herbert Friedman Papers Misc. Coll 113, APS.

11. Carruthers, "A Program of Astrophysical Studies in the Far Ultraviolet Using Image Converters," March 30, 1966, attached to H. Friedman to M. Estella Steele, Property Control Officer NASA, March 31, 1966, folder "NASA Carruthers Proposal, 1966–1968," box 41, HF/APS.

12. Carruthers, "A Program of Astrophysical Studies," See figure 6.

13. Carruthers, 8–10.

14. Friedman to Lyman Spitzer, February 17, 1966, folder "Spitzer," box 2, HF/APS.

15. Carruthers, "A Program of Astrophysical Studies," 16.

16. Both Andrew Millikan and Martin Schwarzschild expressed concerns about manned ballooning as an appropriate vehicle for science in the 1930s and in the 1950s. See DeVorkin, *Race to the Stratosphere* and *Science with a Vengeance*; and DeVorkin, "When to Send Your Telescope Aloft." Committees of scientists advising NASA were also expressing this concern.

17. Carruthers, "A Program of Astrophysical Studies," 18.

18. Carruthers, "A Program," 32–34, 37–38. This capability was achieved only in the Hubble era.

19. Carruthers, "A Program," 39.

20. Carruthers, "A Program," 41.

21. Conrad Swanson, Contracting Officer's Representative, Scientific Payloads Branch MSFC, to Friedman, May 23, 1966, folder "NASA Carruthers Proposal, 1966–1968," box 41, HF/APS.

22. Neufeld, *Von Braun*, 403–404. On the many tensions between centers, see also Newell, *Beyond the Atmosphere*, part 4, "Love-Hate Relationships."

23. Swanson to Friedman, May 23, 1966, folder "NASA Carruthers Proposal, 1966–1968," box 41, HF/APS.

24. Swanson to Friedman, May 23, 1966.

25. Carruthers to Swanson, May 26, 1966, folder "NASA Carruthers Proposal, 1966–1968," box 41, HF/APS.

26. George Carruthers to "Uncle Ben," September 25, 1966, B. F. Carruthers Papers, Schomburg Archives, NYPL MG433, box 1, folder 4. See also Carruthers Papers, National Air and Space Museum, box 1. Carruthers's claim that he had an approved proposal for an Apollo mission has not been verified.

27. Homer Newell to Friedman, August 10, 1966, folder "NASA Carruthers Proposal, 1966–1968," box 41, HF/APS. Walden correspondence was not located.

28. Neufeld, *Von Braun*, 403–404.

29. From a rough set of notes taken during the meetings, attached to: "Summary Minutes: Astronomy Subcommittee of the Space Science Steering Committee," (Meeting Number 2 - FY67) October 24–26, 1966. "Roman Committee October 66" folder. KH/BC." On Henize, https://baas.aas.org/pub/karl-gordon-henize-1926-1993/release/2.

30. Roman, "Nancy Grace Roman and the Dawn of Space Astronomy," 16.

31. Roman, to Carruthers, April 2, 1969, folder "Carruthers Proposals, 1967–1972," box 41, HF/APS.

32. Nancy G. Roman, interview, August 19, 1980, Niels Bohr Library & Archives, American Institute of Physics, College Park, MD, USA, accessed January 3, 2023, www.aip.org/history-programs/niels-bohr-library/oral-histories/4846.

33. Roman, interview, August 19, 1980.

34. Quoted in David DeVorkin, "How Nancy Grace Roman Shaped Hubble," *Physics Today*, accessed May 19, 2021, https://physicstoday.scitation.org/do/10.1063/PT.6.4.20200401c/full/.

35. Nancy Roman, interview with Frank K. Edmondson, November 17, 1982, Roman Papers, AIP, box 4, series II, folder 8, 1976—1982, NR/AIP.

36. See "NASA Resources 1969–1978," *NASA Historical Data Book*, IV, *(*NASA SP-4012), chapter 1.

37. Carruthers and Friedman, Defense Purchase Request H-13241A, October 21, 1966, folder "NASA Carruthers Proposals, 1966–1968," box 41, HF/APS.

38. Carruthers and Friedman, Defense Purchase Request; and Carruthers, Friedman, and Snoddy, "Far UV Spectrographic Sky Survey," n.d., draft clipped to J. H. Schulman, NRL to Director of Grants and Research Contracts, NASA, January 23, 1967, cc: Nancy Roman, "NASA Carruthers Proposals, 1966–1968," box 41, HF/ADS.

39. The NASA Office of Policy and University Affairs was responsible for coordinating the review of all university proposals to NASA.

40. Newell interview with Richard Hirsh, https://www.aip.org/history-programs/niels-bohr-library/oral-histories/4795-2. AIP.

41. Carruthers and Friedman, "Defense Purchase Request," 6.

42. Carruthers, "Status Report: Far U.V. Image Converter for EMR," May 11, 1967, attached to Carruthers to Nancy Roman, May 31, 1967, "NASA Carruthers Proposals, 1966–1968," Friedman papers, box 41. McColl. 113, HF/ADS.

43. Carruthers, "Status Report."

44. George Carruthers, "An Upper Limit."

45. Carruthers to Roman, May 31, 1967, folder "NASA Carruthers Proposals, 1966–1968," box 41, HF/APS.

46. Friedman to Henry J. Smith, June 30, 1967, folder "NASA Carruthers Proposals, 1966–1968," box 41, HF/APS.

47. "OSSA Astronomy Subcommittee" notes on "Proposal of Conceptual Payload Design for Astronomical Experiments," February 8, 1967, "Roman Committee May 1967," folder, KH/BC.

48. Tifft, "Two-Dimensional Area Scanning with Image Dissectors," 137–144; "William G. Tifft," Wikimedia Foundation, https://www.williamgtifft.com/home/.

49. "Astronomy Subcommittee, 8–10 Feb '67" folder, KH/BC. Boggess's comments are from "EMR Ultraviolet Proposals Reviewed at the February 8–10, 1967, Astronomy Subcommittee Meeting," Carruthers, "Far UV Spectrographic Sky Survey," and W. G. Tifft, "Ultraviolet Photographic Survey of Star Fields and Emission Regions," KH/BC. See also "Minutes of Astronomy Subcommittee of the Space Science Steering Committee Meeting No. 3-FY67," February 8–10, 1967, proposal 3–67 G; and Carruthers, Friedman, and Snoddy, "Far UV Spectrographic Sky Survey." RG255, UD 16W Entry 75 Meeting Files, 7/1/1966–6/30/1969, box 3, RG255\NARA. See also Secretary, Astronomy Subcommittee to The File, May 15, 1967, "Far Ultraviolet Proposals Reviewed at the February 8–10, 1967, Astronomy Subcommittee Meeting," box 3, Astro SubCom #2 FY 67, "Minutes/ Astronomy Subcommittee," folder 3, Mtg 3 FY 1967 February 8–10, 1967, Attachment 6, RG255/NARA.

50. Carruthers, "Far UV Spectroscopy from the Lunar Surface," October 14, 1969, Proposal to NASA, p. 17, folder "Carruthers Proposals," box 41. See also Folder "Carruthers Proposals 1966–69," box 19, HF/APS.

51. Charles Brown, interview with DeVorkin and Callahan, 20.

52. Ezell, *NASA Historical Data Book: Volume III*, chapter 2, section 127, and table 3–3. See also Harwit, *Cosmic Discovery*, 48–49, figure 1.11.

53. Homer Newell recalled that "we found it quite distressing when NSF chose to fund rocket astronomy at NRL." Homer Newell to Frank K. Edmondson, February 20, 1981, Roman Papers, AIP, box 4, series II, folder 8, "1976–1982," NR/AIP. The issue of management and control in space astronomy has been addressed by Smith, "Early History of Space Astronomy," 149–161.

54. Fifth magnitude is roughly equivalent to the faintest stars perceptible by the naked eye on a dark night. Herbert Friedman to Associate Administrator, NASA, January 24, 1968, supporting George Carruthers, R. C. Henry, Paul Patterson, Talbot Chubb, and Herb Friedman; "Proposal for Continued Support of NRL's Program of Stellar Ultraviolet Photography and Spectroscopy," table 1, folder "NASA Carruthers Proposals, 1966–1968," box 41, HF/APS.

55. Herbert Friedman to Associate Administrator, NASA, January 24, 1968, "Postscript to Proposal."

56. Henry and Carruthers, "Far-Ultraviolet Photography of Orion," 531.

57. On *Pioneer 10*, see Fimmel, Swindell, and Burgess, *Pioneer Odyssey*.; and Fimmel, van Allen, and Burgess, *Pioneer*.

58. "George Carruthers, interview with Glen Swanson, March 25, 1999," accessed May 12, 2023, https://historycollection.jsc.nasa.gov/JSCHistoryPortal/history/oral_histories/CarruthersGR/CarruthersGR_3-25-99.htm.

59. "The Pioneer Missions," accessed August 17, 2024 The Pioneer Missions - NASA Technical Reports Server (NTRS), accessed August 29, 2024. Pioneer Program (nasa.gov). https://ntrs.nasa.gov/search?q=The%20Pioneer%20missions&page=%7B%22from%22:0,%22size%22:25%7D (Accessed August 29, 2024).

60. "Program Is Proposed for Outer-Planet Trips in 70's," 59. See also Ray Newburn, "The Grand Tour of the Outer Planets," accessed May 12, 2023, https://calteches.library.caltech.edu/2805/1/newburn.pdf.

61. Friedman to AA, Office of University Affairs, NASA Garfield Robinson, November 29, 1968, enclosing Carruthers, "Proposal for Measurement of the Structure and Composition of Jupiter's Atmosphere Using Ultraviolet Photometers on Pioneers," folder "Carruthers Proposals," box 41, HF/APS.

62. Carruthers, "Proposal for Measurement" in Friedman to AA, November 29, 1968, 18.

63. Friedman to AA, Office of University Affairs, NASA Garfield Robinson, November 29, 1968.

64. W. M. Collins to Friedman, October 27, 1969, folder "NASA Carruthers Proposals, 1966–1968," box 41, HF/APS. A search of relevant NASA records at NARA, regarding *Pioneer F & G* deliberations, did not mention Carruthers's proposal. It might not have even reached the relevant committees. Records reviewed included folder "Proposals Reviewed by Members of the Advisory Subcommittees of the Space Science and Applications Steering Committee," box 1, "NASA Minutes/Astronomy Subcommittee 7/1/1967—6/30/1969," WG 255-71A6314, National Archives Identifier: 55317132, RG255/NARA.

65. William Baum, notes of meeting: "Grand Tour Briefing," Pioneer F. and G., November 17, 1970, box 22, folder 9, William Baum Papers/Lowell Observatory Archives, WB/LOA.

66. Carlson and Judge, "The Extreme Ultraviolet Dayglow of Jupiter," 327–343; and Carlson and Judge, "Pioneer 10 Ultraviolet Photometer Observations at Jupiter Encounter," 3623–3633. Carlson and Judge were competitive in this problem area, and their papers from the mission were highly cited.

67. After its report, NASA's Donald Hearth thanked the committee for their work, noting the challenge "became even greater with the large number and the competitive nature of the proposals received." Donald P. Hearth to Chairmen, "Subcommittees of OSSA Steering Committee," April 21, 1969, box 3, "Minutes/Astronomy Committees," folder "NASA SG Meeting 1-FY1969," RG255-71A6314, RG255/NARA.

68. "Planetary Subcommittee of the Space Science and Applications Steering Committee (Meeting No. 2-FY 69)," January 2–4, 1969, 3, William Baum Papers, box 22, folder 9, Lowell Observatory Archives (LOA), WB/LOA. By the time of this

meeting, Carruthers's proposal had already been rejected. On the varying responsibilities of the Principal Investigator, see Newell, *Beyond the Atmosphere*, 124–125. Thanks once again to Barrett Caldwell for perspective.

69. See, for instance, Carruthers, "Magnetically Focused Electronographic Image Converters." Citations are from Astrophysics Data System, Harvard University, accessed December 3, 2022, https://ui.adsabs.harvard.edu/search/q=%20author%3A%22Carruthers%2C%20G.%20R.%22%20year%3A%201959-1971&sort=citation_count%20desc%2C%20bibcode%20desc&p_=0.

70. Capt. J. C. Matheson, NRL to The Washington Academy of Sciences, December 12, 1968, folder "George Carruthers Material 5517, 1963–76," box 17, HF/APS.

## CHAPTER 7

1. Asher and Kaiser, "Broken Promises Line Riot Area Street."

2. Marya Annette McQuirter, "A Brief History of African Americans in Washington, DC," accessed August 19, 2024. https://www.northstarnews.com/userimages/references/Brief%20History_AADC.pdf.

3. Dean, "A Brief History of White People in Southeast."

4. O'Toole, "Black Scientist Develops Apollo Instrument."

5. O'Toole, "Black Scientist Develops Apollo Instrument."

6. Dean, "A Brief History of White People in Southeast"; Paul and Moss, *We Could Not Fail*, 109.

7. Paul and Moss, *We Could Not Fail*, 109.

8. See "Black Population 1970," accessed November 2, 2024. https://www2.census.gov/library/publications/decennial/1970/pc-s1-supplementary-reports/pc-s1-11.pdf, and "This GIF Shows How the D.C. Area's Demographics Have Changed since 1970," January 14, 2020, https://dcist.com/story/20/01/14/this-gif-shows-how-the-d-c-areas-demographics-have-changed-since-1970/. On the primary sites of the rioting, see Asher and Kaiser, "Broken Promises Line Riot Area Street."

9. McCurdy, *Inside NASA*, 13. This characteristic has been described by many researchers and writers. Susan Cain observed that "introverts prefer to work independently, and solitude can be a catalyst to innovation." See Cain, *Quiet*, 74. See also Roe, "A Psychologist Examines 64 Eminent Scientists," 221–225, who observes that these people tend to be firstborn and from a professional middle-class family, and that "the one thing that all of these 64 scientists have in common is their driving absorption in their work. They have worked long hours for many years, frequently with no vacations to speak of, because they would rather be doing their work than anything else."

10. Quoted in Capt. J. C. Matheson, NRL to The Washington Academy of Sciences, December 12, 1968, folder "George Carruthers Material 5517, 1963–76," box 17, HF/APS.

11. Richard Conn Henry to the author, January 4, 2021.

12. Beattie, *Taking Science to the Moon*.

13. Homer Newell, Associate Admin for Space Science and Applications to Director, Manned Spacecraft Center May 9, 1966, folder "Experiments—Proposals," box 1, RG 255 UD-11W 58 72A7952, RG255/NARA.

14. Beattie, *Taking Science to the Moon*, 56. The flight mirror was 32 inches.

15. E. H. Wells, "Optical Astronomy Package Feasibility Study for Apollo Applications Program," Executive Summary Report, TM-53496, MSFC, August 5, 1966.

16. Wells, "Optical Astronomy Package," 3.

17. Wells, 3–5.

18. Wells, 6–13.

19. Wells, 6–13.

20. Sternglass, "Application," 872.

21. Seamans, "Project Apollo," 124.

22. George Carruthers, Grady Hicks, and T. Chubb, "Apollo Far-Ultraviolet Auroral Photography Experiment," April 25, 1969, 4, folder "Carruthers Proposals," box 41, HF/APS.

23. Among many histories, see Sullivan, *Assault on the Unknown*; and Korsmo, "The Genesis of the International Geophysical Year," 38.

24. Carruthers, Hicks, and Chubb, "Apollo Far-Ultraviolet Auroral Photography Experiment," 4.

25. "Ultraviolet Arcs," 133.

26. "OGO 4, 1967-073A," accessed August 19, 2024, NASA, (Accessed August 19, 2024). https://nssdc.gsfc.nasa.gov/nmc/spacecraft/displayTrajectory.action?id=1967-073A.

27. Carruthers, Hicks, and Chubb, "Apollo Far-Ultraviolet Auroral Photography Experiment," 7.

28. The breadth and depth of NRL's activities in upper air physics, rocket spectroscopy, and space science generally can be appreciated in Friedman's annual reports. See, for instance, Friedman, *Annual Report of the E.O. Hulburt Center for Space Research for Space Research, Naval Research Laboratory, for Fiscal Year 1969*.

29. Carruthers, Hicks, and Chubb, "Apollo Far-Ultraviolet Auroral Photography Experiment," 3.

30. Carruthers, Hicks, and Chubb, 19.

31. Carruthers, Hicks, and Chubb, 37, figure 3.

32. Carruthers, Hicks, and Chubb, 22.

33. DeVorkin, *Race to the Stratosphere*; and DeVorkin, "When to Send Your Telescope Aloft."

34. Naugle to Carruthers, October 30, 1969, folder "NASA, Carruthers Proposal, 1969," box 41, HF/APS.

35. NASA, Manned Spacecraft Center, "Announcement," table 3–5.

36. NASA SPD-9P-052, S-169, and S-177. Fastie's instrument, both Moon and Earth related, had goals similar to Carruthers's. His flew on *Apollo 17*.

37. Wishnis, *Optical Technology Experiment System*, figures 3 and 4.

38. Smith, *The Space Telescope*, 72.

39. Nancy Roman to George R. Carruthers, April 2, 1969, and W. M. Collins to Herbert Friedman, October 27, 1969, folder "NASA, Carruthers Proposals, 1969," box 41, HF/APS.

40. "Apollo Far UV Auroral Photography Experiment," April 25, 1969; "Continued Support for Rocket Astronomy," May 20; "Spectroscopic Study of Diffuse UV Background," October 13; "Far-Ultraviolet Spectroscopy of Diffuse Background Radiation, Diffuse Nebulae, and Stars from the Lunar Surface," October 14; "Research and Development of Electronic Image Converters for Advanced Space Astronomy Missions," October 23. Folder "NASA Carruthers Proposals, 1969," box 41, HF/APS.

41. Nancy Roman to George Carruthers, April 2, 1969, folder "Carruthers Proposals, 1967–1972," box 41, HF/APS.

42. Roman, "Exploring the Universe," 507.

43. Herbert Friedman to George E. Mueller, August 28, 1969, folder "NASA Coresp, Folder 2, 1958–1972," box 41, HF/APS.

44. George Carruthers, "Far UV Spectroscopy from the Lunar Surface," and "Spectrographic Study of Diffuse Ultraviolet Background Radiation from the Lunar Surface," P11-69, October 14, 1969, and P12-69, October 13, 1969. "Carruthers Proposals, 1967–1972," box 41, HF/APS.

45. W. C. Hall to Garfield A. Robinson, AA for University Affairs, October 20, 1969, folder "Carruthers Proposals, 1967–1972," box 41, HF/APS.

46. By the time Carruthers submitted his proposals, the nature and properties of quasars were frequently debated in the scientific literature, including *Nature* and *Science* and popular magazines like *Sky & Telescope*. For an overview of their status at the time, see Smith, "Summary Remarks on Quasars."

47. George R. Carruthers, "Proposal for Research and Development of Electronographic Image Converters for Advanced Space Astronomy Missions," P15–69, October 23, 1969, attached to W. C. Hall (NRL) to Garfield Robinson (NASA), October 28, 1969, folder "Carruthers Proposals, 1967–1972," box 41, HF/APS.

48. George Carruthers, "Far UV Spectroscopy of Diffuse Background Radiation . . . from the Lunar Surface," P11–69, October 14, 1969, folder "Carruthers Proposals, 1967–1972," box 41, HF/APS; W. C. Hall to Garfield A. Robinson, AA for University Affairs, October 20, 1969, folder "Carruthers Proposals, 1967–1972," box 41, HF/APS.

49. Carruthers, "Proposal for Research and Development of Electronographic Image Converters."

50. Thornton Page, "Lunar Surface Telescope," attached to Page to Friedman, October 11,1969, folder "NRL, Geo. Carruthers Mat. S-201 Apollo Lunar Surface Camera" 50IT, 1969–1974, box 19, HF/APS.

51. Beattie, *Taking Science to the Moon*, 68.

52. Thornton Page, "Lunar Surface Telescope," September 9, 1969, and "Attachment C Experiment Proposal," October 6, 1969, folder "NRL, Geo. Carruthers Mat. S-201 Apollo Lunar Surface Camera, 50IT, 1969–1974," box 19, HF/APS.

53. Thornton Page to Herbert Friedman, October 11, 1969, folder "NRL, Geo. Carruthers Mat. S-201 Apollo Lunar Surface Camera 50IT, 1969–1974," box 19, HF/APS.

54. Visco, "The Operations Research Office," 24–33. See also "Thornton Leigh Page, 1913—" manuscript autobiography (1983), Thornton Page Papers, MS1983–002 Special Collections, Newman Library, Virginia Tech.

55. Shrader, *History of Operations Research*, 97.

56. Newell, *Beyond the Atmosphere*, 214.

57. Phinney, *Science Training History of the Apollo Astronauts*, 253.

58. See, especially, a well-publicized symposium in 1969 sponsored by the AAAS: Sagan and Page, *UFOs*.

59. Page through Friedman to Alan Berman, "Activities Outside NRL," February 17, 1971, folder "NRL, Geo. Carruthers Mat. S-201 Apollo Lunar Surface Camera 50IT, 1969–1974," box 19, HF/APS. See also Page, "Two Years at NASA Manned Spacecraft Center," Thornton Page Papers, MS1983-002 Special Collections, Newman Library, Virginia Tech, TP/VT. Copy in Thornton Leigh Page Papers, Wesleyan University vertical subject files, 1000-176, Special Collections & Archives, Wesleyan University, TP/WU.

60. Thornton Page to John Naugle, April 13, 1971. "OAO Files," box 15, NASA History Office, A-34041, John E. Naugle Papers Collection, NASA Headquarters Historical Reference Collection, JN/NHO.

61. Carruthers interview with DeVorkin, August 18, 1992.

62. "Two Years at NASA Manned Spacecraft Center," Thornton Leigh Page Papers, Wesleyan University vertical subject files, 1000–176, Special Collections & Archives, Wesleyan University, TP/WU.

63. "Two Years at NASA Manned Spacecraft Center."

64. Allenby, "Lunar Orbital Science," 48.

65. William Baum notes, "Category I Recommendations," undated, attached to "Surface APO Proposals—Discipline Assignments," November 24–26, 1969, William Baum Papers/Archives, Lowell Observatory, box 22, folder 9, WB/LOA.

66. James A. McDivitt (Manager, Apollo Spacecraft program) to R. A. Petrone (NASA Headquarters), January 21, 1970, "*Apollo 16*" folder, box #1 "Program and Project Files Relating to Shuttle Launches," W255-89-0615 RG255/NARA.

67. Page to Capt. Lee Sherer (Code MAL, NASA HQ), April 25, 1970, and to Don Wiseman, July 14, 1970, cc to Friedman and Carruthers. "Preliminary Work on S-201 Experiment for Apollo-17," box 19, folder "NRL, Geo. Carruthers Mat. S-201 Apollo Lunar Surface Camera, 1969–1974," HF/APS. See also Page to Allenby, April 22, 1969, box 2, folder "Astronomy Subcommittee Meeting 3-FY69," Attachment 93, RG255-71A6314 RG255/NARA.

68. Beattie, *Taking Science to the Moon*, chapter 7.

69. Madhulika Guhathakurta interview with Gregory Good, David DeVorkin, and Katie Boyce-Jacino, February 28, 2018, AIP.

70. Page to Capt. Lee Sherer, April 25, 1970, box 19, folder "NRL, Geo. Carruthers Mat. S-201 Apollo Lunar Surface Camera 50IT, 1969–1974," HF/APS.

71. Page to Sherer, April 1970.

72. Page to Friedman, May 5, 1970, box 19, folder "NRL, Geo. Carruthers Mat. S-201 Apollo Lunar Surface Camera 50IT, 1969–1974," HF/APS.

73. Friedman to Page, June 17, 1970, box 19, folder "NRL, Geo. Carruthers Mat. S-201 Apollo Lunar Surface Camera 50IT, 1969–1974," HF/APS.

74. Frank K. Edmondson, *AURA and Its US National Observatories*, 117–122.

75. Carruthers to Mayall, March 26, 1970; Mayall to Carruthers, April 16–17, 1970, box 19, folder "NRL, Geo. Carruthers Mat 1970–1971," HF/APS.

76. Carruthers, Heckathorn, and Opal, "Rocket Ultraviolet Imagery," 346–356.

77. Newell to Edmondson, February 20, 1981, Nancy Roman Papers.

78. “SCOUT” rocket came in many configurations (for the Solid Controlled Orbital Utility Test). It was a multistage solid propellant rocket the used both by the USAF and NASA starting in 1960 for a wide range of applications. See Newell, *Beyond the Atmosphere*, 136–137. See also J. Powell, “Blue Scout,” 22–30.

79. Phillip Mange to Friedman, April 6, 1971, box 19, folder “NRL, Geo. Carruthers Mat 1970–1971,” HF/APS.

80. “C. Opal,” in Carruthers, “Proposal for Study of Modification of Far Ultraviolet Schmidt Camera/Spectrograph for Spacelab Astronomical Observations,” March 31, 1976, 16, box 19, folder “NRL, George Carruthers Mat. 1975–7,” HF/APS.

81. That civilian grade is now equivalent to captain in the Navy. “O-5 Commander,” accessed October 25, 2022, https://www.federalpay.org/military/navy/commander. See also: https://www.mccsokinawa.com/uploadedFiles/MainSite/Content/Marine_and_Family/Marine_and_Family_Programs_-_Resources/Transition_and_Employment_Assistance/Federal-Rank-Equivalency.pdf.

82. Phil Mange to author, June 20, 2020, conveyed by Virginia Mange via email.

83. Page to Conway, July 16, 1970, box 19, folder “NRL, Geo. Carruthers Mat. S-201 Apollo Lunar Surface Camera 50IT, 1969–1974,” HF/APS.

84. Phil Mange to author, June 20, 2020.

85. Page, “Statement of Work: Principal Investigator Services for the Lunar-Surface Ultraviolet Camera (Experiment S-201), Apollo Mission 17 [sic],” July 16, 1970, box 19, folder “NRL, Geo. Carruthers Mat. S-201 Apollo Lunar Surface Camera, 50IT, 1969–1974,” HF/APS.

86. Page, “Statement of Work,” 2.

87. Martin W. Malloy (MAL) to File, “Ultraviolet Camera/Spectrometer Experiment S-201, *Apollo 16*,” November 27, 1970, box 19, folder “NRL, Geo. Carruthers Mat. S-201 Apollo Lunar Surface Camera 50IT, 1969–1974,” HF/APS.

88. N. P. Patterson for the file, December 3, 1970, box 19, folder “NRL, Geo. Carruthers Mat. S-201 Apollo Lunar Surface Camera, 50IT, 1969–1974,” HF/APS.

89. G. T. Orrok (Bellcomm) to R. J. Allenby (NASA/MAL), December 3, 1970, box 19, folder “NRL, Geo. Carruthers Mat. S-201 Apollo Lunar Surface Camera 50IT, 1969–1974,” HF/APS.

90. Orrok to Allenby.

91. Friedman to Page, June 17, 1970, box 19, folder “NRL, Geo. Carruthers Mat. S-201 Apollo Lunar Surface Camera 50IT, 1969–1974,” HF/APS.

92. Page to Don Wiseman, et al., MSC, "Preliminary Work on S-201 Experiment for Apollo-17[sic]," July 14, 1970, box 19, folder "NRL, Geo. Carruthers Mat. S-201 Apollo Lunar Surface Camera 50IT, 1969–1974," HF/APS.

93. G. Carruthers, "Electromagnetic Schmidt/Spectrograph for Far-Ultraviolet Sky Survey," May 3, 1971, attached to: Talbot A. Chubb, "Optical Geophysics Satellite," August 20, 1971, box 19, folder "NRL, Geo. Carruthers Mat. 1970–1971," HF/APS.

94. Carruthers (manuscript), "Further Developments of Magnetically Focused, Internal Optic Image converters," *Fifth Symposium on Photoelectronic Image Devices,* Imperial College, London. September 16, 1971, box 19, folder "NRL, Geo. Carruthers Mat., 1972," HF/APS. On the details of the spectral band options, see Carruthers and Page, "Far UV Camera/Spectrograph," 13–15.

95. P. Mange to Code 7000 (Friedman), April 5, 1971, box 19, folder "NRL, Geo. Carruthers Mat. S-201 Apollo Lunar Surface Camera 50IT, 1969–1974," HF/APS.

96. Carruthers, interview, July 27, 1992.

97. Rocco Petrone to NRL (Carruthers), July 16, 1971, box 19, folder "NRL, Geo. Carruthers Mat. S-201 Apollo Lunar Surface Camera 50IT, 1969–1974," HF/APS.

98. Mange to Codes 4000, 1000, November 9, 1971, box 19, folder "NRL, Geo. Carruthers Mat. S-201 Apollo Lunar Surface Camera 50IT, 1969–1974," HF/APS.

99. "Navy Specialist Probes Deep Space."

100. O'Toole, "Black Scientist Develops Apollo Instrument," box 17, copy in folder "George Carruthers Material 5517, 1963–76," HF/APS.

101. Undated note, found with Carruthers to Distribution, "Newspaper Article on S-201 and P.I." December 10, 1971, box 17, folder "George R. Carruthers Material," HF/APS.

102. Undated note.

103. Rhinelander, "Lunar Observatory is Planned"; Rhinelander, "Wesleyan Astronomer Works on Lunar Telescope Project"; and Rhinelander, "Astronomer's 2½ Year Project to Be the First Moon Telescope."

104. The "Preparation Milestones" are from W. David Woods and Tim Brandt, "*Apollo 16*, Day 1, part 1: Launch and Reaching Earth Orbit." North American Aviation was at that time being merged into Rockwell. *Apollo Flight Journal*, accessed August 18, 2024. https://history.nasa.gov/afj/ap16fj/01_Day1_Pt1.html.

## CHAPTER 8

1. Jones, *Apollo 16 Lunar Surface Journal*, "Loading the Rover," at 119:46:24. See also Swanson, *"Before This Decade Is Out . . ." Personal Reflections on the Apollo Program*, chapter 11, 274–276, https://history.nasa.gov/SP-4223/ch11.htm.

2. Jones, *Apollo 16 Lunar Surface Journal*, "ALSEP Off-load," at 120:47:00.

3. "Thornton Leigh Page, 1913—" manuscript autobiography (1983), Thornton Page Papers, MS1983-002 Special Collections, Newman Library, Virginia Tech, 10.

4. Jones, *Apollo 16 Lunar Surface Journal*, "EVA-1 Closeout," at 125:17:00.

5. Page and Carruthers, "Far Ultra-Violet Observations," 331; Carruthers and Page, "The S201 Far-Ultraviolet Imaging Survey," 447–462.

6. Mission Evaluation Team, *Apollo 16 Mission Report*, section 4.8.

7. Carruthers and Page, "Far UV Camera/Spectrograph," 13–15.

8. Jones, *Apollo 16 Lunar Surface Journal*, "EVA-1 Closeout," at 125:57:00. Duke's humor is noted in various places. See Speakers.com, "Keynote Speaker: Charlie Duke," YouTube video, January 15, 2019, https://www.youtube.com/watch?v=5JWPCpgRgY0.

9. Carruthers interview with Swanson, 1999.

10. Jones, *Apollo 16 Lunar Surface Journal*, "Down the Ladder for EVA-2," at 143:00:00.

11. "Space Camera Takes Special Pix"; "Interplanetary Gases in Lunar UV Photography," 54; "NASA Awards Black Scientist"; "Moon Camera Invented by a Black Man."

12. George Carruthers, "Proposal to NASA: Far UV Spectroscopy from the Lunar Surface," October 14, 1969, box 41, folder "NASA, Carruthers Proposals 1966–69," HF/APS.

13. Thornton Page to Herbert Friedman, May 5, 1970, box 19, folder "NRL, Geo. Carruthers Mat. S-201 Apollo Lunar Surface Camera 50IT, 1969–1974," HF/APS.

14. C. Kraft to Herbert Friedman, May 23, 1972, box 17, folder "George Carruthers Material 5517, 1963–76," HF/APS. On the updated agreement, which never applied to scientific data, only hardware, see *National Air and Space Museum Collections Rationale* (Washington, DC: Smithsonian Institution, 2016), 188.

15. "The Lunar Surface Ultraviolet Camera: Some Observations on the Management Processes Involved in Its Manufacture," unsigned and undated, attached to Herbert Rabin to Herbert Friedman, July 10, 1972, "NRL, Geo. Carruthers Mat. S-201 Apollo Lunar Surface Camera," 50IT, 1969–1974], box 19, MS/APS Coll 113 (IV).

16. "The Lunar Surface Ultraviolet Camera," 3.

17. "The Lunar Surface Ultraviolet Camera," 3

18. W. H. Conway, July 13, 1972, box 19, folder "NRL, Geo. Carruthers Mat. S-201 Apollo Lunar Surface Camera, 50IT, 1969–1974," HF/APS. No record has been found of a response from Carruthers.

19. Mange to Code 7000, "Comment on Mr. Cohen's Critique of the LSUC Program," July 17, 1972, box 19, folder "NRL, Geo. Carruthers Mat. S-201 Apollo Lunar Surface Camera 50IT, 1969–1974," HF/APS.

20. Jones, *Apollo 16 Lunar Surface Journal*, "Loading the Rover," at 119:47:47.

21. Quote from Mission Evaluation Team, *Apollo 16 Mission Report*, 14–93. https://www.nasa.gov/history/alsj/a16/a16mr.html, accessed August 19, 2024. See also Jones, *Apollo 16 Lunar Surface Journal*, "Loading the Rover," at 119:55:00. https://www.nasa.gov/history/alsj/a16/a16.html, accessed August 19, 2024.

22. This error was oddly reminiscent of the scene in the famous 1950 George Pal movie, *Destination Moon*. Greasing any moving parts exposed to space was a no-no, dramatized by Dick Wesson's uninformed technician character greasing the antenna deployment system, which necessitated a harrowing extravehicular maneuver en route to the Moon—a great excuse for drama.

23. NASA, *Apollo 16 Preliminary Science Report*, vii–xi.

24. Friedman, *Annual Report of the E.O. Hulburt Center for Space Research, Naval Research Laboratory, for Fiscal Year 1972*, iv.

25. Friedman, 16. He also reported on continuing Aerobee flights and balloon ascents, of the launch of the solar monitoring satellite *SOLRAD 10 (Explorer 44)*, observations from OSO-7 of the breakup of a solar coronal streamer, and continued development of large solar instruments for the planned Skylab mission. The Radio Division had also been active, mapping water vapor sources using long baseline interferometry. The Division had published some 72 papers that year.

26. Carruthers and Page, "*Apollo 16* Far-Ultraviolet Camera/Spectrograph: Earth Observations," 788, 789. This composition profile had been known for several years. See Miller, Fastie, and Isler, "Rocket Studies of Far-Ultraviolet Radiation in an Aurora," 3353–3365. In 1970, NRL rocket studies had shown that the atomic hydrogen was produced by rocket and missile exhaust plumes. See Hicks, Chubb, and Meier, "Observations of Hydrogen Lyman $\alpha$ Emission from Missile Trails," 10101–10109.

27. Carruthers interview with Glen Swanson, March 25, 1999, "NASA Johnson Space Center Oral History Project," NASA, accessed November 3, 2021, https://historycollection.jsc.nasa.gov/JSCHistoryPortal/history/oral_histories/CarruthersGR/CarruthersGR_3-25-99.htm.

28. [Carruthers, Page, Meier, Chubb, Friedman], [Proposal for] "Additional Analysis of Data Obtained by Far-UV Camera/Spectrograph Apollo-16 Experiment S-201," manuscript attached to Carruthers to Richard Allenby, Code NAL, NASA January 8, 1973, and Carruthers to W. F. Eichelman Code TN4, box 19, folder "NRL Geo Carruthers Mat., 1973," HF/APS.

29. Meier interview. Born in Pittsburgh in 1940, he worked his way through Duquesne University and then earned his PhD in 1966 from the University of Pittsburgh, taking jobs from supermarket clerk to construction worker. As a child, he subscribed to *Popular Science* and had a chemistry set and a microscope. His life changed when he took physics in his senior year in high school. His first impression was, "What the hell's this all about?" and then a "light bulb went off. And I said, 'I love this stuff!'" His advisor introduced him to scientific rocketry: as Meier recalled, one day, he gave him a box of seemingly random parts and asked, "How would you like to fly a rocket experiment for your thesis?"

30. Meier interview, 9.

31. Donahue and Meier, "Distribution of Sodium," 2803–2829. See also Carruthers, et al., "Additional Analysis of Data Obtained by Far-UV Camera/Spectrograph Apollo-16 Experiment S-201," manuscript attached to Carruthers to Richard Allenby, Code NAL, NASA January 8, 1973, and to Carruthers to W. F. Eichelman Code TN4, January 8, 1973, box 19, folder "NRL Geo Carruthers Mat., 1973," HF/APS.

32. As an undergraduate at UCLA in the early 1960s, the author was assigned to serve Page's personal needs when he was in residence and confirms Page's character. On Page, see Osterbrock, "Thornton L. Page, 1913–1996," 1461–1462.

33. Englert et al., "Michelson Interferometer."

34. Bill Anders, "50 Years after 'Earthrise,' a Christmas Eve Message from Its Photographer," Space.com, December 24, 2018, accessed August 20, 2023, https://www.space.com/42848-earthrise-photo-apollo-8-legacy-bill-anders.html.

35. Meier interview with DeVorkin.

36. Carruthers et al., "Far Ultraviolet Wide Field Imaging, and Photometry," 80.

37. On radio stars, see Sullivan, *Cosmic Noise*, chapter 14. And on the importance of optical confirmation, see Dewhirst, "The Optical Identification of Radio Sources," 178–190.

38. David DeVorkin, "New Tools, New Universes: The Great Correlation Era," in press.

39. Heckathorn email to DeVorkin, May 3, 2023. "PDS" is for Pacific Data Systems. Microdensitometry had its challenges. Heckathorn chaired a paper session titled "How to Diagnose a Microdensitometer" at a 1983 NASA workshop calling for "specific examples of hardware failures or the results of unconscious misuse." Heckathorn,

"How to Diagnose a Microdensitometer." For an example of the early use of the 1010, see Barry, Cromwell, and Schoolman, "Spectral Quantification," 462–471.

40. George Carruthers and Thornton Page, "The S201 Far-Ultraviolet Imaging Survey," 459, 462.

41. Cpt. Earl Sapp USN to Chris Kraft, May 15, 1972; Chris Kraft to Cpt Earl Sapp USN, July 10, 1972; Rocco Petrone to NRL, July 16, 1972, folder "NRL, Geo. Carruthers Mat. S-201 Apollo Lunar Surface Camera 50IT, 1969–1974]," box 19, HF/APS.

42. P. Mange to Code 7000, July 18, 1972, folder "NRL, Geo. Carruthers Mat. S-201 Apollo Lunar Surface Camera 50IT, 1969–1974," box 19, HF/APS. Carruthers never gave up hope though. In their 1984 paper, he and Page described their proposed second generation "Mark II . . . Space Schmidt survey telescope" bound for a SPARTAN flight on the shuttle. They predicted that it would match the depth and clarity of the Palomar Schmidt maps.

## CHAPTER 9

1. "New View of the Earth."

2. Paul and Moss, *We Could Not Fail*, 3 and 279, note 104, based upon Mark A. Thompson, "Space Race: African American Newspapers Respond to Sputnik and Apollo 11," master's thesis, University of North Carolina, 2007, https://digital.library.unt.edu/ark:/67531/metadc5115/m1/1/.

3. Spigel, *Welcome to the Dreamhouse*, chapter 3, 142. Spigel describes the reactions of the Black press, which became critical of NASA, but remained positive in features that highlighted Black aerospace workers. Among several books highlighting the achievements of Black people in NASA and the space industry, see *Blacks in Science: Ancient and Modern*, chapters "Space science: the African-American contribution," 238-257, by John Henrik Clarke; and "Blackspace" by James G. Spady, 258-268. See also Paul and Moss, *We Could Not Fail*, 2015.

4. "Earth's Eye on the Moon," 61-63.

5. Lydia R. Thaxton to NRL Public Affairs Office, May 20, 1972, box 17, folder "George Carruthers Material 5517, 1963–76," HF/APS.

6. "Moon Camera Invented by Black Man."

7. Redhead to James Sullivan, Director, Public Relations, NRL, May 23, 1972, box 17, folder "George Carruthers Material 5517, 1963–76," HF/APS.

8. "Caribbean House to Have Guests."

9. See "Grenada coup media reaction," US Department of State unclassified release C17671093, accessed August 20, 2024, https://nsarchive.files.wordpress.com/2012/10/bishop-grenada.pdf.

10. O'Toole, "Black Scientist Develops Apollo Instrument," box 17, George Carruthers Material 5517, 1963–76, Herbert Friedman Papers, APS.

11. Fletcher demoted Harris for her criticisms of NASA's hiring practices, and she left in protest in 1973. See Calvino, "A Phenomenon to Monitor"; and McQuaid, "Racism, Sexism, and Space Ventures," 422–449. See also "Civil Rights Tour," DC Historic Sites, accessed August 20, 2024, https://historicsites.dcpreservation.org /items/show/1007.

12. Francis Redhead to John Donnelly, NASA Asst Administrator for Public Affairs, June 29, 1972, box 17, folder "George Carruthers Material 5517, 1963–76," HF/APS.

13. "NASA Cites Dr. Carruthers." See also *Bay State Banner* (Boston, MA), September 7, 1972.

14. "Black Moon Scientist Is Guest of District 16." See also Dewey Sanders to James Sullivan, Public Affairs, NRL, September 21, 1972, box 17, folder "George Carruthers Material 5517, 1963–76," HF/APS.

15. In his exploration of "race champion-heroes," Fouché observes that the perceived luster of holding a patent in the press and other media should not be the sole measure of success for an inventor. It certainly was not for Carruthers. Fouché, *Black Inventors*, 183–184. However, as Eric Hintz adds, "when a Black inventor earned a U.S. patent—objective, government-issued proof of creativity and ingenuity—it diminished racist charges of intellectual inferiority." Hintz, "Tearing Down the Barriers."

16. Fenrich, "The Gates of Opportunity," 210.

17. Carruthers, interview, August 18, 1992.

18. Fouché, *Black Inventors*, 183–184.

19. Feddoes, "Please Be Seated," May 26, 1973; "Certificate of Marriage," May 19, 1973, No. 98663, box 3, folder 35, GC/NASM.

20. "2 from Jamaica Capture 'Miss Carifta' Crown as Judges Agree," *New York Amsterdam News*. The Caribbean Free Trade Association (CARIFTA) was created in 1968 to link Caribbean English-speaking countries.

21. Sutton, "A Proclamation by the Honorable Percy E. Sutton, President of the Borough of Manhattan." Carruthers Papers, box 1, "Correspondence with Sandra," GC/NASM. Carruthers was also honored that day by the Borough of Brooklyn, emphasizing his "pre-eminence in space technology and a source of pride and inspiration to minority-group youngster everywhere." Sabastian Leone, Bureau of Brooklyn, June 21, 1972. Feddoes, "Please Be Seated" (June 17, 1972).

22. Sandra Redhead to George Carruthers, June 21, 1972, Carruthers Papers, box 1, "Correspondence with Sandra," GC/NASM.

23. George Carruthers to Sandra Redhead, August 22, 1972, Carruthers Papers, box 1, "Correspondence with Sandra," GC/NASM.

24. Nicks, *This Island Earth*. The title echoes the frightening 1955 movie of the same name.

25. George Carruthers to "Dear Sandra," August 27, 1972, Carruthers Papers, box 1, "Correspondence with Sandra," GC/NASM.

26. George Carruthers to "Dear Sandra."

27. Sandra Redhead to George Carruthers, September 13, 1972, Carruthers Papers, box 1, "Correspondence with Sandra," GC/NASM.

28. Sandra Redhead to George Carruthers, September 13, 1972.

29. George Carruthers to Sandra Redhead, September 14, 1972, Carruthers Papers, box 1, "Correspondence with Sandra," GC/NASM.

30. George to Sandra, September 14, 1972, Carruthers Papers, "Correspondence with Sandra," box 1, GC/NASM.

31. Sandra to George, September 13, 1972, Carruthers Papers, "Correspondence with Sandra," box 1, GC/NASM.

32. Sandra to George, September 25, 1972, Carruthers Papers, "Correspondence with Sandra," box 1, GC/NASM.

33. Sandra to George, October 24, 1972, "Correspondence with Sandra," Carruthers Papers, box 1, GC/NASM.

34. Gerald Carruthers interview with DeVorkin; "Certificate of Marriage, City of New York, May 19, 1973," Carruthers Papers, folder 1, box 1, GC/NASM.

35. "My Love" scrapbook, box 7, GC/NASM.

36. "Earth's Eye on the Moon," 61–63.

37. Robert R. Meier interview with DeVorkin and Callahan, AIP/NASM. See also "George Carruthers," The HistoryMakers A2004.112, interview with Racine Tucker Hamilton, July 27, 2004, The HistoryMakers Digital Archive, session 1, tape 1, story 3.

38. George to "Dear Sandra," August 27, 1972, "Correspondence with Sandra," box 1, Carruthers Collection, GC/NASM.

39. Sandra Redhead to Office of Admissions, November 14, 1972, "Corresp. with Sandra," box 1, Carruthers Collection, GC/NASM.

40. Sandra Redhead to Office of Admissions.

41. "Course Notes: Circa Fall 1973," Carruthers Papers, box 2, folder 13, "Misc. notes on the environment," GC/NASM.

42. "Bachelor of Arts" Degree Certificate, Carruthers Papers, box 8, folder 2, GC/NASM.

43. George Carruthers to Sandra Carruthers, July 11, 1973, "Correspondence with Sandra," box 1, Carruthers Collection, GC/NASM. See also Gerald Carruthers interview with DeVorkin; and Thelma Jones to the author, December 28, 2020. According to his brother Gerald, George did have a car in Washington, but it seems that he continued to commute on his bicycle, unless he had plans for offsite work that day. Gerald Carruthers commentary.

44. Francis Redhead ("Dad") to George Carruthers, March 15, 1975, Carruthers Papers, box 3, folder 10, "Miscellaneous Invitations," GC/NASM.

45. Meier interview with DeVorkin and Callahan, 67.

46. Suzanne Phillip and Gwen Ochoa, correspondence with DeVorkin.

47. Simon, telephone conversation with DeVorkin.

48. Phillip and Ochoa, correspondence with DeVorkin.

49. Phillip and Ochoa, correspondence with DeVorkin.

50. Carwell to DeVorkin, email, August 19, 2020.

## CHAPTER 10

1. On *Skylab*, see Compton and Benson, *Living and Working in Space*.

2. Heckathorn interview with DeVorkin.

3. For the full story, see Compton and Benson, *Living and Working in Space*, part 1, chapter 1.

4. A frequently cited phrase. Newkirk, Ertel, Brooks, *Skylab*, part 2C.

5. Belew and Stuhlinger, *Skylab*, chapter 5. See also Compton and Benson, *Living and Working in Space*, appendix D.

6. Alexander, "Big Comet Heads for Sun."

7. O'Toole, "Skylab Likely to See Comet."

8. Herbert Rabin to Chester M. Lee, July 16, 1973, box 19, folder "NRL, Geo. Carruthers Mat. S-201 Apollo Lunar Surface Camera 50IT, 1969–1974," HF/APS.

9. Conway to Carruthers, July 3, 1973, box 8B, "UV Camera for SL4," GC/MSRC.

10. Maran recalls that Carruthers made a presentation to his panel, but it was probably for his Aerobee package. But he also recalls that Maurice Dubin, Chief of Cometary Physics from NASA Headquarters, "wanted the UV camera added to Skylab and ran it by me for 'independent' opinion." Maran to Author, October 28, 2022.

11. Herbert Rabin to William C. Schneider, July 18, 1973, box 19, folder "NRL, Geo. Carruthers Mat. S-201 Apollo Lunar Surface Camera 50IT, 1969–1974," HF/APS.

12. Mange to Code 7000, July 23, 1973, box 8B, "UV Camera for SL4," NTA files from NRL.

13. "Mission Requirements. Third Skylab Mission SL-4 August 27, 1973" III, NASA 1-MRC-001F, addendum n.d., "Flag Sheet S-201/SL-4" included an S-201B "UV Electronographic Camera" as "Change 5" to the program, "UV Camera for SL4," box 8B, GC/MSRC.

14. Herbert Rabin to William C. Schneider, July 18, 1973, box 19, folder "S-201 Apollo Lunar Surface Camera 50IT, 1969–1974," addendum 3, HF/APS.

15. George Carruthers to William Conway and Thornton Page, August 16, 1973, box 8B, NRL/NTA collection, Skylab files.

16. Thornton Page to distribution, April 26, 1973, "Skylab files," box 8B, NRL/NTA collection, Skylab files.

17. "Observation of Ionospheric Perturbations by Skylab S-201," November 9, 1973, box 19, folder "George Carruthers Material 5517, 1963–76," HF/APS. See also George Carruthers, "Far Ultraviolet Imaging and Photometry, Experiment No. 803," April 8, 1977, File 6.2.2-2 "Space Experiments Support Program (SESP Space Flight Request)," box 1, "Project 71A01-63," RG255/NARA. See also Hicks, Chubb, and Meier, "Observations of Hydrogen Lyman α Emission from Missile Trails," 10101–10109.

18. "Timeline," October 22–28, 1973, box 8B, Skylab files, NRL/NTA collection.

19. "Mission Requirements Change 8," December 3, 1973, attached to: "On Kohoutek," pages 3-275–3-309, box 8B, Skylab files, NRL/NTA collection.

20. Page to Carruthers et al., November 20, 1973, Dump Tape 363-09/D-439, box 8B, Skylab files, NRL/NTA collection.

21. Skylab astronaut Dump Tape 345-09 "PLT," Skylab files, box 8B, NRL/NTA collection.

22. Compton and Benson, *Living and Working in Space*, chapter 17.

23. Snoddy and Gary, "Skylab Observations of Comet Kohoutek."

24. Page, "Far-UV Observations of Comet Kohoutek," 299. *Electrography* and *electronography* are interchangeable terms. The barrier membrane is illustrated in Carruthers, "Electromagnetic Cameras for the Vacuum Ultraviolet," 93–113, figure 2.

25. Page to Carruthers, May 10, 1974, Skylab files, box 8B, NRL/NTA collection.

26. Robert Meier to J. A. Blamont, April 17, 1974, Skylab files, box 8B, NRL/NTA collection.

27. Rabin to Kleinknecht (Skylab Office) Johnson Space Center, March 25, 1974, box 19, folder "NRL, Geo. Carruthers Mat. 1974," HF/APS.

28. Carruthers et al., "Lyman-α Imagery of Comet Kohoutek," 526.

29. Opal and Carruthers, "Ultraviolet and Infrared Observations of Comets," 8.

30. Meier et al., "Hydrogen Production Rates," 283–290. In the 1950s, Smithsonian Astrophysical Observatory astronomer Fred Whipple theorized that the heads of comets were composed of clumps of meteoritic stones frozen in ice. This model became popularly known as the "dirty snowball" theory. See Whipple, *The Mystery of Comets*.

31. Page, *Final Report*, 3. *Skylab's* ATM instrument observations of Kohoutek are reviewed in Snoddy and Gary, "Skylab Observations of Comet Kohoutek," 7. Further, Compton and Benson, *Living and Working in Space*, noted that "the most successful experiments of the *Skylab* group were the far-ultraviolet camera . . . and the photometric camera" (352).

32. William M. Jackson to Richard Maulsby, Program Manager, National Medal of Technology and Innovation, March 28, 2001, courtesy Gerald Carruthers.

33. Fastie et al., "Rocket and Spacecraft Studies," 398. There are many explanations for Kohoutek's behavior. See, for instance, Joe Rao, "How the 'Comet of the Century' Became an Astronomical Disappointment," Space.com, January 22, 2020, https://www.space.com/comet-kohoutek-flop-of-the-century.html.

34. Herbert Rabin to Robert Fellows, *Pioneer* Venus Program, NASA, December 14, 1973, folder "NRL Geo Carruthers Mat., 1973," "NRL, Geo. Carruthers Mat. 1974; Rabin to "NASA," January 24, 1974, "NRL, Geo. Carruthers Mat. 1974," box 19, HF/APS.

## CHAPTER 11

1. Karl Henize, University of Texas proposal to NASA, "Deep Sky Survey in Ultraviolet Wavelengths from Space Shuttle," March 22, 1974, attached to letter, Henize to Carruthers, April 4, 1974, box 19, folder "NRL, Geo. Carruthers Mat. 1974," HF/APS.

2. On the first selection, see Hersch, *Inventing the American Astronaut*, chapter 3.

3. Karl Henize proposal.

4. Morrow, et. al, "Accommodating Science and Technology Development."

5. Author's informal discussions with those attending an "Aerobee Wake" at the Goddard Space Flight Center at the time.

6. R. Giacconi to James Jodon, November 11, 1977, folder "Giacconi, Riccardo—1966–1968," box 2, HF/APS. Giacconi led the *Uhuru* ("freedom" in Swahili) mission launched in 1970 from Kenya as the first dedicated all-sky X-ray source mapper. He continued to campaign successfully for larger and larger robotic imaging-X-ray missions.

7. George Carruthers, "Development of Electrographic Camera for Deep UV Sky Survey Telescope," March 22, 1974, figure 4, appendix B, in Henize," Deep Sky Survey," box 19, folder "NRL, Geo. Carruthers Mat. 1974), HF/APS.

8. Carruthers, appendix B, 1.

9. Martin Marietta Technical Summary Document, "Atmospheric, Magnetospheric, and Plasmas in Space (AMPS) Spacelab Payload Definition Study," November 5, 1976, https://ntrs.nasa.gov/api/citations/19770021221/downloads/19770021221.pdf.

10. Carruthers, "Photoelectric Devices in Astronomy," 221.

11. Carruthers, "Proposal for Participation in Scientific Definition of Space Shuttle Missions for Atmospheric, Magnetospheric, And Plasmas-in-Space (AMPS) Payloads," accompanying Rabin to NASA, January 24, 1974, box 19, folder "NRL, Geo. Carruthers Mat. 1974," HF/APS.

12. "Lyman Alpha Explorer Mission," undated, approx. December 1974, box 19, folder "Geo. Carruthers Mat 1970–1971," HF/APS. The *IUE*, first proposed in the early 1960s, finally flew in January 1978 as Explorer 57. It was an international collaboration initiated by the UK Science Research Council and ESA. The *IUE* employed a 16-inch f/15 Ritchey-Chrétien reflecting telescope designed for narrow-field and high-resolution spectroscopic observations and employed two sets of UV-to-visible image converters feeding SEC vidicon tubes. Launched into an eccentric geosynchronous orbit over the Atlantic Ocean, it had the distinction of being directly operated by astronomers in sequence at workstations at NASA's Goddard Space Flight Center and by ESA's facility at the Villafranca Del Castillo in Madrid, Spain. See Boggess et al., "IUE Spacecraft and Instrumentation," 372–377.

13. Carruthers, "Lyman Alpha Explorer Mission."

14. Naugle and Logsdon, "Space Science," 13.

15. Robert Meier to David DeVorkin, October 30, 2020.

16. Seeley interview with DeVorkin and Callahan, 16, 23.

17. Henry interview with DeVorkin, April 15, 2015, 23. Henry also called out Gil Fritz as a collaborator. Johns Hopkins University Oral History Program, S. Leslie.

18. "George Carruthers," The HistoryMakers A2004.112, interviewed by Racine Tucker Hamilton, July 27, 2004, The HistoryMakers Digital Archive, session 1, tape 2, story 7.

19. Jack R. Lister (Personnel Office, NASA JSC) to George R. Carruthers, January 16, 1968; Astronaut (Mission Specialist) Candidate Program, August 2, 1976, FOIA Tracking Number 22-JSC-F-00230, courtesy Robert Young (NASA Government Information Specialist) to the author, March 31, 2022.

20. Shafritz and Atkinson, *The Real Stuff*, 13–14; on the rationale for including minorities, see 135.

21. Shafritz and Atkinson, *The Real Stuff*, 135.

22. N. P. Patterson (Bellcomm) to Distribution, November 17, 1971, "Astronomy on the Space Shuttle," box 46, folder "President's Science Advisory Committee, 1961–1973," HF/APS.

23. Herbert Friedman to Philip E. Culbertson, March 31, 1976, box 17, folder "NASA Correspondence folder 2, 1958–72," HF/APS.

24. McCurdy, *Space and the American Imagination*, 186.

25. George Carruthers to Astronaut (Mission Specialist) Candidate Program, August 2, 1976, FOIA Tracking Number 22-JSC-F-00230, courtesy Robert Young (NASA Government Information Specialist) to the author, March 31, 2022.

26. Carruthers to Astronaut (Mission Specialist) Candidate Program, August 2, 1976.

27. Carruthers to Astronaut (Mission Specialist) Candidate Program, August 2, 1976.

28. See "Eight Scientists Are Candidates for Spacelab 2 Flight," 76–78. On STARLAB, see also Anderson et al., "STARLAB," 613.

29. Carruthers to Astronaut (Mission Specialist) Candidate Program, August 2, 1976.

30. Carruthers to Astronaut (Mission Specialist) Candidate Program, August 2, 1976. FOIA Tracking Number 22-JSC-F-00230, courtesy of Robert Young (NASA Government Information Specialist) to the author, March 31, 2022.

31. "Sixth Group of Shuttle Astronaut Applicants Includes One Woman," 2, courtesy of Jon Uri.

32. See Phelps, *They Had a Dream*.

33. George D. "Pinky" Nelson interview with Jennifer Ross-Nazzal, May 6, 2004, https://historycollection.jsc.nasa.gov/JSCHistoryPortal/history/oral_histories/NelsonGD/NelsonGD_5-6-04.htm.

34. George Nelson, interview, May 18, 2021.

35. George Nelson, interview.

36. "Three at JSC Are Payload Specialist Candidates for First Spacelab Mission," 1, courtesy of Jon Uri. On Giuli, see "R. Thomas Giuli," Space Facts, accessed November 3, 2022, http://www.spacefacts.de/bios/candidates/nasa6/english/giuli_thomas.htm.

37. George Nelson to the author.

38. Shayler and Burgess, *NASA's First Space Shuttle Astronaut Selection*, 2.

39. Shayler and Burgess, 26–28.

40. George Carruthers draft proposal to NASA, "Study of Modification of Far Ultraviolet Schmidt Camera/Spectrograph for Spacelab Astronomical Observations," box 19, March 31, 1976, folder "NRL, George Carruthers Mat. 1975–7," HF/APS.

41. John Uri (JSC-NA141) to the author, May 18, 2021.

42. "Memorandum" attached to Carruthers to Astronaut (Mission Specialist) Candidate Program, August 2, 1976, FOIA Tracking Number 22-JSC-F-00230, courtesy of Robert Young (NASA Government Information Specialist) to the author, March 31, 2022.

43. "Memorandum."

44. Shayler and Burgess, *NASA's First Space Shuttle Astronaut Selection*, 28.

45. Significant budget cuts for the space sciences are noted in Laurence Soderblom, Philip Davis, and Jeffrey Rosendahl, "National Aeronautics and Space Administration Space Science Advisory Committee: Minutes of the October 20–27, 1980, Meeting," unpublished minutes, author's files.

46. Henize response to request from Duane L. Ross, attached to Ross to Henize, October 13, 1977, "Astronaut Candidate Qualifications Inquiry," attached to Rober Young to David DeVorkin, March 31, 2022, FOIA Tracking Number 22-JSC-F-00230, author's files.

47. Spigel, *Welcome to the Dreamhouse*, chapter 3, 148–149.

48. See Ruley, "Revolt of the Scientist-Astronauts," 5–12.

49. See O'Leary, *The Making of an Ex-astronaut*.

50. "Report of the Subcommittee on Scientist Astronauts of the NASA Space Program Advisory Council," September 8, 1975, John E. Naugle Papers Collection, NASA Headquarters Historical Reference Collection, box 36, NASA, JN/NHO. This was an elite committee. Chaired by Homer Newell, it included Hans Mark, then director of NASA's Ames Research Center and soon to become the secretary of the Air Force and then NASA deputy administrator. Also on the panel were astronaut Owen Garriott, Charles A. Berry—known as the "astronauts' doctor" as director of life sciences at NASA—and Leon Silver, a Caltech geologist then at the astrogeology branch of the United States Geological Survey and contracted to work with NASA on Apollo Lunar missions. See Charles A. Berry, Interview with Carol Butler, April 29, 1999, and Leon T. Silver, Interview with Carol Butler, May 5, 2002.

51. Frederick Seitz to George M. Low, September 8, 1975, John E. Naugle Papers Collection, NASA Headquarters Historical Reference Collection, box 36, subject file "S," JSC Oral History Project, NASA History Office, JN/NHO.

52. Carruthers to Nathaniel Cohen, NASA HQ, February 18, 1975, "Appendix 4" 44 in "Report of the Subcommittee on Science Astronauts of the NASA Space Program Advisory Council," September 8, 1975, John E. Naugle Papers Collection, NASA Headquarters Historical Reference Collection, box 36, subject file "S," NASA History Office, JN/NHO.

53. Prinz never flew because of the *Challenger* disaster. She held various administrative and technical positions on missions like Spacelab. Bartoe flew on shuttle mission *STS-51-F* to observe the Sun with instruments on *Spacelab 2* in August 1985.

54. Wray et al., "Space Schmidt Telescope," 141.

55. Carruthers, "High Resolution, Large Format Electrographic Cameras for Space Astronomy," proposal submitted to the Seventh Symposium on Photoelectric Image Devices, Imperial College, London, September 4–8, 1978. Draft copy in the Carruthers Collection, NASM Archives. It was published in *Advances in Electronics and Electron Physics* 52, (1980): 283–294.

56. Henize, "The Role of Surveys in Space Astronomy," *Optical Telescope Technology*. Proceedings of a workshop held at Marshall Space Flight Center, Huntsville, AL, April 29 to May 1, 1969. Prepared by the Office of Space Science and Applications, NASA SP-233, National Aeronautics and Space Administration, 1970, p.17.

57. The best-known example of this practice was the creation of two Schmidt wide-field survey cameras in the 1930s to produce deep sky surveys so that astronomers would know where to point the 200-inch Palomar telescope to find interesting new objects. The result was the Palomar Sky Survey. See Florence, *The Perfect Machine*, 257–261.

58. Carruthers et al., "Far Ultraviolet Wide Field Imaging and Photometry," 112.

59. In Heckathorn's words, "I did mostly ground-based visual kinds of electrography and tried to do things to support George's scientific requirements from the ground. I enjoyed going out to the McDonald Observatory and other places, collecting data of interesting things that he may have seen with his ultraviolet cameras." Heckathorn interview, 20.

60. Heckathorn, born in 1944, trained at Carleton College in Minnesota and then at Northwestern and taught courses at the Adler to support himself. He had a morning paper delivery route and keenly remembers being fascinated by an auroral display: "The skies absolutely lit up. It was the most amazing thing I've ever seen . . . before or after. There were all kinds of auroral rays and curtains and colors and flashes. It was just amazing. So, I really got interested in the sky." Heckathorn interview, 20.

61. Heckathorn interview.

62. Carruthers, "Far Ultraviolet Cameras Experiment on STS-39," 226. See also *NRL Labstracts*, 1, 3.

63. See "Space Shuttle Mission STS-39, Press Kit March 1991," NASA History Collection, accessed November 14, 2022, https://historycollection.jsc.nasa.gov/JSCHistoryPortal/history/shuttle_pk/pk/Flight_040_STS-039_Press_Kit.pdf.

64. Carruthers, "Far Ultraviolet Cameras Experiment on STS-39," 227.

65. William O. Glascoe III, "Summer Research Apprentice Program Final Report for 1985," August 23, 1985, courtesy W. Glascoe. Carruthers's role in the Science and Engineering Apprenticeship Program was mentioned in https://www.biography.com/inventor/george-carruthers.

66. Carruthers to Glascoe, April 14, 1989. Carruthers signed off typically as "George."

67. Roman, "Nancy Grace Roman," 24.

68. Carruthers, "Far Ultraviolet Cameras Experiment on STS-39," 226–231.

69. George Carruthers, The HistoryMakers A2004.112, interviewed by Racine Tucker Hamilton, July 27, 2004, The HistoryMakers Digital Archive, session 1, tape 3, story 5.

70. Brown and Giacconi, "New Directions for Space Astronomy," 617–619.

71. On the ASTRO Observatory, see "NASA Space Science Data Coordinated Archive," NASA, accessed October 29, 2022, https://nssdc.gsfc.nasa.gov/nmc/spacecraft/display.action?id=ASTRO-1.

72. Carruthers, "A High-Resolution Ultraviolet Shuttle Glow Spectrograph," 97–105. See also Carruthers et al., "The Far-Ultraviolet Imaging Spectrograph Experiment," 890.

73. Carruthers et al., "The Far-Ultraviolet Imaging Spectrograph Experiment," 890.

74. Potter and Parker, "Karl Gordon Henize, 1926–1993," 1604–1605.

## CHAPTER 12

1. Carruthers interview with DeVorkin, August 18, 1992.

2. Herbert Friedman to S. L. Mandelshtam, December 7, 1981, box 2, folder "Mandelshtam, S. L. 1967–1986," HF/APS.

3. Herbert Friedman, interview with Hirsh, AIP.

4. Herbert Friedman, interview with Hirsh, AIP.

5. Herbert Friedman, interview with Hirsh, AIP.

6. Phillip Mange, interview with Ron Doel and Fay Korsmo.

7. See, for instance, Herbert Gursky, "Giving Women Their Due," 91.

8. Hirsh, *Glimpsing an Invisible Universe*; Giacconi et al., "Evidence for X Rays from Sources outside the Solar System," 439–443.

9. Phillip Mange, interview with Ron Doel and Fay Korsmo.

10. Private communication, Robert Meier to DeVorkin, May 3, 2023.

11. Heckathorn interview, 27–28.

12. Heckathorn interview.

13. Burke et al., *ONR Research Opportunities in Astronomy and Astrophysics*, iv. Copy courtesy of National Radio Astronomy Observatory/Associated Universities, Inc. Archives, Papers of Bernard F. Burke, folder Councils and Committees Series/ Naval Studies Board Unit. et al., Naval Studies Board, March 17, BB/NRAO.

14. Burke et al., *ONR Research Opportunities in Astronomy and Astrophysics*.

15. Burke et al., *ONR Research Opportunities in Astronomy and Astrophysics*, 25.

16. Burke et al., 2.

17. David Latham, interview.

18. Relevant literature on the migration of detector choice from vacuum tube technologies to solid-state technologies in space astronomy includes Thompson, "The Carnegie Image Tube Committee," and, specific to the Hubble Space Telescope, Smith and Tatarewicz, "Counting on Invention," who emphasize the development time required for the latter to become competitive. See also their "Replacing a Technology."

19. "Far Ultraviolet Imaging and Photometry," box 1, folder 6.0, USS 3–7 Mission (August 1977), Program and Project Files Relating to Shuttle Launches, W255-89-0615, RG255.NARA.

20. Thompson, "The Carnegie Image Tube Committee," chapter 7.

21. Carruthers wrote positively about their promise in "Charge-Coupled Devices," 13–21. Revised from a talk given in 1983.

22. See Thompson, "The Carnegie Image Tube Committee," 162–163. On Hubble specifically, see Smith and Tatarewicz, "Replacing a Technology"; and Smith and Tatarewicz, "Counting on Invention." See also Smith, *The Space Telescope*, 329–331. The classic fictional treatment of scientific obsolescence is McCormmach, *Night Thoughts of a Classical Physicist*, who constructed his story from historical commentary.

23. Carruthers, "Further Developments of Electrographic Image Detectors," 279; Lowrance and Carruthers, "Electron Bombarded Charge-Coupled Device (CCD) Detectors," 112.

24. "Electrographic" is a variation of "electronographic."

25. Griboval, "The U.T. Electrographic Camera: A Powerful Detector."

26. Heckathorn, interview.

27. Heckathorn and Opal, "Optical Identification," 919. See also Opal and Heckathorn, "A Search for 'Unidentified' Stellar Objects," 857.

28. Heckathorn and Opal, "Optical Identification."

29. Heckathorn, "Electrographic Detectors," 164. Morgan, *Photo-Electronic Image Devices*, , 153–165. One salient factor they did not discuss was data recovery. Unlike film, which required physical retrieval, CCD data could be telemetered.

30. Smith and Tatarewicz, "Replacing a Technology," and Smith and Tatarewicz, "Counting on Invention."

31. Heckathorn et al., "Electrographic Detectors," 165.

32. The Hubble Space Telescope's wide field/planetary camera split the incoming light into four physically separated beams to hit single CCDs that would then have their signals combined. It would be another decade before contiguously placed CCD arrays would become dominant in astronomy. See Smith, *The Space Telescope*, 329–331.

33. As argued generally in Edgerton, *The Shock of the Old*.

34. Griboval, "The U.T. Electronographic Camera: Present Status," 305–314.

35. Heckathorn and Carruthers, "An Investigation of Auto Radiographically Intensified Electrographic Imagery," 672.

36. Heckathorn, interview with DeVorkin.

37. Seeley, interview with DeVorkin.

38. Carruthers et al., "Rocket Ultraviolet Observations of Comet Halley," 1155–1167. Carruthers and Opal had hoped to observe the comet from Spacelab on the shuttle, either as guest observers on *ASTRO1* or with their own camera. See B. I. Edelson to Chet B. Opal, April 19, 1984, box 1, folder 6.4, "SCI./INST, 6.4.4 Instruments," W255-89-0615, RG255/NARA. See also EM/Astro Program Manager to JA11/Astro Mission manager, April 24, 1984, "Astro 1 Guest Observers," box 1, folder 6.4, "SCI./INST, 6.4.4 Instruments," W255-89-0615, RG255/NARA. Carruthers was originally a member of the "OSS-4 Halley Comet Science Advisory Committee" to plan for Halley in March 1986. See Ronald B. Felice and Marc Bensimon to Michael J. S. Belton, [n.d. circa September 9, 1981], box 1, folder "Program and Project Files Relating to Shuttle Launches," W255-89-0615, RG255/NARA. Pressures pushed up the flight date from March to January 1986, and it perished on the fateful *Challenger*. On the flight manifest, see "NASA—STS-51L Mission Profile," NASA, accessed December 21, 2022, https://www.nasa.gov/mission_pages/shuttle/shuttlemissions/archives/sts-51L.html. On the committee's charge, see Ronald Felice and Marc Bensimon (noted above), and on the proposal, see Niedner et al., "The ASTRO1 Mission and Halley's Comet," 900–901.

39. Seeley, interview. Born in Victorville, California, in 1956, Seely came from a peripatetic military family that moved around the United States as well as France and England, where Seeley completed high school. When they returned to the United States, Seeley took various odd jobs including a "gas jockey and mechanic."

College was at first not affordable, so he became a laborer in Baltimore, excavating the "twin tunnels" for Penn Station. He was finally able to enter the University of Maryland, supported by a half-time co-op program where his work exposed him to electronics. This caused him to shift to electrical engineering and eventually to a co-op position at NRL, working in Carruthers's laboratory in 1984 in his junior year. By then, he was married, and his wife Alicia also helped him through college.

40. Carruthers, et al., "The *SPARTAN-281* Far Ultraviolet Imaging Spectrograph," 87C.

41. Seeley, interview.

42. Gerald Carruthers phone call, August 5, 2023; Gerald Carruthers interview. August 5, 2023.

43. Seeley interview.

44. Meier interview with DeVorkin and Callahan, 75–76.

45. Smith and Tatarewicz, "Counting on Invention"; and Smith and Tatarewicz, "Replacing a Technology."

46. Janesick and Blouke, "The Sky on a Chip," 238–242.

47. Seeley interview with DeVorkin, 19. Back thinning is a process of chemical etching to thin the wafer to improve quantum efficiency. See "Spectroscopy," Laser Focus World, accessed September 28, 2021, https://www.laserfocusworld.com/test-measurement/spectroscopy/article/16555243/spectroscopy-backilluminated-ccds-enable-advanced-spectroscopy-instrumentation.

48. Conway interview with DeVorkin and Callahan, 17.

49. Meier, "Ultraviolet Spectroscopy and Remote Sensing," 1–185.

50. Charles Brown interview with DeVorkin and Callahan. See also Conway, Prinz, and Mount, "Middle Atmosphere High Resolution Spectrograph," 50–60.

51. Conway interview with DeVorkin, 11.

52. Tousey's large group included veteran analyst Charlotte E. Moore Sitterly, whom he recruited from NIST after she retired. Starting at Princeton in the 1920s, Sitterly compiled data for what became, in 1945, her invaluable resource for astronomical spectroscopy, *A Multiplet Table of Astrophysical Interest*. Tousey's group used it as a guide. She also advised NRL staff like Brown on its best uses and prepared him for his next assignments. Spectra of the elements occur in families called "multiplets."

53. Brown interview with DeVorkin and Callahan.

54. Conway interview, 18.

55. Meier letter to DeVorkin, May 5, 2023.

56. Conway interview, 13–14.

57. Conway interview with DeVorkin and Callahan, 20.

58. Conway interview.

59. Seeley, Conway, and Carruthers, "The Middle Atmosphere High Resolution Spectrograph Investigation," 41–48.

60. Stevens et al., "MAHRSI Observations," 3213–3216.

61. Englert interview with DeVorkin and Callahan, and Meier email to DeVorkin, May 3, 2023. See also Englert and Harlander, "Flatfielding in Spatial Heterodyne Spectroscopy," 4583.

62. Englert interview with DeVorkin and Callahan.

63. See "ARGOS (Satellite)," Wikimedia Foundation, accessed November 8, 2024, https://en.wikipedia.org/wiki/ARGOS_(satellite). See also "ARGOS Satellite Serves as Platform for Leading-Edge Technology and Research," accessed November 8, 2024, https://boeing.mediaroom.com/1999-01-06-ARGOS-Satellite-Serves-as-Platform-for-Leading-Edge-Technology-and-Research.

64. An NRL historical retrospective linked the success of the *Apollo 16* imaging of the geocorona to the development of GIMI: "Such a detection of UV features associated with the ionosphere from the distance of the Moon made clear that it would be possible to image or monitor the ionosphere steadily from a platform at geosynchronous altitude." Wolff and Dymond, "90's.3IV: ARGOS PART IV," 49

65. Garland Dixon interview; Dixon to DeVorkin, January 22, 2021.

66. Dixon interview.

67. On informal education in the space sciences, see Hawkins, "Summary of Informal Education Discussions," 147–154. And within the same volume, Thomas, Carruthers, and Takamura, "Collaborative Support for Solar Eclipse 2001 Activities," 187–190.

68. Carruthers and Seeley, "Global Imaging Monitor of the Ionosphere," 22–333. See also Carruthers, Seeley, and Dymond, "Electron-Bombarded CCD Sensors," 1–8.

69. Carruthers and Seeley, "Global Imaging Monitor of the Ionosphere."

70. Dymond, et al. "High-Resolution Airglow." See also Thonnard et al., "Results from the High Resolution Airglow."

71. Nuth, Lowrance, and Carruthers, "NEOCAM," 495–504.

72. Meier email to DeVorkin, May 3, 2023.

73. Brown interview with DeVorkin and Callahan, 20.

74. Howard interview with DeVorkin and Callahan.

75. Meier email to DeVorkin.

76. Englert interview with DeVorkin and Callahan.

77. Englert interview with DeVorkin and Callahan. *ICON* was air launched by a Pegasus rocket in 2019.

78. See "ICON: Ionic Connection Explorer," NASA, accessed April 26, 2023, https://www.nasa.gov/sites/default/files/atoms/files/icon_presskit_oct2019.pdf. Englert, who was the principal investigator, told the author that the Michelson Interferometer for Global High-Resolution Thermospheric Imaging "met all mission success criteria." Englert to DeVorkin, email, April 26, 2023. See Englert et al., "Michelson Interferometer."

79. Brown interview with DeVorkin and Callahan. Giovane interview.

80. NRL Record Reports April 1976, in conjunction with the Air Force Avionics Laboratory; Report 625162, Record #168484 ("Final report"), July 1, 1993; Quote from "An Atlas of Far-Ultraviolet Observations of Solid Rocket Plumes and Their Far-UV Brightness Distributions," Abstracts only, NRL Archives.

81. Seeley interview with DeVorkin and Callahan, 3.

82. Meier interview with DeVorkin and Callahan. Meier later recalled that he and others were upset when part of Carruthers's office space was converted into a lunchroom, not laboratory space. Meier email to DeVorkin May 3, 2023.

83. Meier interview.

84. Giovane interview.

85. Meier interview.

86. Giovane to DeVorkin, email, January 10, 2023.

87. "Senior Executive Service," Department of the Navy, accessed May 6, 2021, https://www.secnav.navy.mil/donhr/About/Senior-Executives/Pages/Default.aspx.

88. Heckathorn interview with DeVorkin. Meier has added that Carruthers "was always in trouble with the NRL Safety Team. For example, he had cabinets full of unmarked containers of chemicals." Meier email to DeVorkin, May 3, 2023.

89. Meier interview, 51. See also Meier talk outline "George Carruthers," September 7, 2006, courtesy of Robert Meier.

90. Conway interview with DeVorkin and Callahan, 14.

91. Meier interview, 55.

92. Kenneth Dymond to author, December 29, 2021.

93. Meier to DeVorkin, March 1, 2021.

94. Meier interview, 68.

95. Brown interview with DeVorkin and Callahan. See also Laurence Soderblom, Philip Davis, and Jeffrey Rosendahl, "National Aeronautics and Space

Administration Space Science Advisory Committee: Minutes of the October 20–27, 1980, Meeting," unpublished minutes, author's files.

96. Giovane interview with DeVorkin.

97. Gursky to the Space Telescope Science Institute, n.d. (November 1998), NTA Files.

98. Conway interview with DeVorkin and Callahan, 20.

99. Callahan comment in Brown interview.

100. Giovane interview with DeVorkin.

101. Interview with anonymous NRL staffer.

102. Callahan comment in Howard interview with DeVorkin and Callahan.

103. Englert interview 35–36.

104. Meier interview, added commentary, September 19, 2020, AIP.

105. Anonymous staffer interview with DeVorkin and Callahan, AIP.

106. Giovane interview, March 13, 2021, 10, AIP.

107. Jessye Bemley Talley interview with DeVorkin.

## CHAPTER 13

1. George Carruthers, The HistoryMakers A2004.112, interview with Racine Tucker Hamilton, July 27, 2004, The HistoryMakers Digital Archive, session 1, tape 3, story 5.

2. George Carruthers, "Speech, Career Conference," rough draft, April 3, 1960, Carruthers Papers, NASM Archives Division, box 3, folder 12, GC/NASM.

3. George Carruthers, "Speech, Career Conference."

4. George Carruthers to "Uncle Ben," September 22, 1970, Benjamin Carruthers Papers, Schomberg MG433 MG433_b1_f14, BC/SA. His uncle received an award from that magazine earlier that year.

5. George Carruthers to "Uncle Ben." No record has been found of a published article.

6. "NASA Awards Black Scientist"; Fellows, "Please Be Seated," June 17, 1972.

7. "Report on Black Inventors and Innovators: New Perspectives," The Lemelson Center for the History of Invention and Innovation (2020–21), 24, accessed July 20, 2023, https://invention.si.edu/sites/default/files/Lemelson-Center-Black-Inventors-Innovators-Report-updated-2021-10.pdf.

8. See, for instance, NRL *Bulletin,* April 18, 1958, no. 30; *Labstracts,* January 17, 1962, 6; and March 1, 1962, 2, courtesy of Angelina Callahan.

9. NRL, *Labstracts*, October 17, 1986, 5, courtesy of Angelina Callahan. On the importance of exposure to role models, see Bell et al., "Who Becomes an Inventor in America?," 647–713.

10. NRL, *Labstracts,* "News and Views from Personnel," December 1974, 3, courtesy of Angelina Callahan.

11. Responses from Heckathorn, Brown, Seeley, and Meier to a March 21, 2021, author's query.

12. Henry Pickard to the author, March 27, 2021. Quoted from Bradby, "The Naval Research Laboratory," 32.

13. This included STEP (Student Temporary Employment Program)—short- and long-term employment, flexible, and very popular. It was absorbed into SCEP (Student Career Experience Program), the Navy's co-op program, which then became part of SEAP (Science & Engineering Apprenticeship Program), a popular program for high school students. There also was NREIP (Naval Research Enterprise Internship Program), the higher education equivalent of SEAP. Descriptions adapted from Henry Pickard to the author, March 27, 2021.

14. Henry Pickard to the author, March 27, 2021.

15. NRL, *Labstracts,* August 12, 1966, courtesy of Angelina Callahan.

16. Rouse, "Ionization Equilibrium Equation of State," 599.

17. Anonymous NRL staffer interview with DeVorkin and Callahan. Carruthers was aware of his casual dress, but it did not differ significantly from dress habits of others in the laboratory. Still, it is not clear if he was singled out for his dress or his race. On dress habits of experimenters, see Traweek, *Beamtimes and Lifetimes,* 24–25.

18. NASA Ames, *The Astrogram*, 2.

19. Calvino, "A Phenomenon to Monitor," 1.

20. Paul and Moss, *We Could Not Fail*, 116.

21. Calvino, "A Phenomenon to Monitor," 1.

22. Calvino, "A Phenomenon to Monitor," 33, ref. 73.

23. NASA Ames, *The Astrogram*, 2. Monroe had been a science teacher for some thirteen years where he participated in NASA's "Spacemobile program."

24. Carwell, "Roscoe Monroe Discovered Dr. Carruthers for NASA," 12.

25. Thomas interview, 12; "Appendix B: Resumes of Investigators, Valerie L. Thomas," courtesy of Valerie L. Thomas.

26. Yette, "Give NASA Credit."

27. On exploitation, see Komanduri, Roebuck, and McCamey, "Race and Class Exploitation."

28. George Carruthers interview with DeVorkin, August 18, 1992.

29. George Carruthers interview with DeVorkin, August 18, 1992. See also Carruthers, "Outreach Programs," 29–30.

30. Ambrose, "The National Technical Association;" Cunningham, "Charles Sumner Duke (1879–1952)."

31. Carwell to DeVorkin, August 8, 2020.

32. She grew up in a supportive family and community in Ashland, Virginia and attended Bennett College for Women in Greensboro, North Carolina, graduating in chemistry and biology in 1971 and then obtaining a master's at Rutgers University a year later in health physics. By 1994, she had a leading role at DOE's Berkeley Office.

33. Hattie Carwell to George Carruthers, August 21, 2011, reprinted in *A Tribute to the Career of Dr. George Carruthers*, September 14, 2011, provided by V. Thomas. See also documents between 1983 and 1987 in "NTA Journal Records," box 5, NTA files, Howard University, GC/MSRC. Carruthers served some thirty years in the NTA by then, so twenty thousand hours amounts to one-third time! See also Carwell to Carruthers, August 15, 2011, courtesy of Hattie Carwell.

34. Bradby, "The Naval Research Laboratory," 32–36.

35. Box 6,"NTA Corr 1984" folder, GC/MSRC.

36. Carruthers to Phillip Manuel, Black Collegiate Services, Inc., June 28, 1986, box 7, GC/MSRC.

37. Carruthers to Stinson, August 16, 1993, box 4, GC/MSRC.

38. Carruthers manuscript, "Introduction to Technical Papers Issue and Call for Papers," *NTA Journal*, spring 1993, box 4, GC/MSRC.

39. Carruthers, "Emerging Space Applications," *NTA Journal*, winter 1994, 11, box 4, GC/MSRC.

40. Carruthers, *NTA Journal*, 1991, 65 #2: 20, box 4, GC/MSRC.

41. George Carruthers, "Charge-Coupled Devices," 13–21. Revised from talk given in 1983. On the early limitations with CCDs, see Smith, *The Space Telescope*, 329–331.

42. Carruthers to Board of Directors, NTA February 6, 1985, box 2, GC/MSRC. Both Gregory and Bolden remained active in the upper echelons of NASA—Gregory as deputy and then acting NASA administrator in 1985 and Bolden as NASA administrator in 2009.

43. Carwell to Carruthers, August 15, 2011, attached to Carwell to DeVorkin, January 18, 2023.

44. George R. Carruthers, "Ultraviolet Remote Sensing of the Middle and Upper Atmosphere." Draft copy of talk presented at an NTA Conference, Cincinnati,

Ohio, 1992, "Stepping into Tomorrow . . . Building on our Technical Tradition," Carruthers Papers, box 4, GC/MSRC. A comprehensive technical review of the field at the time is Meier, "Ultraviolet Spectroscopy and Remote Sensing," 1–185.

45. Draft copy of George R. Carruthers, "Ultraviolet Remote Sensing of the Middle and Upper Atmosphere," 1992, to be presented at NTA Conference, Cincinnati, Ohio 1992, "Stepping into Tomorrow . . . Building on our Technical Tradition," 26–27, box 4, GC/MSRC.

46. Carruthers, George R. "Current Developments and Future Prospects in Space Transportation," *NTA Journal*, April 1992. See also draft copy of George R. Carruthers, "Ultraviolet Remote Sensing of the Middle and Upper Atmosphere," NTA Conference, Cincinnati, Ohio, 1992, "Stepping into Tomorrow . . . Building on our Technical Tradition," box 4, GC/MSRC.

47. "NTA Housing Fund," June 1988, box 2, GC/MSRC.

48. "Board Meeting," correspondence and notes, G. A. Haynes to Directors of the NTA, April 4, 1986, box 3, GC/MSRC.

49. Mike Chapman (NASA Langley) to Carruthers, April 9, 1997, box 9, GC/MSRC.

50. "Executive Director's Recommendations for a Near Term Plan," n.d. circa September 2000; see also "NTA Profile (2000)," n.d., box 1, GC/MSRC.

51. *New York Times* advertisement by Citibank, March 25, 1990, box 10, NTA files at Howard; Sonya Stinson to Carruthers, May 18, 1993, box 4, GC/MSRC.

52. George Carruthers, July 22, 1985. Handwritten report in file "NTA 57th Annual Conference" Houston TX "Career Development Guide," box 1, GC/MSRC.

53. Riggs, "Training Matters!," 224.

54. Hattie Carwell, "MAAT Science Village: A New Approach," NTA 71st Annual Conference (November 1999), 11, GC/MSRC.

55. General program—NTA 71st Annual Conference, November 3, 1999, "Creative Beginning for the New Millennium," box 1, MSRC.

56. George Carruthers interview with DeVorkin, August 18, 1992, 1; Valerie L. Thomas interview with DeVorkin, 24.

57. "NTA Brochure," n.d., box 2, GC/MSRC.

58. For a full profile of Dr. Thomas, see "78-Year-Old Valerie Thomas," *Oprah Daily*, accessed May 7, 2023, https://www.oprahdaily.com/life/a36674183/valerie-thomas-nasa-scientist-interview/.

59. Thomas interview, 14. Modified in Thomas to DeVorkin, March 19, 2023.

60. Thomas interview, 14.

61. Thomas interview.

62. George Carruthers interview, 1992.

63. George Carruthers interview.

64. George Carruthers interview.

65. "S.M.A.R.T. Day at the National Air and Space Museum," courtesy Valerie L. Thomas.

66. On the "BDPA Education and Technology Foundation" scholarships, see: https://bdpa.org/about-us/ (accessed December 7, 2024).

67. "S.M.A.R.T. Day at the National Air and Space Museum," courtesy Valerie L. Thomas. See also Carruthers, "Outreach Programs," 29.

68. See Pollyanna Williams, "The Adventures of Astronomy," in Carruthers, "Outreach Programs," 30.

69. Sawyer, "Comet Chunk"; Sawyer, "More Chunks of Comet Hit Jupiter"; Friedlander, "Sky Watch."

70. Peery interview, with DeVorkin. AIP.

71. S.M.A.R.T., "Compendium of Abstracts." This was likely a preliminary schedule as subsequent announcements indicated different days and speakers.

72. R. Berendzen to NASA Office of Space Science, June 25, 2003, Re: Minority University and College Education, "Earth Science Proposal 9/04," box 7, GC/MSRC. Berendzen headed the DC Space Grant Consortium at the time.

73. Box 9, "Pyramids to Planets" folder, GC/MSRC.

74. Carruthers and Thomas, "Space Science Education," 257. PowerPoint courtesy of V. Thomas.

75. Carruthers and Thomas, "Sun-Earth Connection Education."

76. "GSA Partners with DC Link and Learn New Technology Center to Prepare DC Youths and Adults for High Paying Jobs," US General Services Administration, November 23, 1998.

77. V. Thomas interview, 24.

78. V. Thomas interview, 18.

79. V. Thomas to the author, February 21, 2020.

80. Henry Pickard to the author, March 27, 2021. Quoted from Bradby, "The Naval Research Laboratory," 32.

81. McKinney, "Naval Research Laboratory Voted a Best Diversity Company 2010," NRL News and Press Releases, dated July 7, 2010, https://www.nrl.navy.mil/Media/News/Article/2577239/naval-research-laboratory-voted-a-best-diversity-company-2010/.

82. "Carruthers, Johnson EEO Award Recipients," NRL *Labstracts,* October 17, 1986, 3, courtesy Angelina Callahan.

83. "Racial Stereotype Busters," 133–134.

84. "Time Life Releases First Volume of Books on Influential Blacks, *Miami Times,* October 21, 1993.

85. Dahlburg et. al., *NRL SSD Research Achievements, IV*, 49.

86. "U.S. Naval Research Laboratory (NRL) Personnel Management Demonstration Project," 33970–34046.

87. "U.S. Naval Research Laboratory (NRL) Personnel Management Demonstration Project," 34016.

88. "Science & Engineering Professionals" Performance Standards (December 17, 1998) Element 2, Level IV, "NRL Personnel" folder, box 1, Carruthers Papers, GC/NASM.

89. Draft copy of "NRL Personnel Management Demonstration Project Requirements Document," November 16, 2000. Carruthers identified as "Incumbent." "NRL Personnel" folder, box 1, Carruthers Papers, GC/NASM.

90. "Military Pay," USAGov, accessed May 20, 2021, https://www.usa.gov/military-pay.

91. Meier to DeVorkin, notes, March 12, 2021.

92. George Carruthers, SF50 Part E.1 "Reasons for Resignation, June 2002 (date uncertain), box 1, folder "NRL Personnel," Carruthers Papers, GC/NASM.

93. On annuitants in the Navy, see "Employment of Federal Civilian Annuitants," Secretary of the Navy (accessed August 30, 2024). https://www.secnav.navy.mil/donhr/Documents/CivilianJobs/FedCivAnnuitants.pdf.

94. "Special Report," 3.

95. Carruthers, "Outreach Programs," 30; Carruthers and Washington, "Space Science Education," 941.

96. "George Carruthers," *Biography*, accessed April 9, 2023, https://www.biography.com/inventor/george-carruthers. See also Carruthers to S. Danagoulian, August 18, 2004, box 7, folder "N.C.A.T. University Invited Lecture 9/23/04 4," and folder "Earth Science Proposal 9/04," GC/MSRC; Prabhakar Misra, "Remembering Dr. George Robert Carruthers," dated January 30, 2021, Private Memorial packet.

97. Banks, "The First Spectrum of Titanium."

98. Banks was born in Atlantic City, New Jersey, but grew up in Washington, DC. He specialized in electronics and wrote a thesis on the spectrum of titanium. See "Astronomers of the African Diaspora: Harvey Washington Banks," University

at Buffalo, Department of Mathematics, accessed August 30, 2024, http://math.buffalo.edu/mad/physics/banks_harveyw.html. For a helpful overview, see also Fikes, "From Banneker to Best," published as "Careers of African Americans in Academic Astronomy," *Journal of Blacks in Higher Education* (Autumn 2000): 132–134. See also Patterson, "Professor Banks Dies of Heart Attack."

99. Peery interview, AIP.

100. Peery interview, AIP.

101. Cowley, "Benjamin Franklin Peery Jr. (1922–2010)," 28. Peery interview, with DeVorkin.

102. Cowley, 28.

103. Peery interview.

104. Peery interview.

105. Peery interview.

106. He was one of several dozen to be interviewed in the series that year. See "Full Cast & Crew," accessed November 20, 2022, https://www.imdb.com/name/nm1352733/?ref_=tt_cl_i_17.

107. "1987–89: The Howard University Catalog," Howard University Catalogs, January 1, 1987, 266, https://dh.howard.edu/hucatalogs/73/.

108. Morris et al., "The Howard University Program," 45–53. Misra led the expansion, along with other colleagues (notably Demetrius Venable, Gregory Jenkins, Vernon Morris, and Belay Demoz), first establishing a PhD program in the atmospheric sciences. See "The Howard University Program in Atmospheric Sciences," Taylor & Francis, accessed July 4, 2022, https://www.tandfonline.com/doi/full/10.5408/10-180.1. Perspective can also be gained from an oral history of Demetrius Venable, interviewed with David Zierler, May 12, 2021, AIP: "Demetrius Venable," AIP, accessed December 28, 2022, https://www.aip.org/history-programs/niels-bohr-library/oral-histories/47015.

109. Carruthers to Beth Brown (Department of Physics and Astronomy, Howard), September 15, 1989, box 10, GC/MSRC.

110. Misra, "Remembering Dr. George Robert Carruthers," comments at Carruthers's funeral, January 30, 2021, Private Memorial packet, author's files.

111. Misra, "Remembering Dr. George Robert Carruthers."

112. George Carruthers to Vernon Morris, Howard University, December 27, 2004, email, GC/MSRC.

113. Misra, Carruthers, and Jenkins, "Development of an Earth and Space Science-Focused Education Program," 339–345.

114. Misra, Carruthers, and Jenkins, 339–345.

115. Misra, "Remembering Dr. George Robert Carruthers."

116. Misra, Carruthers, and Jenkins 339–345.

117. Padgett interview.

118. "2009 APS March Meeting," Bulletin of the American Physical Society, accessed December 20, 2022, http://meetings.aps.org/Meeting/MAR09/Session/L4. And on the ADS, see "ADS Abstract Service," Harvard, accessed December 20, 2022, https://adsabs.harvard.edu/ads_abstracts.html.

119. Carruthers interview, August 18, 1992.

## CHAPTER 14

1. "NRL's Not-So Hidden Figure," US Naval Research Laboratory, accessed March 20, 2021, https://www.nrl.navy.mil/Media/News/Article/2516211/nrls-not-so-hidden-figure/.

2. Gerald Carruthers to author, email, December 20, 2020.

3. "In Loving Memory of Sophia Barbara Martin," February 23, 1990, Carruthers Papers, box 1, folder 2, GC/NASM.

4. Phillip interview.

5. Phillip and Ochoa correspondence with DeVorkin.

6. Phillip and Ochoa correspondence.

7. Phone call, Gerald Carruthers to DeVorkin, December 15, 2022.

8. "Appendix B: Resumes of Investigators: George R. Carruthers," provided by Valerie Thomas.

9. "Appendix B: Resumes of Investigators: George R. Carruthers," provided by Valerie Thomas.

10. See photos and postcards in box 5, GC/NASM.

11. Paul S. Allen (lawyer) to George Carruthers, "Application for Alien Labor Certification for: Ana Teresa Coplan," March 30, 1987, box 1, folder 8, Carruthers Papers, GC/NASM.

12. George Carruthers to Christine James, July 6, 1991, "Job Description," George Carruthers Papers, box 1, folder 3, GC/NASM.

13. Phillip and Ochoa interview with DeVorkin.

14. George Carruthers to Department of Health and Human Services, Washington, DC, August 18, 1997, Carruthers Papers, box 3, folder 35, GC/NASM. In 1995, Sandra was functional enough to help George determine what her mother needed to obtain permanent residency. See correspondence in box 3, folder 35, GC/NASM.

15. Thelma Jones to the author, December 28, 2020.

16. Deborah Carruthers, telephone interview, May 19, 2021. Gerald Carruthers recalled that George found her at home already dead. Gerald Carruthers to the author, December 14, 2020.

17. Gerald Carruthers interview.

18. "Program to Honor an American Living Legend—Dr. George Carruthers, Astrophysicist/Inventor," 2011, private circulation copies courtesy of Valerie L. Thomas and Gerald Carruthers.

19. "Gullah/Geechee Nation," accessed November 7, 2024, https://gullahgeechee nation.com/gullahgeechee-sea-island-coalition/.

20. "1969—Coast Guard Deploys a Detachment to Greece to Provide HU-16 Pilot Training for the Hellenic Air Force," United States Coast Guard Aviation History, accessed May 19, 2021, https://cgaviationhistory.org/1969-coast-guard -deploys-a-detachment-to-greece/.

21. Debra Carruthers, telephone interview, May 19, 2021.

22. Debra Carruthers telephone interview, February 18, 2021.

23. Debra Carruthers telephone interview l, February 18, 2021.

24. Appointment calendar, 2011, box 7, folder 3, GC/NASM.

25. Debra Thomas to George Carruthers, January 7, 2011, box 3, folder 33, GC/ NASM.

26. George Carruthers, The HistoryMakers A2004.112, interviewed by Larry Crowe, August 27, 2012, The HistoryMakers Digital Archive, session 2, tape 1, story 1, part 2; and session 2, tape 3, story 9. George Carruthers talks about his family, accessed May 20, 2021, https://da.thehistorymakers.org/story/29205, https://da.thehistorymakers.org/story/29234, and https://www.thehistorymakers .org/biography/george-carruthers-41.

27. Hattie Carwell, "2011 Nomination Form," National Medal of Technology and Innovation, copies courtesy of Carwell, John Palafoutas, Department of Commerce, and Gerald Carruthers.

28. William M. Jackson to Richard Mauslby, March 28, 2011, courtesy of Valerie L. Thomas and Gerald Carruthers.

29. Dymond to NMTI Nomination Evaluation Committee, n.d., courtesy of Gerald Carruthers.

30. Dymond to NMTI.

31. Dymond to NMTI.

32. Moos to NMTI Nomination Evaluation Committee, March 22, 2011, courtesy Gerald Carruthers.

33. Tate, "Brown, Political Economy," 157.

34. "2011 Laureates," United States Patent and Trademark Office, accessed February 26, 2023, https://www.uspto.gov/learning-and-resources/ip-programs-and-awards/national-medal-technology-and-innovation/recipients/2011.

35. Meier to Richard Maulsby, program manager, NMTI Nomination Evaluations Committee, n.d., courtesy of R. Meier.

36. The Obama White House, "President Obama Honors the Country's Top Innovators and Scientists of 2011," YouTube video, dated February 1, 2013, https://www.youtube.com/watch?v=yZX6GI2sLWU&t=21m36s.

37. Major General (retired) Charles F. Bolden, "Eulogy," "Celebration of Life," Facebook, January 30, 2021, 2 hrs. 31 min, https://www.facebook.com/388296547997017/videos/246561643584468.

38. Vigil, "Tribute," "Celebration of Life," January 30, 2021, Facebook, 1 hr. 15 min, https://www.facebook.com/388296547997017/videos/246561643584468.

39. Vigil, "Tribute." NASA's Heliospheric missions include "Ionospheric Connection Explorer" (launched 2019), "Global-Scale Observations of the Limb and Disk" (launched January 2018), and SWARM, a constellation of ESA satellites (launched November 2013–March 2018, from the Plesetsk Cosmodrome).

40. NASA's Heliospheric missions: "Ionospheric Connection Explorer" (launched 2019), "Global-Scale Observations of the Limb and Disk" (launched January 2018), and SWARM, a constellation of ESA satellites (launched November 2013–March 2018, from the Plesetsk Cosmodrome).

41. Four of his six highest cited papers have this characteristic, like his 1970 paper on molecular hydrogen: "Rocket Observation of Interstellar Molecular Hydrogen," accessed August 26, 2024, and his 1972 paper on the geocorona. https://articles.adsabs.harvard.edu/pdf/1970ApJ . . . 161L . . 81C.

42. Tessarotto, "Evolution and Recent Development," 278–286.

## APPENDIX A

1. Jessye Bemley Talley interview with DeVorkin, 1–3. All interviews cited here are with DeVorkin.

2. Seeley interview, 26.

3. Talley interview, 6–8.

4. Talley interview, 10.

5. Talley interview, 15.

6. Padgett interview, n.p.

7. Padgett interview, n.p.

8. It is now a public charter middle school on the campus. See Howard Middle School, accessed August 9, 2021, https://hu-ms2.org/.

9. Padgett interview.

10. Padgett interview.

11. Padgett, from a speech at the Italian Community Center, Inc., https://drive.google.com/file/d/1uZGDsjZA6LfJUPJJE-kuo1uUy3dhquak/view.

12. Padgett to DeVorkin, December 27, 2022.

13. Padgett remarks, "Celebration of Life," January 30, 2021, Facebook, https://www.facebook.com/100051800476855/videos/246561643584468.

14. Glascoe interview, 21 November 2020, 6, amended January 20, 2023.

15. Glascoe interview, 9.

16. Glascoe interview, 9.

17. Glascoe interview, 9.

18. Glascoe interview, 35.

19. Glascoe interview, 35.

20. Glascoe interview.

21. Dixon interview, 5.

22. Dixon interview, 7.

23. Dixon interview, 24.

24. Dixon interview, 24.

## APPENDIX B

1. For the details about the conservation process and analyzing the corrosion at an NRL laboratory, see O'Hern and DeVorkin, "What's That Smell?"

2. See O'Hern and DeVorkin, "What's That Smell?"

# BIBLIOGRAPHY


"A Study of NASA University Programs," NASA SP-185, 1968.

Adler Planetarium and Astronomical Museum. *Courses for 1960–1961*. Chicago Park District, 1960.

Alexander, George. "Big Comet Heads for Sun and a Christmastime Show." *Los Angeles Times*, April 8, 1973.

Allenby, R. J. "Lunar Orbital Science." *Space Science Reviews* 11 (1970): 5–53.

Allison, David. *New Eye for the Navy: The Origin of Radar at the Naval Research Laboratory*. NRL Report 8466, 1981.

Anderson, C., K. Henize, E. Jenkins, R. O'Connell, A. Smith, B. Smith, and T. Stecher. "STARLAB: A Spacelab-Based One-Meter General Purpose Telescope." Abstract 29.02.01 151st AAS Meeting, Austin, TX, *Bulletin of the American Astronomical Society* 90, 1977.

Armstrong, E. B. "Electron Image-Forming Devices in Astronomy." *Irish Astronomical Journal* 2, no. 5 (1953): 147–151.

Asher Robert L., and Robert C. Kaiser. "Broken Promises Line Riot Area Street." *Washington Post*, December 29, 1965.

Baldwin, Richard R. *Mission Science Planning Document Apollo Mission J-2 (Apollo 16)*. Rev. ed. NASA Marshall Space Flight Center, 1972.

Banks, Harvey Washington. "The First Spectrum of Titanium from 6000 to 3000 Angstroms." PhD diss., Georgetown University, 1961.

Barry, D. C., R. H. Cromwell, and S. A. Schoolman, "Spectral Quantification." *The Astrophysical Journal* 212 (1977): 462–471.

Baum, William A. "Electronic Photography of Stars." *Scientific American* 194 (1956): 81–92.

Beattie, Donald A. *Taking Science to the Moon: Lunar Experiments and the Apollo Program*. Johns Hopkins University Press, 2001.

Belew, Leland F., and Ernst Stuhlinger. *Skylab: A Guidebook*. 1973.

Bell, Alex, Raj Chetty, Xavier Jaravel, Neviana Petkova, and John Van Reenen. "Who Becomes an Inventor in America? The Importance of Exposure to Innovation." Quarterly Journal of Economics 134, no. 2 (2019): 647–713.

Bell, Thomas. "D.C. Knights of Pythias Hope to Eliminate Color Barrier." *Washington Post*, February 21, 1990. https://www.washingtonpost.com/archive/local/1990/02/22/dc-knights-of-pythias-hope-to-eliminate-color-barrier/8be38101-6d31-4696-a1ea-2f015a940f9a/.

"Benjamin J. Carruthers." *Chicago Defender (National Edition)*, November 11, 1939.

Bennet, Cynthia. "Science Service and the Origins of Science Journalism, 1919–1950." PhD diss., Iowa State University, 2013. https://dr.lib.iastate.edu/handle/20.500.12876/27268.

"Black Moon Scientist Is Guest of District 16." *New York Amsterdam News*, December 9, 1972.

Blue, John T., Jr. "The Black Mr. Hearst." *Journal of Negro Education* 25, no. 2 (1956): 149–151.

Boggess, Albert, F. A. Carr, D. C. Evans, D. Fischel, H. R. Freeman, C. F. Fuechsel, D. A. Klinglesmith, V. L. Krueger, G. W. Longanecker, and J. V. Moore. "IUE Spacecraft and Instrumentation." *Nature* 275, no. 5679 (1978): 372–377.

Booker, H. G. "Fifty Years of the Ionosphere. The Early Years–Electromagnetic Theory." *Journal of Atmospheric and Terrestrial Physics* 36, no. 12 (1974): 2113–2136.

Bowyer, S., E. T. Byram, T. A. Chubb, and H. Friedman. "Observational Results of X-Ray Astronomy." *Annales d'Astrophysique* 28 (1965): 791–802.

Bradby, Marie. "The Naval Research Laboratory." *US Black Engineer* 11 (1987): 32–36.

Brown, Robert A., and Riccardo Giacconi. "New Directions for Space Astronomy." *Science* 238, no. 4827 (1987): 617–619.

Brown, Sanborn C. "A Short History of Gaseous Electronics." In *Gaseous Electronics*, edited by M. N. Hirsh and H. J. Oskam, 1–18. Academic Press, 1978.

Burke, Bernard F., Arthur Davidson, Giovanni Fazio, Carl Fichtel, George D. Gatewood, Riccardo Giacconi, Jonathan E. Grindlay, and Charles H. Townes. *Research Opportunities in Astronomy and Astrophysics*. National Academy Press, 1988.

Burke, Bernard F., Arthur Davidson, Giovanni Fazio, Carl Fichtel, George D. Gatewood, Riccardo Giacconi, Jonathan E. Grindlay, and Charles H. Townes. *ONR Research Opportunities in Astronomy and Astrophysics*. National Academy Press, 1991.

Cain, Susan. *Quiet: The Power of Introverts in a World That Can't Stop Talking*. Crown, 2013.

Calvino, Ruth Joy. "A Phenomenon to Monitor: Racial Discrimination at NASA, 1974–1985." PhD diss., Clemson University, 2020. https://tigerprints.clemson.edu/all_theses/3281.

"Caribbean House to Have Guests." *New York Amsterdam News*, October 4, 1969.

Carlson, R. W., and D. L. Judge. "The Extreme Ultraviolet Dayglow of Jupiter." *Planetary and Space Science* 19, no. 3 (1971): 327–343.

Carlson, R. W., and D. L. Judge. "Pioneer 10 Ultraviolet Photometer Observations at Jupiter Encounter." *Journal of Geophysical Research* 79, no. 25 (1974): 3623–3633.

Carruthers, George. "Far-Ultraviolet Spectroscopy and Photometry of Some Early-Type Stars." *The Astrophysical Journal* 151 (1968): 269–284.

Carruthers, George. "Magnetically Focused Image Converters with Internal Reflecting Optics." *Report on NRL Progress* (1966): 7–13. Abstract in *Scientific and Technical Aerospace Reports* 4, no. 21 (1966).

Carruthers, George. "Outreach Programs for African-American Students in Washington, D.C." *Mercury* 24, no. 3 (1995): 29–30.

Carruthers, George. "An Upper Limit on the Concentration of Molecular Hydrogen in Interstellar Space." *The Astrophysical Journal* 148 (1967): L141–L142.

Carruthers, George, Harry Heckathorn, Chet Opal, Adolf Witt, and Karl Henize. "Far Ultraviolet Wide Field Imaging, and Photometry: Spartan-202 Mark II Far Ultraviolet Camera." *Proceedings of SPIE* 932 (1988): 61–88.

Carruthers, George R. "Atomic and Molecular Hydrogen in Interstellar Space." *Space Science Reviews* 10, no. 4 (1970): 459–482.

Carruthers, George R. "Charge-Coupled Devices (CCD) Arrays for Electronic Imaging." *Journal of the NTA* 60, no. 2 (1986): 13–21.

Carruthers, George R. "Electronic Imaging Devices in Astronomy." *Astrophysics and Space Science* 14, no. 2 (1971): 332–377.

Carruthers, George R. "Far Ultraviolet Cameras Experiment on STS-39: Observations of the Far-UV Space Environment." *NRL Review* (1992): 226–231.

Carruthers, George R. "Far UV Spectroscopy from the Lunar Surface." Proposals to NASA. October 14, 1969. Box 41, HF/APS, 113.

Carruthers, George R. "Further Developments of Electrographic Image Detectors." *Proceedings of SPIE* 279 (1981): 112.

Carruthers, George R. "A High-Resolution Ultraviolet Shuttle Glow Spectrograph." In *The 1993 Shuttle Small Payloads Symposium*, 97–105. NASA, 1993.

Carruthers, George R. Image converter for detecting electromagnetic radiation especially in short wave lengths. US Patent 3,478,216, filed July 27, 1966, and issued November 11, 1969. https://patentimages.storage.googleapis.com/c2/41/70/635c0324de0303/US3478216.pdf.

Carruthers, George R. "Magnetically Focused Electronographic Image Converters for Space Astronomy Applications." *Applied Optics* 8, no. 3 (1969): 633–638.

Carruthers, George R. "Observation of the Lyman Alpha Interstellar Absorption Line in Theta Orionis." *Astrophysical Journal* 156 (1969): L97–L100.

Carruthers, George R. "Photoelectric Devices in Astronomy and Space Research. Part I; Part II." *Physics Teacher* 12 (1974): 135–143, 221–227.

Carruthers, George R. "Rocket Observations of Interstellar Molecular Hydrogen." *The Astrophysical Journal* 161 (1970): L81–L85.

Carruthers, George R. "Space Astronomy in the Far Ultraviolet." *Science Teacher* 40 (1973): 36–39.

Carruthers, George R., R. J. Dufour, J. C. Raymond, and Adolf Witt. "The Far-Ultraviolet Imaging Spectrograph Experiment on Spartan-204 /STS-63." *Bulletin of the American Astronomical Society* 27 (1995): 890.

Carruthers, George R., Harry M. Heckathorn, Reginald J. Dufour, Chet B. Opal, John C. Raymond, and Adolf N. Witt. "The Spartan-281 Far Ultraviolet Imaging Spectrograph." *Proceedings of SPIE* 932 (1988): 87C.

Carruthers, George R., Harry M. Heckathorn, and Chet B. Opal. "Rocket Ultraviolet Imagery of the Andromeda Galaxy." *The Astrophysical Journal* 225 (1978): 346–356.

Carruthers, George R., Robert P. McCoy, Thomas N. Woods, Paul D. Feldman, and Chet B. Opal. "Rocket Ultraviolet Observations of Comet Halley." In *Aerospace Century XXI: Space Sciences, Applications, and Commercial Developments; Proceedings of the Thirty-Third Annual AAS International Conference*, 1155–1167. Univelt, Inc., 1987.

Carruthers, George R., Chet B. Opal, Thornton L. Page, R. R. Meier, and D. K. Prinz. "Lyman-α Imagery of Comet Kohoutek." *Icarus* 23 (1974): 526–537.

Carruthers, George R., and Thornton Page. "*Apollo 16* Far-Ultraviolet Camera/Spectrograph: Earth Observations." *Science* 177 (1972): 788–791.

Carruthers, George R., and Thornton Page. "Far UV Camera/Spectrograph." In *Apollo 16 Preliminary Science Report*, 13–15. Washington, D.C 1972.

Carruthers, George R., and Thornton Page. "The S201 Far-Ultraviolet Imaging Survey: A Summary of Results and Implications for Future Surveys." *Publications of the Astronomical Society of the Pacific* 96 (1984): 447–462.

Carruthers, George R., and Timothy D. Seeley. "Global Imaging Monitor of the Ionosphere: An Ultraviolet Ionospheric Imaging Experiment for the ARGOS Satellite." *Proceedings of SPIE* 1745 (1992): 322–333.

Carruthers, George R., Timothy D. Seeley, and K. F. Dymond. "Electron-Bombarded CCD Sensors for Far-Ultraviolet Measurements of the Upper Atmosphere." In *Photoelectronic Image Devices 1991: Proceedings of the 10th Symposium*, edited by B. L. Moran, 1–8. Institute of Physics Conference Series 121. Bristol, England, 1992.

Carruthers, George R., and M. L. Washington. "Space Science Education and Public Outreach in the Washington, DC Area." *Bulletin of the American Astronomical Society* 31 (1999): 941.

Carruthers, George Robert. "Experimental Investigations of Atomic Nitrogen Recombination." PhD diss., University of Illinois, 1964. Carter, Luther J. "Space Budget: Congress Is in a Critical, Cutting Mood." *Science* 157 (1967): 170–173.

Carwell, Hattie. *Blacks in Science*. Exposition Press, 1977, 1981.

Carwell, Hattie, Valerie L. Thomas, J. Leonard Anthony, Harold Goodridge, Anna Pratt, and Linda Jackson. *A Tribute to the Career of Dr. George R. Carruthers*. Privately Printed, 2011.

Cervenka, Adolph. "One Approach to the Engineering Design of the Advanced Orbiting Solar Observatory." In *The Observatory Generation of Satellites*, SP-30, 62 pages, p. 31. NASA Office of Scientific and Technical Information, 1963.

"Chicago Rocket Society Juniors Shoot for the Sky." *Chicago Daily Tribune*, September 13, 1959.

Christe, Steven, Ben Zeiger, Rob Pfaff, and Michael Garcia. "Introduction to the Special Issue on Sounding Rockets and Instrumentation." *Journal of Astronomical Instrumentation* 5, no. 1 (2016).

Chubb, Talbot, and E. T. Byram. "Stellar Brightness Measurement at 1314 and 1427 A: Observation of the OI Twilight Airglow." *Astrophysical Journal* 138 (1963): 617–630.

Compton, W. David. *Where No Man Has Gone Before: A History of Apollo Lunar Exploration Missions*. NASA, 1989. Accessed October 27, 2024. https://www.hq.nasa.gov/office/pao/History/SP-4214/cover.html.

Compton, W. David, and Charles D. Benson. *Living and Working in Space*. NASA, 1983. https://history.nasa.gov/SP-4208/ch1.htm.

Conway, R. R., D. K. Prinz, and G. H. Mount. "Middle Atmosphere High Resolution Spectrograph." *Proceedings of SPIE* 932 (1988), 50–60.

Corliss, William R. *NASA Sounding Rockets, 1958–1968: A Historical Summary*. NASA, 1971. Accessed October 27, 2024.

Corn, Joseph, and Brian Horrigan. *Yesterday's Tomorrows: Past Visions of the American Future*. Smithsonian Press, 1984.

Cowley, Charles, "Benjamin Franklin Peery Jr. (1922–2010)." *Bulletin of the American Astronomical Society* 43, (2011): 28.

Cunningham, J. "Charles Sumner Duke (1879–1952)." Encyclopedia of Arkansas. Accessed August 15, 2024. https://encyclopediaofarkansas.net/entries/charles-sumner-duke-7794/.

Dahlburg, Jill, George Doschek, et al., eds. *NRL SSD Research Achievements I, II, III, IV, V*. Washington DC: US Naval Research Laboratory, 2015. Accessed June 14, 2023.

Vol 1: "1960–1970" https://apps.dtic.mil/dtic/tr/fulltext/u2/1000471.pdf;

Vol 2: "1970–1980" https://apps.dtic.mil/dtic/tr/fulltext/u2/1000472.pdf;

Vol 3: "1980–1990" https://apps.dtic.mil/dtic/tr/fulltext/u2/1000473.pdf;

Vol 4: "1990–2000" https://apps.dtic.mil/dtic/tr/fulltext/u2/1000474.pdf;

Vol 5: "2000–2010" https://apps.dtic.mil/dtic/tr/fulltext/u2/1000475.pdf.

Daniel, Pete. *Dispossession: Discrimination Against African American Farmers in the Age of Civil Rights*. Chapel Hill: University of North Carolina Press, 2013.

Dean, Eddie. "A Brief History of White People in Southeast." *Washington City Paper*, October 16, 1998. Accessed October 27, 2024. https://washingtoncitypaper.com/article/275849/a-brief-history-of-white-people-in-southeast/.

DeVorkin, David H. *Fred Whipple's Empire: The Smithsonian Astrophysical Observatory, 1955–1973*. Smithsonian Institution Scholarly Press, 2018.

DeVorkin. "Photomultiplier." In *Instruments of Science: An Historical Encyclopedia*, edited by Robert Bud and Deborah Jean Warner, 458–460. Science Museum; NMAH; and Garland, 1997.

DeVorkin. *Race to the Stratosphere: Manned Scientific Ballooning in America*. New York: Springer-Verlag, 1989.

DeVorkin. *Science with a Vengeance*. Springer-Verlag, 1992.

DeVorkin. "When to Send Your Telescope Aloft." *Journal for the History of Astronomy* 50, no. 3 (2019): 265–305.

DeVorkin. "Who Speaks for Astronomy? How Astronomers Responded to Government Funding after World War II." *Historical Studies in the Physical and Biological Sciences* 31, no. 1 (2000): 55–92.

Dewhirst, D. W. "The Optical Identification of Radio Sources." In *Radio Astronomy Today, Talks Given at the Jodrell Bank Summer School, 1962*, edited by H. P. Palmer and M. I. Large, 178–190. Manchester University Press, 1963.

DeYoung, D. J. *National Security and the U.S. Naval Research Laboratory*. Naval Research Laboratory, 1994.

Donahue, T. M., and R. R. Meier. "Distribution of Sodium in the Daytime Upper Atmosphere as Measured by a Rocket Experiment." *Journal of Geophysical Research* 72 (1967): 2803–2829.

Dupree, A. Hunter. *Science in the Federal Government: A History of Politics and Activities to 1940*. Harvard University Press, 1957.

Dymond, Kenneth F., Scott A. Budzien, George R. Carruthers, and Robert P. McCoy. "High-Resolution Airglow and Aurora Spectrograph (HIRAAS) Sounding Rocket Experiment." *Proceedings of SPIE* 3818 (1999). https://doi.org/10.1117/12.364148.

"Earth's Eye on the Moon." *Ebony Magazine*, 28, no. 12, October 1973. 61-63.

Edgerton, David. *The Shock of the Old: Technology and Global History since 1900*. Oxford University Press, 2007.

Edmondson, Frank K. *AURA and Its US National Observatories*. Cambridge University Press, 1997.

"Eight Scientists Are Candidates for Spacelab 2 Flight." *COSPAR Information Bulletin* 82 (1978): 76–78.

Englert, Christoph, and John Harlander. "Flatfielding in Spatial Heterodyne Spectroscopy." *Applied Optics* 45, no. 19 (2005): 4583–4590.

Englert, Christoph R., John M. Harlander, Kenneth Marr, Brian Harding, Jonathan J. Makela, Tori Fae, Charles M. Brown, M. Venkat Ratnam, S. Vijaya Bhaskara Rao, and Thomas J. Immel. "Michelson Interferometer for Global High-Resolution Thermospheric Imaging (MIGHTI) On-Orbit Wind Observations: Data Analysis and Instrument Performance." *Space Science Reviews* 27, no. 3 (2023).

Engstrom, Ralph W. *Photomultiplier Handbook*. Burle Technologies, Inc., 1989.

Ezell, Linda Neuman. *Historical Data Book, Volume III: Programs and Projects 1969–1978*. NASA, 2009.

Fastie, William. G., H. Warren Moos, Richard C. Henry, and P. D. Feldman. "Rocket and Spacecraft Studies of Ultraviolet Emissions from Astrophysical Targets." *Philosophical Transactions of the Royal Society of London. Series A, Mathematical and Physical Sciences* 279, no. 1289 (1975): 391–400.

Feddoes, Sadie. "Please Be Seated." *New York Amsterdam News*, May 26, 1973.

Feddoes, Sadie. "Please Be Seated." *New York Amsterdam News*, June 17, 1972.

Fenrich, Eric. "The Gates of Opportunity: NASA, Black Activism, and Education Access." In *NASA and the Long Civil Rights Movement*, edited by Brian C. Odom and Stephen P. Waring, 206–218. University Press of Florida, 2019.

Fikes, Robert. "From Banneker to Best: Some Stellar Careers in Astronomy and Astrophysics." University at Buffalo Department of Mathematics. Accessed November 12, 2022. http://www.math.buffalo.edu/mad/special/Black.Astronomers-Fikes.pdf.

Fimmel, R. O., W. Swindell, and E. Burgess. *Pioneer Odyssey*. NASA, 1974.

Fimmel, R. O., James Van Allen, and Eric Burgess. *Pioneer: First to Jupiter, Saturn and Beyond*. NASA, 1980.

"5 Boys Suffer Burns in Two S. Side Blasts: Pal Supplies Two with Explosives." *Chicago Daily Tribune*, July 5, 1958.

Florence, Ronald. *The Perfect Machine: Building the Palomar Telescope*. HarperCollins Publishers, 1994.

Fouché, Rayvon. *Black Inventors in the Age of Segregation*. Johns Hopkins University Press, 2003.

Fox, Philip. *Adler Planetarium and Astronomical Museum of Chicago*. Lakeside Press, 1933.

Friedlander, Blaine, Jr. "Sky Watch: Those Jovian Spots." *Washington Post*, July 27, 1994.

Friedman, Herbert. *Annual Report of the E.O. Hulburt Center for Space Research, Naval Research Laboratory, for Fiscal Year 1969*. Naval Research Laboratory, 1969. https://ntrs.nasa.gov/api/citations/19690028912/downloads/19690028912.pdf.

Friedman, Herbert. *Annual Report of the E.O. Hulburt Center for Space Research, Naval Research Laboratory, for Fiscal Year 1972*. Naval Research Laboratory, 1972.

Friedman, Herbert. "N.R.L. Equipment and Plans Report." In *Les spectres des astres dans l'ultraviolet lointain: communications presentees au dixieme colloque international d'astrophysique tenu a Liege, les 11, 12, 13 et 14 juillet 1960*, 15–24. Institute d'Astrophysique cointe-sclessin, Paris, 1961.

Friedman, Herbert. "Reports of Observatories: The E. O. Hulburt Center for Space Research." *Bulletin of the American Astronomical Society* 1 (1969): 220–228.

Friedman, Herbert. "Reports of Observatories: The E. O. Hulburt Center for Space Research, U. S. Naval Research Laboratory, Washington, D.C." *Astronomical Journal* 70 (1965): 765–777.

Friedman, Herbert. "Reports of Observatories: The E. O. Hulburt Center for Space Research, U. S. Naval Research Laboratory, Washington, D.C., Report 1969–1970." *Bulletin of the Astronomical Society* 3 (1971): 90–101.

Friedman, Herbert, Luther B. Lockhart, and Irving H. Blifford. "Detecting the Soviet Bomb: Joe-1 in a Rain Barrel." *Physics Today* 49, no. 11 (1996): 38–41.

Gatewood, Willard B., Jr. "Aristocrats of Color: South and North the Black Elite, 1880–1920." *Journal of Southern History* 54, no. 1 (1988): 3–20.

"George Archer Carruthers." *Chicago Daily Tribune*, September 27, 1952.

"George Carruthers Wins Science Contest" and "Public School Star Seniors Are Selected." *Chicago Daily Tribune*, February 24, 1957.

Giacconi, Riccardo, Herbert Gursky, Frank Paolini, and Bruno Rossi. "Evidence for X Rays from Sources outside the Solar System." *Physical Review Letters* 9 (1962): 439–443.

Goldstein, L., K. V. Narasinga Rao, and J. T. Verdeyen. "Interaction of Microwaves in Gaseous Plasmas Immersed in Magnetic Fields." In *Electromagnetics and Fluid Dynamics of Gaseous Plasma*, 121. Polytechnic Press, Polytechnic Institute of Brooklyn, 1962.

Goodstein, Anita S. "Slavery." In *Tennessee Encyclopedia*. Accessed October 4, 2022. http://tennesseeencyclopedia.net/entries/slavery/.

Graves, Curtis, and Ivan Van Sertima. "Space Science: The African-American Contribution." In *Blacks in Science: Ancient and Modern*, edited by Ivan Van Sertima, 238–257. Transaction Publishers, 1983.

Green, Constance M., and Milton Lomask. *Vanguard, A History*. NASA, 1970. https://www.nasa.gov/wp-content/uploads/2023/03/sp-4202.pdf

Greenberg, D. S. "Post-Sputnik: Relations between Science, Government Now Passing into More Settled, Mature Stage." *Science* 145, no. 3638 (1964): 1281–1283.

Griboval, Paul J. "The U.T. Electronographic Camera: A Powerful Detector for Ground Based and Space Astronomy: Description and Astronomical Performance." *Proceedings of SPIE* 172 (1979). https://doi.org/10.1117/12.957101.

Griboval, Paul J. "The U.T. Electronographic Camera: Present Status, Astronomical Performance and Future Developments." *Advances in Electronics and Electron Physics* 52 (1980): 305–314.

"Group Builds Experimental Rocket Motor." *Chicago Daily Tribune*, August 2, 1959.

Gursky, Herbert. "Giving Women Their Due." *Physics Today* 43, no. 8 (1990): 91.

Gursky, Herbert. "Herbert Friedman." In *Biographical Memoirs*, 91–110. Vol. 88. National Academy of Sciences Press, 2006.

Hallam, F. K. L. "Astronomical Ultraviolet Radiation." In *The Middle Ultraviolet: Its Science and Technology*, edited by A. E. S. Green, 219. Wiley Series in Pure and Applied Optics. John Wiley & Sons, Inc., 1966.

Harwit, Martin. *Cosmic Discovery: The Search, Scope, and Heritage of Astronomy*. Basic Books, 1981.

Hawkins, Isabel. "Summary of Informal Education Discussions." In *NASA Office of Space Science Education and Public Outreach Conference 2002*, 147–154. ASP Conference Series, vol. 319. 2004.

Heckathorn, Harry. "George Robert Carruthers." *Physics Today* 74, no. 5 (2021): 64.

Heckathorn, Harry M. "How to Diagnose a Microdensitometer." In *Astronomical Microdensitometry Conference*, edited by Daniel A. Klinglesmith, 1–97. NASA, 1984.

Heckathorn, Harry M., and George R. Carruthers. "An Investigation of Auto Radiographically Intensified Electrographic Imagery." *Publications of the Astronomical Society of the Pacific* 93 (1981): 672.

Heckathorn, Harry M., and C. B. Opal. "Optical Identification and Spectra of Visually Faint, High-Latitude S-201 UV-Stars." *Bulletin of the American Astronomical Society* 14 (1982): 919.

Heckathorn, Harry. C. B. Opal, P. Seitzer, E. M. Green, and E. P. Bozynan. "Electrographic Detectors versus Charge Coupled Devices: A Comparison of Two Quality Panoramic Detectors for Stellar Photometry." *Advances in Electronics and Electron Physics* 64A (1985): 153–165.

Heiles, Carl. "Physical Conditions and Chemical Constitution of Dark Clouds." *Annual Review of Astronomy and Astrophysics* 9 (1971): 293–322.

Henry, Richard C., F. Meekins, H. Friedman, E. T. Byram, G. Fritz. "Possible Detection of a Dense Intergalactic Plasma." *The Astrophysical Journal* 153 (1968) L11-L16.Henry, Richard C., and George R. Carruthers. "Far-Ultraviolet Photography of Orion: Interstellar Dust." *Science* 170 (1970): 527–531.

Henry, Richard C., G. Fritz, J. F. Meekins, H. Friedman, and E. T. Byram. "Possible Detection of a Dense Intergalactic Plasma." *The Astrophysical Journal* 153 (1968): L11.

Hembree, Ray. "Summary of British Image Tube Symposium." In *Proceedings of the Image Intensifier Symposium (October 24–26, 1961): Fort Belvoir, Virginia*. NASA, 1961. https://ntrs.nasa.gov/citations/19620004873.

Hersch, Matthew. *Inventing the American Astronaut*. London: Palgrave MacMillan, 2012.

Hevly, Bruce William. "Basic Research within a Military Context: The Naval Research Laboratory and the Foundations of Extreme Ultraviolet and X-Ray Astronomy." PhD diss., Johns Hopkins University, 1987.

Hevly, Bruce William. "Building a Washington Network for Atmospheric Research." In *The Heavens and the Carnegie Institution of Washington*. Series: History of Geophysics 5, edited by Greg Good, 143–148. History of Geophysics, vol. 5. ISBN: 0-87590-279-0. American Geophysical Union, 1994.

Hicks, G. T., T. A. Chubb, and R. R. Meier, "Observations of Hydrogen Lyman [Alpha] Emission from Missile Trails." *Journal of Geophysical Research* 104, no. A5 (1999): 10101–10109.

Hintz, Eric S. "Tearing Down the Barriers for Black Inventors Begins with Honoring Their Historic Breakthroughs." *Smithsonian Magazine*, March 1, 2022. https://www.smithsonianmag.com/smithsonian-institution/tearing-down-barriers-black-inventors-honoring-historic-breakthroughs-180979652/.

Hirsh, Richard. *Glimpsing an Invisible Universe: The Emergence of X-Ray Astronomy*. Cambridge University Press, 1983.

Hoidale, Marjorie McLardie. *Meteorological Data Report*. White Sands Missile Range, NM: Atmospheric Sciences Laboratory, 1966. https://ntrs.nasa.gov/api/citations/19660025511/downloads/19660025511.pdf.

Holloway, Jonathan B. "NRL's Not-So Hidden Figure." US Naval Research Laboratory. February 27, 2017. https://www.nrl.navy.mil/Media/News/Article/2516211/nrls-not-so-hidden-figure.

Holmes, Keith C. *Black Inventors: Crafting over 200 Years of Success*. Global Black Inventors Research Projects, Inc., 2008.

Hughes, Langston, and Benjamin Carruthers. *Cuba Libre*. The Ward Ritchie Press, 1948.

Imes, William Lloyd. "The Legal Status of Free Negroes and Slaves in Tennessee." *Journal of Negro History* 4, no. 2 (1919): 254–272.

"Interplanetary Gases in Lunar UV Photography." *Science News* 102, no. 4 (1972).

Janesick, James and Morley Blouke. "The Sky on a Chip." *Sky & Telescope* 74 (1987): 238–242.

Jenkins, Dennis. *Space Shuttle: The History of Developing the National Space Transportation System*. Walsworth Publishing Co., 1996.

Jerald, Ambrose, Jr. "The National Technical Association: A Hallmark for Access and Success." American Geophysical Union, Fall Meeting 2017.

Jones, Eric M., ed. *Apollo 16 Lunar Surface Journal*. Revised March 5, 2016. https://www.nasa.gov/history/alsj/a16/a16.html.

Komanduri, S. Murty, Julian B. Roebuck, and Jimmy D. McCamey, Jr. "Race and Class Exploitation: A Study of Black Male Student Athletes (BSAS) on White Campuses." *Race, Gender & Class* 21 (2014): 156–173.

Kondo, Yoji, and James E. Kupperian. "Interaction of Neutral Hydrogen and Electrons and Protons in the Radiation Belts: And the Consequent Lyman-Alpha Emission." *Astronomical Journal* 71 (1966): 860–861.

Korsmo, Fay L. "The Genesis of the International Geophysical Year." *Physics Today* 60, no. 7 (2007): 38.

Lallemand, Andre. "Application de l'optiqué électronique à la photographie." *Comptes rendus de l'Académie des Sciences* 203 (1936): 243.

Lallemand, Andre. "L'optiqué électronique et les telescopes." *L'Astronomie, 49* (1945): 26–33.

Laney, Monique. *German Rocketeers in the Heart of Dixie*. Yale University Press, 2015.

Latour, Bruno. "Give Me a Laboratory and I Will Raise the World." In *Science Observed*, edited by K. Knorr-Cetina and M. Mulkay, 141–170. Sage Publishing, 1983.

Latour, Bruno, and Steve Woolgar. *Laboratory Life: The Construction of Scientific Facts*. Princeton, NJ: Princeton University Press, 1986.

Launius, Roger D. *NACA to NASA to NOW*. NASA, 2022.

Launius, Roger D. "An Unintended Consequence of the IGY: Eisenhower, Sputnik, the Founding of NASA." *Acta Astronautica* 67, no. 1–2, (2010): 254–263.

Launius, Roger D., John M. Logsdon, and Robert W. Smith, eds. *Reconsidering Sputnik: Forty Years since the Soviet Satellite*. Harwood Academic Press, 2000.

Leak, William M. *Rocket Research at NRL*. Naval Research Laboratory, 1954.

Livingston, William C. "Image-Tube Systems." *Advances in Astronomy and Astrophysics* 11, (1973): 95–114.

Livingston, William C. "Properties and Limitations of Image Intensifiers Used in Astronomy." *Advances in Electronics and Electron Physics* 23 (1967): 347–384.

Logsdon, John, editor. "Space Research: Directions for the Future—Report of a Study by the Space Science Board." In *Exploring the Unknown*, edited by John M. Logsdon, Amy Page Snyder, Roger D. Launius, Stephen J. Garber, and Regan Anne Newport, 579–585. Vol. 5. NASA, 2001.

Lowrance, J. L., and George Carruthers. "Electron Bombarded Charge-Coupled Device (CCD) Detectors for the Vacuum Ultraviolet." *Proceedings of SPIE* 279 (1981): 123–128.

Lubkin, Gloria. "Charge-Coupled Devices Would Be Cheap, Compact." *Physics Today* 23, (1970): 17–18.

MacQueen, R. M., C. L. Ross, and Thomas K. Mattingly. "Solar Corona Photography." In *Apollo 16 Preliminary Science Report*. NASA, 1972. Chapter 31, 3–4.

McBride, David, and Monroe H. Little. "The Afro-American Elite, 1930–1940: A Historical and Statistical Profile." *Phylon* 42, no. 2 (1981): 105–119.

McCormmach, Russell. *Night Thoughts of a Classical Physicist*. Harvard University Press, 1991.

McCray, W. Patrick. *Keep Watching the Skies*. Princeton University Press, 2008.

McCurdy, Howard E. *Inside NASA: High Technology and Organizational Change in the U.S. Space Program*. Johns Hopkins University Press, 1993.

McCurdy, Howard E. *Space and the American Imagination*. Washington, DC: Smithsonian Institution Press, 1997. Reprint, Johns Hopkins University Press, 2011.

McGuire, Brett A. "2021 Census of Interstellar, Circumstellar, Extragalactic, Protoplanetary Disk, and Exoplanetary Molecules." *The Astrophysical Journal Supplement Series* 259, Issue 2, (2022), 51.

McQuaid, Kim. "'Racism, Sexism, and Space Ventures': Civil Rights at NASA in the Nixon Era and Beyond." In *Impact of Spaceflight*, edited by Steven J. Dick and Roger Launius, 422–449. NASA Office of External Relations, History Division, 2007.

McWhorter, Diane. *Carry Me Home: Birmingham, Alabama, the Climactic Battle of the Civil Rights Revolution*. Simon & Schuster, 2001.

Meier, Robert R. "Philip W. Mange." *Physics Today*, October 16, 2020. https://physicstoday.scitation.org/do/10.1063/pt.6.4o.20201016b/full/.

Meier, Robert R. "Ultraviolet Spectroscopy and Remote Sensing of the Upper Atmosphere." *Space Science Reviews* 58, no. 1 (1991): 1–185.

Meier, Robert R., C. B. Opal, H. U. Keller, T. L. Page, and George R. Carruthers. "Hydrogen Production Rates from Lyman-Alpha Images of Comet Kohoutek (1973 XII)." *Astronomy and Astrophysics* 52, no. 2 (1976): 283–290.

Mercer, R. D., L. Dunkelman, and Thomas K. Mattingly. "Gum Nebula, Galactic Cluster, and Zodiacal Light Photography." In *Apollo 16 Preliminary Science Report*. NASA, 1972. Chapter 31, 1-2.

Miller, Joseph S. "Astronomical Instrumentation Acquires Pedagogical Tools." *Physics Today* 42 (1989): 97–98.

Miller, Joseph S. "Optical Instrumentation for Large Telescopes in the Next Decade." *Bulletin of the American Astronomical Society* 21 (1989): 1146–1147.

Miller, Ron. "The Archaeology of Space Art." *Leonardo* 29 (1996): 139–143.

Miller, R. E., W. G. Fastie, and R. S. Isler. "Rocket Studies of Far-Ultraviolet Radiation in an Aurora." *Journal of Geophysical Research: Space Physics* 73, no. 11 (1968): 3353–3365.

Misra, Prabhakar, George Carruthers, and Gregory S. Jenkins. "Development of an Earth and Space Science-Focused Education Program at Howard University." *Journal of Geoscience Education* 54, no. 3 (2006): 339–345.

Mission Evaluation Team. *Apollo 16 Mission Report*. Houston: Manned Spacecraft Center, 1972. Accessed August 15, 2024, https://www.nasa.gov/history/alsj/a16/a16mr.html.

Mitton, Jacqueline, Simon Mitton, and Jocelyn Bell-Burnell. *Vera Rubin: A Life*. Harvard University Press, 2021.

"Moon Camera Invented by a Black Man." *New York Amsterdam News*, December 16, 1972.

Moore, Charlotte E. *A Multiplet Table of Astrophysical Interest*. United States Department of Commerce National Bureau of Standards, 1945.

Morgan, B. L., *Photo-Electronic Image Devices*, 64, part A, (1986): 153–165.

Morris, Vernon R., Everette Joseph, Sonya Smith, and Tsann-wang Yu. "The Howard University Program in Atmospheric Sciences (HUPAS): A Program Exemplifying Diversity and Opportunity." *Journal of Geoscience Education* 60 (2012): 45–53. https://www.tandfonline.com/doi/full/10.5408/10-180.1.

Morrow, Robert C., John P. Wetzel, Robert C. Richter, Kathy Benzin, and Christopher Allison. "Accommodating Science and Technology Development Sortie Missions in the Post Space Shuttle Era." 48th International Conference on Environmental Systems, July 8-12, 2018. https://ttu-ir.tdl.org/bitstream/handle/2346/74243/ICES_2018_304.pdf?sequence=1&isAllowed=y.

Morton, David L., Jr., and Joseph Gabriel. *Electronics: The Life Story of a Technology*. Johns Hopkins University Press, 2007.

Morton, Donald C., and Lyman Spitzer, Jr. "Line Spectra of Delta and Pi Scorpii in the Far-Ultraviolet." *Astrophysical Journal* 144 (1966): 1–12.

Moss, Steven L. "NASA and Racial Equality in the South, 1961–1968." Master's thesis, Texas Tech University, 1997.

Mumm, Susan. "AE Alum to Be Presented University's Highest Honor." The Grainger College of Engineering. April 7, 2014. https://aerospace.illinois.edu/news/ae-alum-be-presented-universitys-highest-honor.

NASA. *Apollo 16 Preliminary Science Report*. NASA, 1972.

NASA. Manned Spacecraft Center "Announcement" No: 69-61; September 30, 1969. No. 69–138.

NASA. SPD-9P-052 "Apollo Lunar Exploration Missions." Manned Spacecraft Center, Houston, November 1, 1969.

"NASA Awards Black Scientist." *The Bay State Banner*, August 10, 1972.

"NASA Cites Dr. Carruthers." *Tri-State Defender*, August 5, 1972.

National Research Council. *Sounding Rockets: Their Role in Space Research*. National Academies Press, 1969.

Naugle, John E., and John M. Logsdon. "Space Science: Origins, Evolution, and Organization." In *Exploring the Unknown*, edited by John M. Logsdon, Amy Page Snyder, Roger D. Launius, Stephen J. Garber, and Regan Anne Newport, 1–15. Vol. 5. NASA, 2001.

"Navy Specialist Probes Deep Space." *Chicago Daily Defender*, June 27, 1970.

Neufeld, Michael. *Von Braun: Dreamer of Space, Engineer of War*. Knopf, 2007.

Newell, Homer E. *Beyond the Atmosphere: Early Years of Space Science*. NASA, 1980.

Newell, Homer E. *Sounding Rockets*. McGraw-Hill, 1959.

Newkirk, Roland W., Ivan D. Ertel, and Courtney Brooks. *Skylab: A Chronology*. Washington, DC: NASA, 1977. https://history.nasa.gov/SP-4011/key.htm.

"New View of the Earth." *Washington Post, Times Herald*, June 9, 1972.

Nicks, Oran, ed. *This Island Earth*. NASA, 1970.

Niedner, M. B., M. F. A'Hearn, J. C. Brandt, A. D. Code, A. F. Davidsen, B. Donn, P. Feldman et al. "The ASTRO-1 Mission and Halley's Comet." *Publications of the Astronomical Society of the Pacific* 97 (1985): 900–901.

Nuth, J. A., III, J. L. Lowrance, and George R. Carruthers. "NEOCAM: The Near Earth Object Chemical Analysis Mission." *Earth, Moon, and Planets* 102 (2008): 495–504. https://doi.org/10.1007/s11038-007-9178-y.

Odom, Brian C., and Stephen P. Waring, eds. *NASA and the Long Civil Rights Movement*. University Press of Florida: 2019.

O'Hern, Robin, and David DeVorkin. "What's That Smell? Collaboration and Conservation of a Film Transport from *Apollo 16*." *National Air and Space Museum* (blog). September 6, 2016. https://airandspace.si.edu/stories/editorial/what%E2%80%99s-smell-conserving-apollo-16-film-transport.

O'Leary, Brian T. *The Making of an Ex-astronaut*. Houghton Mifflin, 1970.

Opal, C. B, and George R. Carruthers. "Ultraviolet and Infrared Observations of Comets: Recent Results and Prospects for the Shuttle ERA." In *American Institute of Aeronautics and Astronautics, 14th Aerospace Sciences Meeting*. 1976.

Opal, C. B., and Harry M. Heckathorn. "A Search for 'Unidentified' Stellar Objects." *Bulletin of the American Astronomical Society* 13 (1981): 857.

Osterbrock, D. "Thornton L. Page, 1913–1996." *Bulletin of the American Astronomical Society* 28 (1996), 1462–1462.

O'Toole, Thomas. "Black Scientist Develops Apollo Instrument—Telescope to Give Views of Earth's Poles." *Washington Post*, November 15, 1971.

O'Toole, Thomas. "Skylab Likely to See Comet." *Washington Post, Times Herald*, July 24, 1973.

Owen, Tobias B. "The Mid-60's at NRL." In *Report of NRL Progress*, 109–112. Naval Research Laboratory, 1973.

Page, Thornton. "Far-UV Observations of Comet Kohoutek and Other Targets with the S201 Electronographic Camera on Skylab 4." In *Electrography and Astronomical Applications, Proceedings of the Conference*, 297–305. University of Texas, 1974. https://ui.adsabs.harvard.edu/abs/1974eaa..conf..297P/abstract.

Page, Thornton. *Final Report: Measurements of Far-Ultraviolet Photographs from Skylab IV and Apollo 16*. Washington, DC: NASA Center for Aerospace Information, 1977.

Page, Thornton, and George Carruthers. "Far Ultra-Violet Observations of the Large Magellanic Cloud." *Bulletin of the American Astronomical Society* 4 (1972): 331.

Page, Thornton, and Lou Williams Page. *Space Science and Astronomy: Escape from Earth*. Macmillan, 1976.

Panel on Astronomical Facilities. *Ground-Based Astronomy, A Ten-Year Program*. National Academy of Sciences, 1964.

Patterson, Gregory A. "Professor Banks Dies of Heart Attack." *Hilltop News*, January 26, 1979.

Paul, Richard, and Steven Moss. *We Could Not Fail: The First African Americans in the Space Program*. University of Texas Press, 2015. Reprinted in 2016.

Pedrick, Alexis C. "Percy Julian and the False Promise of Exceptionalism." *Distillations* 2 (2023): 2–5.

Petrosian, Vahé. "Arthur B. C. Walker (1936–2001)." *Bulletin of the American Astronomical Society* 33, no. 4 (2001). https://baas.aas.org/pub/arthur-b-c-walker-1936-2001/release/1.

Phelps, J. Alfred. *They Had a Dream: The Story of African-American Astronauts*. Presidio Press, 1994.

Phinney, William C. *Science Training History of the Apollo Astronauts*. NASA, 2015.

Plummer, Brenda. "The Newest South: Race and Space on the Dixie Frontier." In *NASA and the Long Civil Rights Movement*, edited by Brian C. Odom and Stephen P. Waring. University Press of Florida, 2019, 61–108.

Powell, J. "Blue Scout—Military Research Rocket." *British Interplanetary Society Journal* 35, (1982): 22–30.

Potter, Andrew, and Robert Parker. "Karl Gordon Henize, 1926–1993." *Bulletin of the American Astronomical Society* 26 (1994): 1604–1605.

Pratt, Anna. *Program to Honor an American Living Legend: George R. Carruthers, Astrophysicist/Inventor*. Carruthers Collection, NASM. Privately printed, 2011.

President's Science Advisory Committee, Joint Space Panels. *The Space Program in the Post-Apollo Period*. The White House, 1967. https://books.google.com/books?id=9me95yllziUC&pg=PR1.

*Proceedings of the Image Intensifier Symposium, October 24–26, 1961, Fort Belvoir, Virginia*. NASA. https://ntrs.nasa.gov/citations/19620004873.

"Program Is Proposed for Outer-Planet Trips in 70's." *Physics Today* 22, no. 10 (1969): 59.

"Racial Stereotype Busters: Black Scientists Who Made a Difference." *Journal of Blacks in Higher Education*, no. 25 (1999): 133–134. https://www.jstor.org/stable/2999414.

"Reunion Dinner Served Chi Delta Sigma Sorority." *Chicago Defender*, March 30, 1935.

Rhinelander, David H. "Astronomer's 2½ Year Project to Be the First Moon Telescope." *Hartford Courant*, April 20, 1972.

Rhinelander, David H. "Lunar Observatory Is Planned." *Hartford Courant*, July 21, 1969.

Rhinelander, David H. "Wesleyan Astronomer Works on Lunar Telescope Project." *Hartford Courant*, November 23, 1969.

Riggs, Blake. "Training Matters! Narrative from a Black Scientist." *Molecular Biology of the Cell* 32 (2021): 223–225.

Roe, Ann. "A Psychologist Examines 64 Eminent Scientists." *Scientific American* 185 (1952): 221–225. https://www.gwern.net/docs/iq/high/anne-roe/1952-roe.pdf.

Roman, Nancy. "Nancy Grace Roman and the Dawn of Space Astronomy." *Annual Review of Astronomy and Astrophysics* 57 (2019): 1–34. https://www.annualreviews.org/doi/full/10.1146/annurev-astro-091918-104446.

Roman, Nancy Grace. "Exploring the Universe: Space-Based Astronomy and Astrophysics." In *Exploring the Unknown*, edited by John M. Logsdon, Amy Page Snyder, Roger D. Launius, Stephen J. Garber, and Regan Anne Newport, 579–585. Vol. 5. NASA, 2001.

Rosen, Milton. *Viking Rocket Story*. Harper, 1955.

Rouse, Carl A. "Ionization Equilibrium Equation of State." *Astrophysical Journal* 135 (1962): 599.

Ruley, John. "Revolt of the Scientist-Astronauts." *Quest* 29 (2022): 5–12.

Ryan, Cornelius, ed. "Man Will Conquer Space Soon!" *Collier's Weekly*, March 22, 1952, through April 30, 1954.

Sagan, Carl, and Thornton Page, eds. *UFOs: A Scientific Debate*. Cornell University Press, 1972.

Sapolsky, Harvey M. *Science and the Navy: The History of the Office of Naval Research*. Princeton University Press, 1990.

Sawyer, Cathy. "Comet Chunk Leaves Larger-than-Earth Scar on Jupiter." *Washington Post*, July 19, 1994.

Sawyer, Cathy. "More Chunks of Comet Hit Jupiter, Giving Astronomers a Striking View." *Washington Post*, July 18, 1994.

Seamans, Robert C., Jr. *Project Apollo: The Tough Decisions*. Monographs in Aerospace History. No. 37. NASA, 2005.

Seeley, T. D., R. R. Conway, and George R. Carruthers. "The Middle Atmosphere High Resolution Spectrograph Investigation." In *Photoelectric Image Devices, the McGee Symposium. Proceedings of the 10th Symposium on Photoelectric Image Devices*, edited by B. L. Morgan, 41–48. Institute of Physics, 1991.

Selme, Pierre. "L'optique électronique et les télescopes." *L'Astronomie* 64 (1945): 26–33.

Shafritz, Jay M., and Joseph D. Atkinson. *The Real Stuff*. Praeger, 1985.

Shayler, David J., and Colin Burgess. *NASA's First Space Shuttle Astronaut Selection: Redefining the Right Stuff*. Springer Praxis Books, 2020.

Shetterly, Margot Lee. *Hidden Figures: The Story of the African-American Women Who Helped Win the Space Race*. HarperCollins, 2017.

Shrader, Charles R. *History of Operations Research in the United States Army*. Vol 1. Washington, DC: Office of the Deputy Under Secretary of the Army for Operations Research, 2006. https://history.army.mil/html/books/hist_op_research/CMH_70-102-1.pdf.

Shull, J. M., and S. Beckwith. "Interstellar Molecular Hydrogen." *Annual Review of Astronomy and Astrophysics* 20 (1982): 163–190.

"Sixth Group of Shuttle Astronaut Applicants Includes One Woman." *JSC Roundup* 16, no. 221 (1977).

Sluby, Patricia Carter. *The Inventive Spirit of African Americans: Patented Ingenuity*. Praeger, 2004.

S.M.A.R.T. "Compendium of Abstracts. Saturday Workshops in Aerospace Science and Technology," July 1992–June 1998. Courtesy Valerie L. Thomas; copy also in box 9, NTA files at NASM.

"S.M.A.R.T. Day at the National Air and Space Museum." *The S.M.A.R.T. Times*, spring 1993.

Smith, Harlan J. "Summary Remarks on Quasars." In *Quasars and High-Energy Astronomy, Proceedings of the 2nd Texas Symposium on Relativistic Astrophysics, Held in Austin, December 15–19, 1964*, edited by K. N. Douglas et al., 181–182. Gordon & Breach, 1969.

Smith, Robert W. "Early History of Space Astronomy: Issues of Patronage, Management and Control." *Experimental Astronomy* 26 (2009): 149–161.

Smith, Robert W. *The Space Telescope: A Study of NASA, Science, Technology, and Politics*. Cambridge University Press, 1989.

Smith, R. W., and J. H. Tatarewicz. "Counting on Invention: Devices and Black Boxes in Very Big Science." *Osiris* 9 (1994): 101–123.

Smith, R. W., and J. H. Tatarewicz. "Replacing a Technology: The Large Space Telescope and CCDs." *IEEE Proceedings* 73 (1985): 1221–1235.

Snoddy, William C., and G. Allen Gary. "Skylab Observations of Comet Kohoutek." In *Conference on Scientific Experiments on Skylab*. American Institute of Aeronautics and Astronautics, 1974.

"Space Camera Takes Special Pix." *Daily Defender*, July 18, 1972.

Space General. "Fine Attitude Control System Final Report July 1966. NASA-CR-75392.

Spady, James G. "Blackspace." In *Blacks in Science: Ancient and Modern*, edited by Ivan Van Sertima, 258–268. New Brunswick, NJ: Transaction Publishers, 1983.

"Special Report: American Minorities in Astronomy." *Mercury* 24, no. 3 (1995): 3.

Spigel, Lynn. *Welcome to the Dreamhouse*. Duke University Press, 2001.

Spitzer, Lyman, Jr. "Behavior of Matter in Space." *Astrophysical Journal* 120 (1954): 1–17.

Spitzer, Lyman, Jr. *Diffuse Matter in Space*. Wiley Interscience, 1968.

Spitzer, Lyman, Jr., and Franklin R. Zabriskie. "Interstellar Research with a Spectroscopic Satellite." *Publications of the Astronomical Society of the Pacific* 71 (1959): 412–420.

Stebbins, Joel. "The Law of Diminishing Returns." *Sky & Telescope* 3, no. 4 (1944): 5–8.

Stecher, T. P., and J. E. Milligan. "An Interim Interstellar Radiation Field." *Annales d'Astrophysique* 25 (1962): 268–270.

Sternglass, E. J., "Application of secondary electron image and storage tubes to space astronomy," *Astronomical Journal* 71 (1966): 872–873.

Stevens, Michael H., Robert R. Conway, Joel G. Cardon, and James M. Russell, III. "MAHRSI Observations of Nitric Oxide in the Mesosphere and Lower Thermosphere." *Geophysical Research Letters* 24 (1997): 3213–3216.

Strong, John, H. Victor Neher, Albert E. Whitford, C. Hawley Cartwright, and Roger Hayward. "Glassblowing as a Central Practice in Experimental Physics." In *Procedures in Experimental Physics*. Prentice-Hall, 1938, 1–38.

Sullivan, Walter. *Assault on the Unknown: The International Geophysical Year*. McGraw-Hill, 1961.

Sullivan, Walter. "Ninety per Cent of the Universe Found 'Missing' by Astronomer." *The New York Times*, December 29, 1960.

Sullivan, Woodruff T., III. *Cosmic Noise: A History of Radio Astronomy*. Cambridge University Press, 2009.

Swanson, Glen E., ed. *Before This Decade Is Out: Personal Reflections on the Apollo Program*. NASA History Series. NASA, 1999.

Tate, William F., IV. "Brown, Political Economy, and the Scientific Education of African Americans." *Review of Research in Education* 28 (2004): 47–184.

Teeter, Ron, and Bob Reynolds. *Final Report on the NASA Suborbital Program: A Status Review*. Battelle Columbus Laboratories, 1983.

Terebish, V. Yu. "Two-Mirror Schwarzschild Aplanats. Basic Relations." *Astronomy Letters* 31, no. 2 (2005): 129–139. https://arxiv.org/abs/astro-ph/0502121v1.

Tessarotto, Fulvio. "Evolution and Recent Development of the Gaseous Photon Detectors Technologies." *Nuclear Instruments and Methods in Physics Research, A* 912 (2018): 278–286.

Thomas, Valerie L., George R. Carruthers, and E. Takamura. *NASA Office of Space Science Education and Public Outreach Conference 2002*. ASP Conference Series, vol. 319. 2004.

Thompson, Samantha Michelle. "The Carnegie Image Tube Committee and the Development of Electronic Imaging Devices in Astronomy, 1953–1976." PhD diss., Arizona State University, 2019.

Thonnard, S. E., K. D. Wolfram, Kenneth F. Dymond, C. B. Fortna, S. Budzien, A. Nicholas, and R. McCoy. "Results from the High Resolution Airglow and Aurora Spectroscopy (HIRAAS) Experiment (August 2001)." In *AIAA Space 2001 Conference and Exposition*. American Institute of Aeronautics and Astronautics, 2001. https://doi.org/10.2514/6.2001-4663.

"Three at JSC Are Payload Specialist Candidates for First Spacelab Mission." *JSC Roundup* 16, no. 221 (1977): 1.

Tifft, William G. "Two-Dimensional Area Scanning with Image Dissectors." *Publications of the Astronomical Society of the Pacific* 84, no. 497 (1972): 137–144.

"Time Life Releases First Volume of Books on Influential Blacks." *Miami Times*, October 21, 1993.

Tolchin, Martin. "Teen-Age Rocket Groups Need a Parental Brake." *The New York Times*, December 12, 1957.

Tousey, Richard. "Rocketry." *Smithsonian Contributions to Astrophysics* 1 (1956): 39–44.

Traweek, Sharon. *Beamtimes and Lifetimes: The World of High Energy Physicists*. Harvard University Press, 1988.

"2 From Jamaica Capture 'Miss Carifta' Crown as Judges Agree." *New York Amsterdam News*, March 8, 1969.

"Ultraviolet Arcs in the Equatorial Ionosphere," *1972 Review: Naval Research Laboratory*, 133

"U.S. Naval Research Laboratory (NRL) Personnel Management Demonstration Project; Department of the Navy (DON), Washington, DC." *Federal Register* 64, no. 121 (1999): 33970–34046.

Van Sertima, Ivan. *Blacks in Science: Ancient and Modern*. Transaction Publishers, 1983.

Verdeyen, Joseph T. "The Gaseous Electronics Laboratory." Accessed November 15, 2022. https://web.archive.org/web/20071017121448/http://lope.ece.uiuc.edu/history.htm.

Visco, Eugene P. "The Operations Research Office." *Army History* 38 (1996): 24–33.

Wakelam, Valentine, Emeric Bron, Stephanie Cazaux, Francois Dulieu, Cecile Gry, Pierre Guillard, Emilie Habart, et al. "H2 Formation on Interstellar Dust Grains:

The Viewpoints of Theory, Experiments, Models and Observations." *Molecular Astrophysics* 9 (2017): 1–36.

[Wallops Island Missile Range]. *Sounding Rocket Program Handbook*. Wallops Island, VA: National Aeronautics and Space Administration-Goddard Space Flight Center-Wallops Flight Facility, 2001. Accessed August 15, 2024. https://www.nasa.gov/wp-content/uploads/2023/09/sounding-rocket-program-handbook.pdf.

Walsh, John, "News and Comment: Graduate Education. Navy Program in Rocket Astronomy Opens New Horizons to University Scientists." *Science* 141 (1963): 788–790.

Weitekamp, Margaret A. *Space Craze: America's Enduring Fascination with Real and Imagined Spaceflight*. Smithsonian Institution Press, 2022.

Weitekamp, Margaret A. "Space History Matures—and Reaches a Crossroads." In *NASA and the Long Civil Rights Movement*, edited by Brian C. Odom and Stephen P. Waring, 11–28. University Press of Florida, 2019.

Wells, E. H. *Optical Astronomy Package Feasibility Study for Apollo Applications Program Executive Summary Report*. George C. Marshall Space Flight Center, 1966.

Whipple, F. L., *The Mystery of Comets*. Cambridge University Press, 1985.

Whyte, William H., Jr. *The Organization Man*. Doubleday Anchor Books, 1957.

Wilkerson, Isabel. *The Warmth of Other Suns: The Epic Story of America's Great Migration*. New York: Vintage, 2010.

Winder, Alvin. "Residential Invasion and Racial Antagonism in Chicago." *Phylon* 12 (1951): 239–241.

Wishnis, Herbert F. *Optical Technology Experiment System*. Vol. 1. Perkin-Elmer Corporation, 1967.

Wolff, Michael T. and Kenneth F. Dymond. "ARGOS PART IV: The Global Imaging Monitor of the Ionosphere (GIMI) Experiment." In *NRL SSD Research Achievements: 1990–2000*, Vol. 4, 49. Naval Research Laboratory, October 30, 2015. https://apps.dtic.mil/sti/tr/pdf/AD1000474.pdf

Wood, Frank Bradshaw. *The Present and Future of the Telescope of Moderate Size*. Philadelphia: University of Pennsylvania Press, 1958.

Wray, James D., Harlan J. Smith, Karl G. Henize, and George R. Carruthers. "Space Schmidt Telescope." *Proceedings of SPIE* 332 (1982): 141–150.

Yeager, Ashley Jean. *Bright Galaxies, Dark Matter, and Beyond: The Life of Astronomer Vera Rubin*. MIT Press: 2021.

Yette, Samuel F. "Give NASA Credit for Recognizing Courageous Blacks in Space." *Philadelphia Tribune*, July 18, 1997.

"Young Set Enjoys Gay House Party." *Chicago Defender*, August 14, 1937.

# INDEX

GRC is George Robert Carruthers; other abbreviations can be found in the List of Abbreviations on page xvii. Page numbers in *italics* indicate an image. Notes are indicated by the page number, the letter n, and the note number. For example, 332n20 is note 20 on page 332. For people who helped the author with this book, please go to the Acknowledgments.